International Conference on

Elastic and Diffractive Scattering

σ

Proceedings of the VIIIth Blois Workshop

International Conference on

Elastic and Diffractive Scattering

σ

Protvino, Russia 28 June – 2 July 1999

Editors

V. A. Petrov

A. V. Prokudin

Institute for High Energy Physics,
Protvino, Russia

World Scientific
Singapore • New Jersey • London • Hong Kong

Published by

World Scientific Publishing Co. Pte. Ltd.

P O Box 128, Farrer Road, Singapore 912805

USA office: Suite 1B, 1060 Main Street, River Edge, NJ 07661

UK office: 57 Shelton Street, Covent Garden, London WC2H 9HE

British Library Cataloguing-in-Publication Data
A catalogue record for this book is available from the British Library.

International Conference on Elastic and Diffractive Scattering
BLOIS VIII

ISBN 981-02-4118-6

This book is printed on acid-free paper.

Printed in Singapore by World Scientific Printers

Preface

These Proceedings are supposed to reflect the state of affaires in the study of hadron diffraction phenomena, one of the most difficult subjects of modern high-energy physics, by July 1999 or, in more inflated style, by the end of the XXth century.

One may hope that all principal "schools of thought" in the domain of strong interactions at high energies are represented more or less fully.

I would be very happy if the content of this book became obsolete by the time it has been accessible to readers. Long experience shows, however, that it will be hardly the case. This mainly concerns theory where the progress is slow and even tiny steps (not always clear forth or back, if any) trigger numerous (often imitative) publications. It is an experiment that steadily supplies us with new, sometimes quite suprising, physical effects. This can be clearly seen clearly in the talks presented by HERA, LEP and Tevatron in this volume.

From many keywords and concepts characterizing the Conference I would distinguish "hard diffraction", an interesting phenomenon embracing both short- and large-distance dynamics underlying particle collisions.

The reader will find several theoretical proposals to deal with "hard diffraction"and to explain and describe its spectacular peculiarities.

We hope that at least some of them will be helpful for better understanding of strong interactions at high energies.

I have to add that Blois Workshop was for the first time hosted by Russia and it would be hardly possible without financial help and other assistance of Russian Foundation of Basic Research, Ministry of Science and Technology of Russian Federation and Directorate of IHEP.

Organizing Committee is very much obliged to Elena Nevezhina, Irina Filimonova and Olga Laskovaya whose efforts made activity of the Working Group exclusively efficient.

Vladimir A. Petrov
Chairman of the Organizing
Committee

CONTENTS

Laura Iannotti

INCLUSIVE DIFFRACTION AND LEADING BARYON PRODUCTION AT HERA

Laura Iannotti
(on behalf of the ZEUS and H1 Collaborations)
Pennsylvania State University and Tel Aviv University
E-mail: iannotti@desy.de

A review of ZEUS and H1 data on diffractive deep inelastic scattering and leading baryon production is presented.

1 Introduction

Diffractive processes have been studied for more than thirty years in hadronic interactions [1].

At HERA, diffractive interactions can be thought as a beam of almost-real or virtual photons fluctuating into a hadronic or partonic configuration ($\gamma \to$ vector meson, $q\bar{q}, q\bar{q}g, \ldots$), which diffracts off the proton [2].

The final state (see figure) is characterised subsequently by a hadronic system X of mass M_X originating from the diffractive dissociation of the photon, while the proton either remains intact or is excited to a low mass baryonic state Y. Because of the nature of the colourless exchange, hereafter referred to as *pomeron* (*IP*) exchange, the two systems are separated in rapidity.

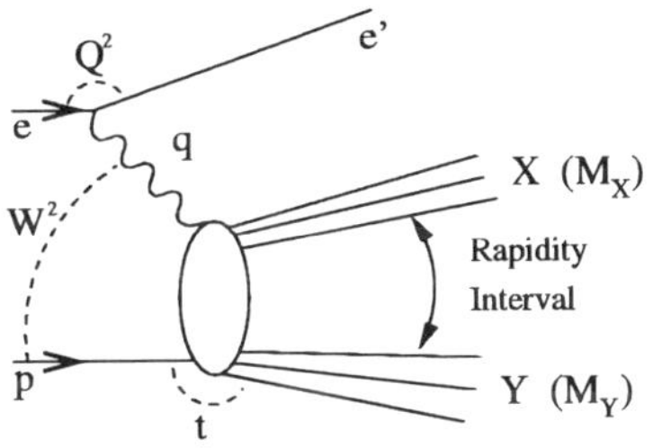

In addition to the virtual photon-pomeron interaction one also expects meson exchanges to contribute to the total deep inelastic scattering cross section and to the production of nucleons carrying a large fraction of the incoming proton momentum (leading baryons). Assuming a simple Regge expansion, the contribution of a colour singlet exchange to the production of a leading baryon with a fraction of proton beam momentum x_L is of the form $\sim (1-x_L)^{1-2\alpha(t)}$. Thus *IP* exchange is expected to dominate the production cross section for protons with energy close to the proton beam energy, whereas reggeon and pion exchange are of increasing importance at lower values of x_L. Neutron production can only occur through the exchange of an isovector particle, notably the π^+ [3].

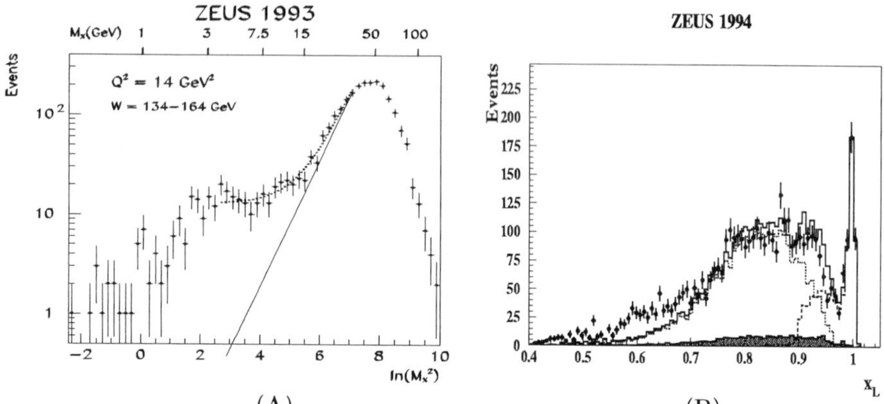

(A) (B)

Figure 1: **(A)**: *Distribution of ln M_X^2 for W and Q^2 as denoted in the figure. The solid line shows the exponential fall-off of the non-diffractive contribution.* **(B)**: *Observed spectrum of the fraction x_L of the proton beam momentum carried by the scattered proton. The full line is the result of fitting this distribution with a sum of contributions from Monte Carlo generators for double dissociation (shaded area), for pion exchange (dotted line) and for photon diffractive dissociation (dashed line).*

2 Inclusive Diffractive Photon Dissociation

Inclusive photon diffractive dissociation includes all pomeron mediated reactions of the kind $\gamma p \to XY$, where Y is either the proton (single diffractive dissociation) or an excited state (double diffractive dissociation).
Diffractive dissociation events are characterised by a gap in rapidity between the systems X and Y which is not exponentially suppressed as a function of the gap width, as opposed to non-diffractive interactions, where the colour string connecting the struck parton and the proton remnant fills the rapidity interval during the hadronisation process. Also, single diffractive dissociation events are characterised by a proton emerging from the interaction with small transverse momentum and carrying a large fraction (x_L or z) of the beam momentum ($x_L \gtrsim 0.95$).
Based on these simple features, both ZEUS and H1 have developed selection methods exploiting the main features of their own detectors.
The event selection used by H1 is based on the presence of a gap in rapidity and relies on the ability to measure hadronic activity in the central detector up to $\eta \sim 3.4$ [a] and to veto any activity in the region $3.4 < \eta < 7.5$ [4]. The contribution of events in which the proton has also dissociated is estimated to

[a] The pseudo-rapidity η is defined as $-\ln(\tan(\theta/2))$, where θ is the polar angle measured with respect to the proton direction.

be $\lesssim$ 10% and is restricted to events where the proton dissociates into a mass $M_Y \lesssim$ 1.6 GeV and escapes the forward tagging devices.
The selection method developed by ZEUS (called the M_X *method* in the following) is based on the observation that the invariant mass M_X measured in the calorimeter consists of two components with very different behaviours [5]; the diffractive contribution is identified as the excess of events at small M_X above the exponential fall-off of the non-diffractive contribution with decreasing $\ln M_X^2$ (see fig. 1(A)). Given the size of the non-diffractive subtraction, this method limits the measurements to small masses ($M_X \lesssim 15$ GeV). The contribution of double dissociation is $\sim$ 30% and is restricted to $M_Y \lesssim 5.5$ GeV.
An alternative method developed at ZEUS consists of tagging the scattered proton in the final state using a proton spectrometer (LPS). The spectrum of the scattered proton fractional momentum x_L for a DIS sample is shown in fig. 1(B). A clean selection of diffractive events is obtained by requiring $x_L > 0.97$, where the background contamination from processes such as π trajectory exchange or proton dissociation remains below 3%. The use of this method is limited by statistics due to the low spectrometer acceptance ($\sim$ 6%) [6].

2.1 The diffractive structure function

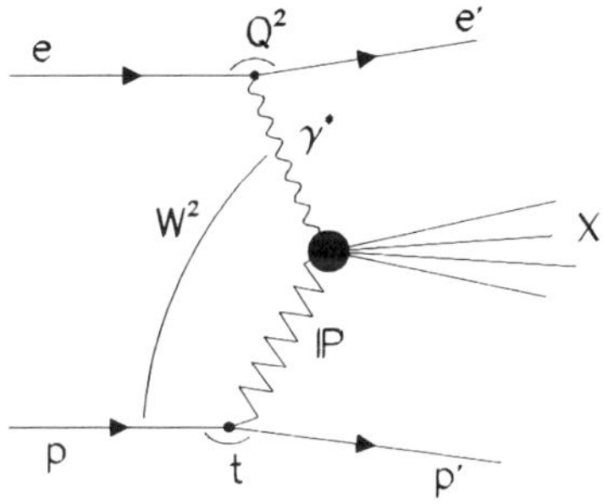

In the Ingelman-Schlein model [7], diffractive DIS is depicted as a two step process: first a pomeron, modeled as a colorless object made of partons, is emitted by the proton, then the photon interacts with a pomeron constituent (see figure). This approach, where the virtual photon probes directly the partonic structure of the pomeron, led to the concept of diffractive structure function F_2^D.
In addition to the variables used to describe inclusive DIS such as Q^2, W, x and y, new variables need to be introduced to account for the additional vertex in the diagram. These are: $x_{I\!P} = \frac{q\cdot(p-p')}{q\cdot p} \simeq \frac{M_X^2+Q^2}{W^2+Q^2}$ and $t = (p-p')^2 \simeq -p_T^2/x_L$, where M_X is the mass of the photon dissociated system, p_T and x_L are the transverse momentum and fractional momentum, respectively, of the scattered proton. With the above picture in mind, $x_{I\!P}$ can be interpreted as the fraction of the proton momentum carried by the pomeron and a scaling variable equivalent to Bjorken-x but relative to the pomeron momentum is defined as $\beta = \frac{Q^2}{2q\cdot(p-p')} = \frac{x}{x_{I\!P}} \simeq \frac{Q^2}{Q^2+M_X^2}$.

The diffractive structure function F_2^D is related to the four-fold differential

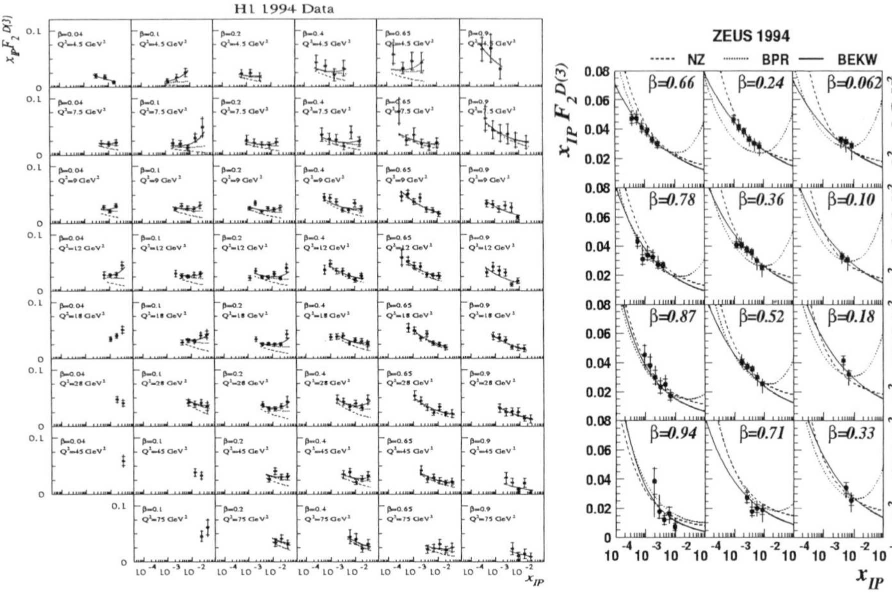

(A) (B)

Figure 2: **(A)**: *the diffractive structure function* $F_2^{D(3)}$ *multiplied by* $x_{I\!P}$ *as a function of* $x_{I\!P}$ *for fixed values of* β *and* Q^2 *as measured using H1 data. The solid line represents a fit in which the pomeron and the reggeon exchange contribute with maximum interference. The pomeron contribution alone is shown as a dashed line, the interference term is shown as a dotted line.* **(B)**: *the diffractive structure function* $F_2^{D(3)}$ *multiplied by* $x_{I\!P}$, *as a function of* $x_{I\!P}$ *for fixed values of* β *and* Q^2, *measured using the ZEUS data. The data are compared with the predictions of some of pQCD inspired models*[9].

cross section

$$\frac{d^4\sigma^D_{(ep\to e'Xp')}}{d\beta dQ^2 dx_{I\!P} dt} = \frac{4\pi\alpha^2}{\beta Q^2}(1 - y + \frac{y^2}{2})F_2^{D(4)}(\beta, Q^2, x_{I\!P}, t) \quad ,$$

where contributions from the longitudinal structure function and from Z° exchange have been neglected. In the Ingelman-Schlein approach, the diffractive structure function is expressed as the product of two terms $F_2^{D(4)} = f_{I\!P/p}(x_{I\!P}, t)F_2^{I\!P}(\beta, Q^2)$, where the first term depends on $x_{I\!P}$ and t and is called the *pomeron flux factor*. In the simple Regge model, the flux factor takes the same form as in hadron-hadron interactions and is usually expressed as

$f_{I\!P/p}(x_{I\!P},t) = N\beta^2_{I\!P/p}(t) x_{I\!P}{}^{1-2\alpha_{I\!P}(t)}$, where $\beta(t)$ is the function describing the coupling of the pomeron trajectory to the proton. We refer to this assumption as *Regge factorisation* at the proton vertex. The second term ($F_2^{I\!P}$) represents the *structure function of the pomeron.*

In most of the analyses the proton is not detected and t is not measured, so the t-integrated diffractive structure function $F_2^{D(3)} \propto \frac{1}{x_{I\!P}{}^a} F_2^{I\!P}(\beta, Q^2)$ is used. The exponent a is related to the the pomeron trajectory *averaged over t* through the relation $\overline{\alpha_{I\!P}} = 1 + a/2$. If Regge factorisation holds, a single exponent a independent of β and Q^2 should be able to describe the $x_{I\!P}$ dependence of the diffractive structure function.

The measurements of H1 [4], based on the data collected in 1994 (integrated luminosity of 1.96 pb^{-1}), cover the kinematic region $4.5 < Q^2 < 75$ GeV^2, $0.0002 < x_{I\!P} < 0.04$ and $0.04 < \beta < 0.9$ and are shown in fig. 2 (A) as $x_{I\!P} \cdot F_2^{D(3)}$, as a function of $x_{I\!P}$, for different values of β and Q^2. As shown in the figure, in the low β, high $x_{I\!P}$ region a breaking of factorisation is observed. The factorisation breaking is predicted by various models based on pQCD [9]. However, the data are well described by including the exchange of a reggeon in addition to the pomeron. A fit to the data yields $\alpha_{I\!P}(0) = 1.203 \pm 0.02(\text{stat.}) \pm 0.013(\text{syst.})^{+0.03}_{-0.035}(\text{mod.})$ and for the intercept of the secondary trajectory $\alpha_{I\!R}(0) = 0.50 \pm 0.11(\text{stat.}) \pm 0.11(\text{syst.})^{+0.09}_{-0.010}(\text{mod.})$.

The diffractive structure function $F_2^{D(3)}$ has also been measured by ZEUS, with the full 1994 DIS statistics (corresponding to 2.61 pb^{-1}) in the range $7< Q^2 <140$ GeV^2, $60< W <200$ GeV and for small masses ($M_X <15$ GeV) using the M_X method and the results are shown in fig. 2 (B).

In the small $x_{I\!P}$ range covered by the data ($x_{I\!P} \lesssim 0.01$), a good fit of the $x_{I\!P}$ dependence of $F_2^{D(3)}$ was obtained assuming a single power a in each β bin [8], giving support to the Regge factorisation at low $x_{I\!P}$ values corresponding to a t-averaged pomeron trajectory of $\overline{\alpha_{I\!P}} = 1.127 \pm 0.009(\text{stat.})^{+0.039}_{-0.012}(\text{sys.})$.

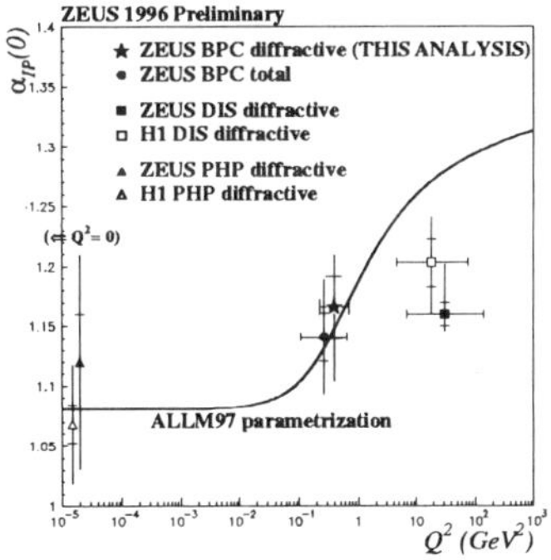

Figure 3: *Pomeron intercept as a function of Q^2 measured by the ZEUS and H1 experiments* [4 8 10 11 12]. *The curve is obtained from the ALLM97 parametrisation* [13]

Most recently the value $\overline{\alpha_{I\!P}}$ has been measured using ZEUS data in the kinematic region $0.22< Q^2 <0.7$ GeV^2 and $90< W <220$ GeV, yelding $\overline{\alpha_{I\!P}} = 1.133 \pm 0.026(\text{stat.})^{+0.043}_{-0.062}(\text{sys.})$ [10].

The average over the t-distribution gives $\alpha_{I\!P}(0) \sim \overline{\alpha_{I\!P}} + 0.03$. The values of $\alpha_{I\!P}(0)$ as a function of Q^2 measured at HERA [4 8 10 11 12] are shown in fig. 3; for $Q^2 > 0.2$ GeV^2 the value of the pomeron intercept appears larger than the value $\alpha_{I\!P}(0) \sim 1.08$ obtained from soft hadron-hadron interactions [14].

2.2 The partonic structure of the pomeron

The validity of the QCD factorisation theorem [15] was extended to diffractive DIS [16]. As a consequence, the use of the DGLAP evolution equation [17] to describe the scaling violations of $F_2^{D(3)}$ is justified and *diffractive* parton distributions of the proton can be derived in a way similar to the ordinary parton distributions of the proton are derived from the inclusive F_2. If, in addition, Regge factorisation is assumed, the pomeron structure function $F_2^{I\!P}$ can be expressed at LO in terms of parton densities *in the pomeron.* No momentum sum rule should be imposed because of the theoretical uncertainties on the normalisation of the pomeron flux factor and because it is not clear that such a rule is appropriate for parton distributions of a virtual exchange.

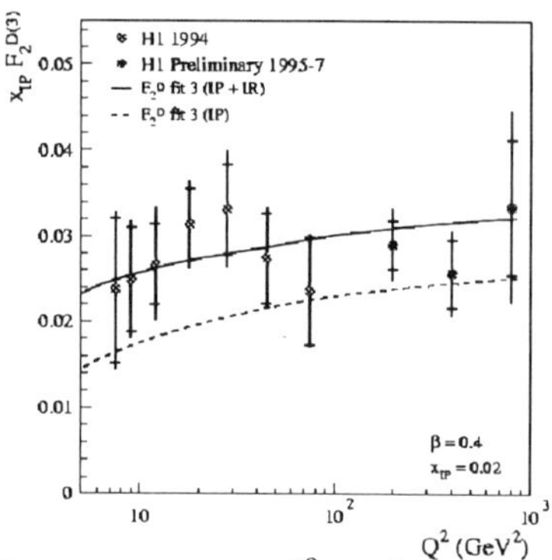

Figure 4: *Q^2 dependence of $x_{I\!P} F_2^{D(3)}$ at $\beta = 0.4$ and $x_{I\!P} = 0.02$. The results of a QCD fit to the data in which both quarks and gluons contribute are also shown.*

H1 has performed a QCD fit to $F_2^{D(3)}$ assuming both pomeron and reggeon flux factors as determined from the data in the $4.5 < Q^2 < 75$ GeV^2 range [4].

The measurement of $x_{I\!P} F_2^{D(3)}$ as a function of Q^2 for fixed $x_{I\!P} = 0.02$ and for fixed $\beta = 0.4$ values, compared with the expectations of the fit, is shown in fig. 4. The low Q^2 fit also describes the high Q^2 data [18].

The scaling violation observed [4] cannot be described by parametrisations containing only quarks; the best fit is obtained with a gluon distribution peaking at high values of the pomeron fractional momentum. The fraction of the pomeron momentum carried by gluons is found to be $> 80\%$ [4 18].

3 Leading Proton and Neutron Production

The production of leading baryons (LB) has been studied by ZEUS [19 20] and H1 [21] in a wide x_L range ($0.2 < x_L < 1$). These processes allow the study of the proton fragmentation region for low transverse momenta (p_T) carried by the LB ($p_T < 0.9$ GeV). Two different approaches exist. In the Regge models, the LB production is mediated by a colour singlet (pomeron, pion or reggeon) between the virtual photon and the incoming proton. If this mechanism is valid, the structure of the exchanged object may be studied. In the fragmentation approach, the LB simply emerge from the target fragmentation and tests of different fragmentation models can be performed.

The production rate of LB has been measured by ZEUS and is shown in fig. 5 [19].The production of LB in the range $0.2 < x_L < 0.9$ and scattering angle $\theta < 0.8$ mrad contributes to $\sim 12\%$ of the inclusive cross section. In contrast to the expectation of proton fragmentation models, the production of leading neutrons (LN) is significantly lower than that of leading protons (LP). As shown in fig. 5, the standard fragmentation models, like the colour dipole model, as implemented in the ARIADNE program [22], cannot describe the data in both shape and normalisation.

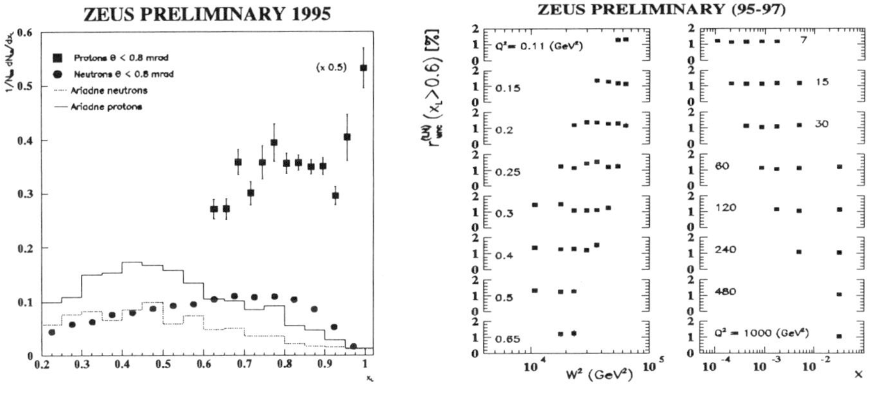

Figure 5: *Left: x_L spectrum for protons and neutrons with scattering angle $\theta < 0.8$ mrad compared to the predictions of the ARIADNE MC. Right: Uncorrected ratio of events with a LN and all inclusive DIS events for $x_L > 0.6$ and $\theta < 0.8$ mrad as a function of x and W^2 at fixed values of Q^2.*

The uncorrected ratio of events with a LN in the range $x_L > 0.6$ and $\theta < 0.8$ mrad and all inclusive DIS events (r_{unc}) as a function of W^2 and x in different Q^2 bins as measured by ZEUS is shown in fig. 5. The ratios are flat, showing that the LN production mechanism exhibits the same x and Q^2 behaviour as the inclusive DIS cross section. Similar conclusions can be obtained from the measurement of LP [20] made by ZEUS and from the measurement of LB structure functions performed by H1 [21].

3.1 The LB structure function

In analogy with the inclusive structure function F_2 the four-fold differential cross section for LB production may be written in terms of the semi-inclusive structure function $F_2^{LB(4)}$ as follows:

$$\frac{d^4\sigma}{dx dQ^2 dz dp_T^2} = 2\pi\alpha^2 x Q^2 (1 - y + \frac{y^2}{2}) F_2^{LB(4)}(x, Q^2, z, p_T^2).$$

where the fraction of the proton beam momentum carried by the LB is indicated as z. The LB structure function has been measured using the H1 data in

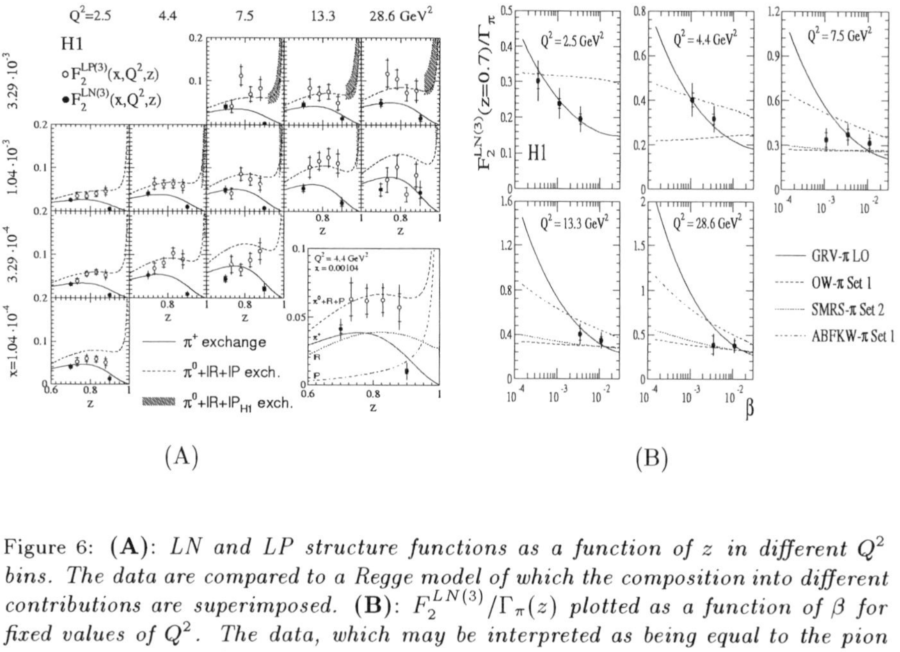

(A) (B)

Figure 6: **(A)**: *LN and LP structure functions as a function of z in different Q^2 bins. The data are compared to a Regge model of which the composition into different contributions are superimposed.* **(B)**: *$F_2^{LN(3)}/\Gamma_\pi(z)$ plotted as a function of β for fixed values of Q^2. The data, which may be interpreted as being equal to the pion structure function F_2^π, are compared to different parametrisations.*

the region $z \geq 0.7$ and $p_T < 0.2$ GeV [21] and compared to the predictions of a Regge model based on $I\!P$, $I\!R$ and π exchange. For LP production π°, $I\!P$ and f_2 exchange have been considered, while for LN only π^+ are assumed [3]. For F_2^π, the GRV-LO [23] parametrisation has been used and it is assumed that [24]

$F_2^{I\!R} = F_2^{\pi}$ and $F_2^{I\!P} = 0.22 F_2^{I\!R}$ at low β. The comparison of the structure function with the different contributions is shown in fig. 6 (A). The LN data can be described by π^+ exchange alone, whereas for the LP data, the *IR* exchange appears to contribute approximately 2/3 of the cross section; *IP* exchange appears to became important only in the diffractive peak.

Since within the Regge model the LN cross section is saturated by π-exchange only, the measurement of F_2^{LN} at $z = 0.7$ and the p_T-integrated π flux factor $\Gamma_\pi(z) = \int dp_T f_{\pi/p}(z, p_T)$ can be used to estimate the π structure function F_2^{π} at low β values: $F_2^{\pi} \sim F_2^{LN(3)}(\beta, Q^2, z = 0.7)/\Gamma_\pi(z)$.

The estimated F_2^{π} within this model is shown in fig. 6 (B) in a previously unexplored low β region. The GRV-LO parametrisation [23] is consistent with the data.

4 Summary

Diffractive interactions have recently attracted a lot of interest, as the soft pomeron no longer describes all the diffractive data measured at HERA. Regge factorisation appears not to hold; this seems to be explained as the effect of additional subleading exchanges. The resulting pomeron intercept is higher than that in soft hadronic interactions starting from $Q^2 > 0.2$ GeV^2.

While pomeron exchange is a well-defined concept in Regge phenomenology, its picture in terms of partons is still an open question. The HERA results show that it is possible to describe inclusive diffraction in DIS with a factorisable model in which a pomeron structure function evolves with Q^2 according to the DGLAP evolution equations. Although the precise shape of the gluon distribution is not well constrained by the present measurements, the data exhibit some preference for a substantial gluon component in the pomeron.

The semi-inclusive structure function of LB shows the same x, Q^2 and W dependence as the inclusive structure function indicating the projectile-target factorisation. Moreover the standard fragmentation Monte Carlo models are not able to describe the data. The LB production in the non-diffractive region seems to be described by Regge-inspired models. Within this framework, the pion structure function may be measured in the low β region from the measured F_2^{LN}.

References

1. See e.g.: K. Goulianos, *Phys. Rep.* **101**, 1969 (1983);
   G. Alberi and G. Goggi, *Phys. Rep.* **74**, 1 (1981).
2. N.N. Nikolaev and B.G. Zakharov, *Z. Phys.* C **49**, 607 (1991).

3. K. Golec-Biernat, J.Kwiecinski and A. Szczurek, *Phys. Rev.* D **56**, 1955 (1997).
4. H1 Collab., C.Adloff *et al.*, *Z. Phys.* C **76**, 613 (1997).
5. ZEUS Collab., M.Derrick *et al.*, *Z. Phys.* C **70**, 391 (1996).
6. ZEUS Collab., J.Breitweg *et al.*, *Eur. Phys. Jou.* C **1**, 81 (1998).
7. G. Ingelman and P. Schlein,*Phys. Lett.* B **152**, 256 (1985).
8. ZEUS Collab., J.Breitweg *et al.*, *Eur. Phys. Jou.* C **6**, 43 (1999).
9. N.N. Nikolaev and B.G. Zakharov, *Z. Phys.* C **53**, 331 (1992);
M. Genovese, N.N. Nikolaev and B.G. Zakharov, *JETP* **81**, 625 (1995);
J. Bartels, J. Ellis, H. Kowalski and M. Wüsthoff, *Eur. Phys. Jou.* C **7**, 443 (1999);
A. Bialas, R. Peschanski and C. Royon, *Phys.Rev.* D **57**, 6899 (1998).
10. ZEUS Collab., J.Breitweg *et al.*, Conf. Paper **500**, Inter. Eur. Conf. on HEP, Tampere (1999).
11. ZEUS Collab., M.Derrick *et al.*, *Z. Phys.* C **63**, 391 (1994);
H1 Collab., S. Aid *et al.*, *Z. Phys.* C **69**, 17 (1995).
12. ZEUS Collab., J.Breitweg *et al.*, *Eur. Phys. Jou.* C **7**, 609 (1999).
13. H. Abramowicz and A. Levy, DESY 97-251.
14. A. Donnachie and P.V. Landshoff, *Phys. Lett.* B **296**, 227 (1992).
15. J.C. Collins, D.E. Soper and G. Sterman, *Nucl. Phys.* B **261**, 104 (1985);
J.C. Collins, D.E. Soper and G. Sterman, *Nucl. Phys.* B **308**, 833 (1988);
G.T. Bodwin, *Phys. Rev.* D **31**, 2616 (1985) and D **34**, 3932 (1986).
16. J.C. Collins, *Phys.Rev.* D **57**, 3051 (1998) and references therein.
17. Yu.L. Dokshitzer, *JETP* **46**, 641 (1977);
V.N. Gribov and L.N. Lipatov, *Sov. Journ. Nucl. Phys.* **14**, 78 (1972);
G. Altarelli and G. Parisi, *Nucl. Phys.* B **126**, 298 (1977).
18. H1 Collab., C.Adloff *et al.*, Conf. Paper **571**, 29th Inter. Conf. on HEP, Vancouver (1998).
19. ZEUS Collab., J.Breitweg *et al.*, Conf. Paper **789**, 29th Inter. Conf. on HEP, Vancouver (1998).
20. ZEUS Collab., J.Breitweg *et al.*, Conf. Paper **501**, Inter. Eur. Conf. on HEP, Tampere (1999).
21. H1 Collab., C. Adloff *et al.*, *Eur. Phys. Jou.* C **6**, 587 (1999).
22. L. Lönnblad, *Comput. Phys. Comm.* **71**, 15 (1992).
23. M. Glück, E. Reya and A. Vogt, *Z. Phys.* C **53**, 651 (1992).
24. A. Szczurek, N.N Nikolaev and J. Speth, *Phys. Lett.* B **428**, 383 (1998).

STATUS OF MEASUREMENTS AND INTERPRETATION OF TOTAL REAL AND VIRTUAL PHOTON-PROTON CROSS SECTIONS

CH. AMELUNG
(for the ZEUS and H1 Collaborations)
Physikalisches Institut der Universität Bonn, Nußallee 12,
53115 Bonn, Germany
E-mail: amelung@physik.uni-bonn.de

The status of measurements and interpretation of the proton structure function F_2 is summarized. The measurements are subjected to DGLAP fits, from which the gluon density is extracted and a comparison to measurements of F_2^c is performed. The longitudinal structure function, F_L, is extracted via an extrapolation of F_2. At low Q^2, the transition from virtual to real photon exchange is studied.

1 Introduction

The proton structure functions carry information about the dynamics of quarks and gluons inside the proton, and they can be used to test the theory of the strong interaction. The measurement of the structure functions is an important part of the physics program at the *ep* collider HERA, and has been performed by the ZEUS and H1 collaborations.

1.1 Kinematics of Deep Inelastic Scattering

Inelastic positron-proton scattering, $e^+p \to e^+X$, can be described in terms of two kinematic variables, x and Q^2, where x is the Bjorken scaling variable and Q^2 the negative squared four-momentum transfer. In the absence of initial- and final-state radiation, $Q^2 = -q^2 = -(k-k')^2$ and $x = Q^2/(2P \cdot q)$, where k and P are the four-momenta of the incoming positron and proton, respectively, and k' is the four-momentum of the scattered positron. The fractional energy transfer to the proton in its rest frame, y, can be related to x and Q^2 by $Q^2 = sxy$, where $s = 4E_eE_p$ is the square of the positron-proton center-of-mass energy. Here, E_e (27.5 GeV) and E_p (820 GeV) are the positron and proton beam energies, respectively. For $Q^2 \ll M_Z^2$, positron-proton scattering is characterized by the exchange of a photon. The photon-proton center-of-mass energy squared is defined as $W^2 = Q^2(1-x)/x + m_p^2$.

The kinematic variables are closely related to the energy, E_e', and angle, θ, of the scattered positron, where θ is measured with respect to the positron

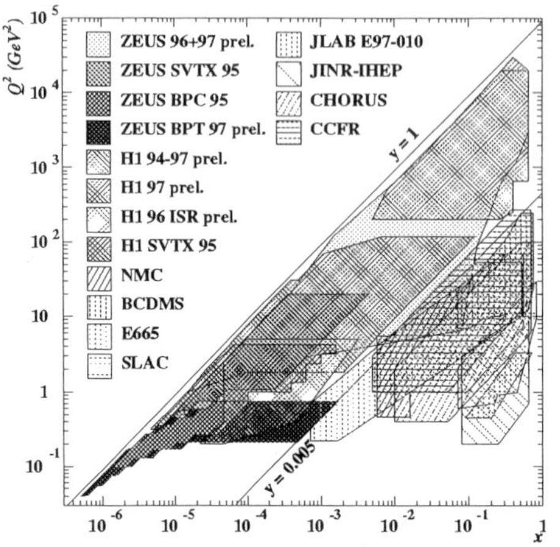

Figure 1: The range in the kinematic plane Q^2 vs. x covered by the HERA experiments and by various fixed-target experiments. In most parts of the Q^2 range, ZEUS and H1 data overlap with the fixed-target data. At low Q^2, the transition from $\gamma^* p$ to γp scattering can be studied. Measurements at medium Q^2 are used for studies and tests of pQCD. At high Q^2, electroweak effects become important and searches for new physics are performed.

beam direction: $y = 1 - E'_e/2E_e(1 + \cos\theta)$ and $Q^2 = 2E_e E'_e(1 - \cos\theta)$.

HERA has extended the previously accessible range in the kinematic variables x and Q^2 by two orders of magnitude, as shown in Fig. 1.

1.2 Cross Sections and Structure Functions

The double-differential cross section for inelastic $e^\pm p \to e^\pm X$ scattering can be expressed in terms of structure functions as

$$\frac{d^2\sigma(e^\pm p \to e^\pm X)}{dy\, dQ^2} = \frac{2\pi\alpha^2}{yQ^4}\left(Y_+ F_2 - y^2 F_L \mp Y_- xF_3\right)(1 + \delta_r), \tag{1}$$

where $Y_\pm = 1 \pm (1-y)^2$, and the radiative correction to the Born cross section, δ_r, is a function of y and Q^2, but to a good approximation independent of F_2 and F_L. From the measured quantity, the double-differential ep cross section, F_2 is extracted by making suitable assumptions about F_L and xF_3.

At $Q^2 \ll M_Z^2$, where xF_3 can be neglected, the ep cross section can also be expressed in terms of total cross sections for the absorption of transversely and longitudinally polarized virtual photons,

$$\frac{d^2\sigma(e^\pm p \to e^\pm X)}{dy\, dQ^2} = \Gamma\left(\sigma_T^{\gamma^* p} + \epsilon\, \sigma_L^{\gamma^* p}\right)(1 + \delta_r), \tag{2}$$

where $\Gamma = \alpha Y_+/(2\pi y Q^2)$, and $\epsilon = 2(1-y)/Y_+$. At low x, the photon-proton cross sections are related to the structure functions by

$$\sigma_{\rm tot}^{\gamma^* p} = \sigma_T^{\gamma^* p} + \sigma_L^{\gamma^* p} \approx 4\pi^2\alpha/Q^2 \cdot F_2 \quad \text{and} \quad \sigma_L^{\gamma^* p} \approx 4\pi^2\alpha/Q^2 \cdot F_L. \tag{3}$$

The so-called reduced cross section

$$\sigma_r = F_2 - \frac{y^2}{Y_+} F_L \mp \frac{Y_-}{Y_+} x F_3 \tag{4}$$

can also be defined, which differs from the double-differential cross section only by a kinematic factor, and can thus be extracted without assumptions about F_L and xF_3. In most of the kinematic region covered by the present analyses, $y^2 F_L$ and $Y_- xF_3$ are negligible, and thus $\sigma_r \approx F_2$.

In lowest-order QCD, F_2 can be expressed as the charge-weighted sum of quark and antiquark densities in the proton. While fixed-target experiments probe predominantly the valence-quark content of the proton, the HERA experiments are mainly sensitive to the sea contribution arising from gluon radiation off quarks and subsequent gluon splitting into quark-antiquark pairs. Only at very high Q^2 do the HERA experiments also probe the valence quarks.

2 Structure Functions at High and Medium Q^2

The measurement of F_2 is an inclusive cross-section measurement using the angle and energy of the detected scattered positron. Good detection efficiency and purity require both tracking and calorimetric energy measurements. The ZEUS F_2 measurement at medium and high Q^2 used the uranium-scintillator calorimeter and the central tracking detector. The acceptance of the latter, however, is limited to $16^\circ < \theta < 165^\circ$. The H1 collaboration upgraded their detector in 1997 to improve the positron detection at small scattering angles by adding four rear discs to their microvertex detector (Backward Silicon Tracker, BST), which extend the tracking acceptance down to 3°.

2.1 Measurement of F_2

Figure 2 shows an overview of F_2 measurements at low and medium Q^2. In selected kinematic regions, ZEUS and H1 have achieved F_2 results with a precision of 1% (stat.) and 3% (sys.), comparable to the typical precision of fixed-target experiments.

The measured F_2 can be well described by a form $F_2 \propto x^{-\lambda}$. The quantity λ decreases as Q^2 decreases, corresponding to the rise of F_2 becoming less steep. As can be seen from Eq. (3), the rise of F_2 with x corresponds to an energy dependence of $\sigma_{\rm tot}^{\gamma^* p}$ of $\sigma_{\rm tot}^{\gamma^* p} \propto W^{2\lambda}$.

Two alternative and complementary theoretical frameworks are used to interpret these data. At medium to high Q^2, i.e. in hard interactions, pQCD is applicable, because α_s is small. At low Q^2, i.e. in soft interactions, α_s is

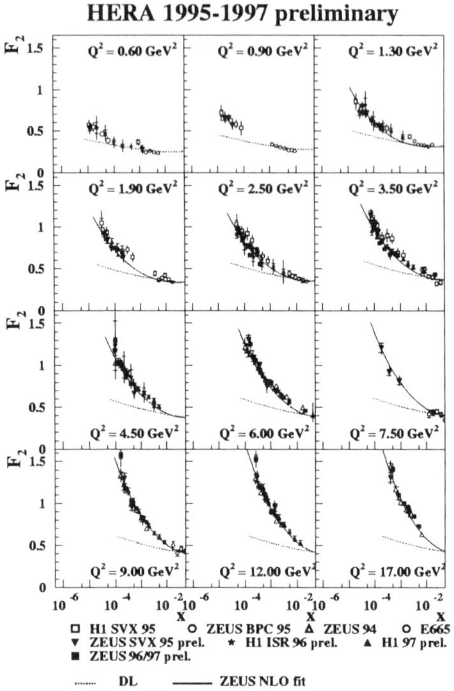

Figure 2: Overview of F_2 measurements at low and medium Q^2. The symbols denote measurements from ZEUS, H1 and E665 in different parts of the kinematic range and by special analysis techniques (SVX: shifted vertex, ISR: events with initial state radiation, BPC: using a special calorimeter close to the beam line). The solid line is the result of a DGLAP fit, the dotted line a Regge-type parameterization by Donnachie and Landshoff [1].

large, and perturbative calculation techniques break down. This is the domain of Regge theory.

2.2 DGLAP Fits

The DGLAP equations [2] predict the evolution of parton densities with Q^2 (but make no statement about the x dependence). They relate the derivative of gluon and quark densities with Q^2 to the convolution of gluon and quark densities with the splitting functions P_{ij}. The latter are calculable in pQCD and are a measure of the probability to find a parton i radiated by a parton j as a function of Q^2.

In order to compare the Q^2 evolution of parton densities as predicted by the DGLAP equations to that of the measured F_2 data, the gluon, sea quark and valence quark densities are parameterized by a suitable form at a starting scale Q_0^2. The parton densities are evolved to other values of Q^2, and F_2 is calculated from the quark densities. The free parameters of the parton density parameterizations are adjusted such that the resulting F_2 parameterization fits

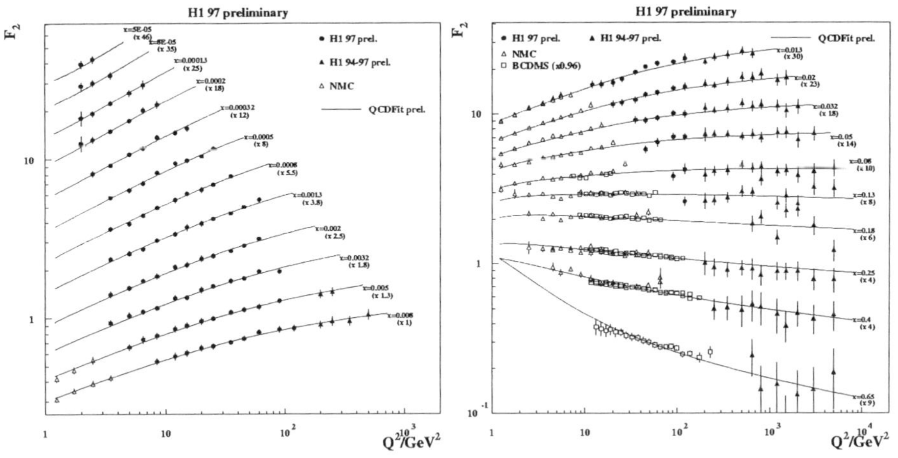

Figure 3: The scaling violations of F_2 as measured by H1 and the fixed-target experiments NMC and BCDMS, displayed as F_2 vs. Q^2. The solid line is the result of a DGLAP fit, which is able to describe F_2 over four orders of magnitude in Q^2.

the data above a cut-off value of $Q^2_{\rm min}$.

Figure 3 shows the result of such a fit to F_2 data from H1. The value of F_2 increases at small x when increasing Q^2, while it decreases simultaneously at large x. The latter is a consequence of valence quarks radiating gluons and thus losing momentum, while the former results from the radiated gluons splitting into low-momentum quark-antiquark pairs. Scaling, i.e. Q^2-independence of F_2, can be seen only at $x \approx 0.1$. The fit describes the measured F_2 values well over four orders of magnitude in Q^2, which is an impressive confirmation of QCD. The fit does not deviate from the data down to the (surprisingly low) value of $Q^2 = 1\,\mathrm{GeV}^2$, where the application of pQCD is somewhat questionable.

2.3 The Gluon Density

The gluon density can be extracted from a DGLAP fit to measured F_2 data. The result from ZEUS is shown in Fig. 4. Much effort has been put into a reliable determination of the precision of this extraction, which is 10%–15%. A comparison of the value of $xg(x, Q^2)$ to the value of $F_2(x, Q^2)$ at low x shows that the overwhelming fraction of partons in the proton at low x are gluons.

Like the quark densities, the gluon density increases with Q^2. However, when going down to Q^2 values as low as $1\,\mathrm{GeV}^2$, the gluon density from the fit becomes compatible with zero at $x < 10^{-2}$, while the sea quark density

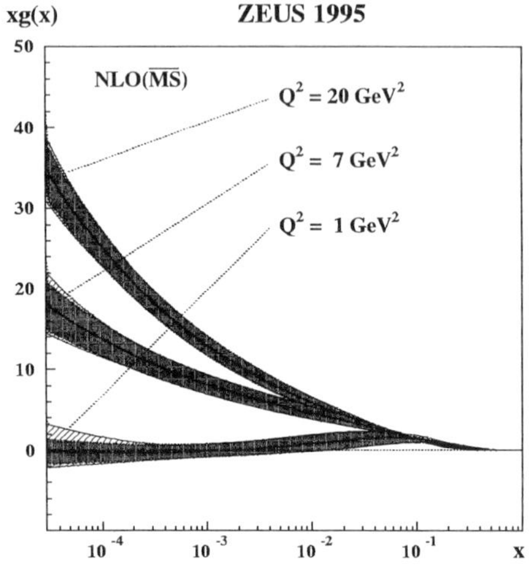

Figure 4: The gluon density resulting from a DGLAP fit to F_2 data from ZEUS, for three different values of Q^2. The shaded bands show the uncertainty of $xg(x, Q^2)$. At $Q^2 = 1\,\text{GeV}^2$, the gluon density as extracted from the fit becomes compatible with zero at $x < 10^{-2}$.

remains positive and rises up to the smallest accessible values of x. This seems to be in contradiction to the concept that quarks at low x are produced by gluon splitting, and may indicate a problem in applying pQCD at such low values of Q^2.

2.4 Measurement of F_2^c

The charm structure functions F_i^c are defined in analogy to (1) with the requirement that the final state contains a $c\bar{c}$ pair:

$$\frac{d^2\sigma(e^\pm p \to e^\pm c\bar{c}X)}{dy\,dQ^2} = \frac{2\pi\alpha^2}{yQ^4}\left(Y_+F_2^c - y^2F_L^c \mp Y_-xF_3^c\right)(1+\delta_r). \quad (5)$$

At HERA, charm is generated predominantly in the photon-gluon fusion process $\gamma g \to c\bar{c}$, and is detected by reconstructing the fragmentation products of the c quark, usually in the channel $D^* \to D^0\pi_{\text{slow}}$, where the D^0 decays further to $K\pi$, $K\pi\pi\pi$, or semi-leptonically. At the current level of precision, F_L^c and F_3^c can safely be neglected. In contrast to the fully inclusive F_2 measurements, further input is needed, e.g. the fragmentation function for $c \to D^*$ and the $D^* \to D^0\pi$ branching ratio. In addition, the event selection efficiency is quite low, because experimentally only a limited kinematic region of the D^* is accessible. Therefore, a Monte Carlo simulation (based on the measured inclusive F_2) has to be used to extrapolate to the full kinematic region.

The result of the extraction of F_2^c by ZEUS is shown in Fig. 5. The charm structure function F_2^c exhibits a steeper rise toward low x than the fully

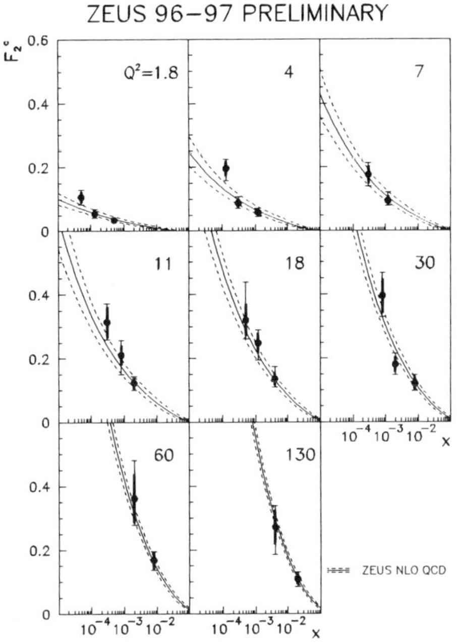

Figure 5: The charm structure function F_2^c as a function of x in bins of Q^2, measured by ZEUS. The solid line shows the central value of F_2^c as calculated from a DGLAP fit to F_2 data, which did not include the F_2^c measurements shown here. The dashed lines indicate the uncertainty of the fit result. In principle, the observed agreement of the measured F_2^c with that extracted from a DGLAP fit to the inclusive F_2 is an important consistency check of the DGLAP formalism; however, the significance of this observation is reduced here due to the extrapolation technique making use of the measured inclusive F_2.

inclusive F_2, and stronger scaling violations. The charm content of the proton, F_2^c/F_2, rises from about 10% at $Q^2 = 1.8\,\text{GeV}^2$ to 25% at $130\,\text{GeV}^2$.

2.5 Extraction of F_L

The most direct method to measure F_L requires a variation of the beam energies, which has not been done yet. However, F_L can be extracted from a comparison of the measured reduced cross section σ_r and an assumption about the behavior of F_2, which is derived, for example, from a DGLAP fit to σ_r in the region of small y (where $\sigma_r \approx F_2$, because the influence of F_L is negligible), and subsequent extrapolation to higher values of y. Then, F_L is obtained from $F_L = Y_+/y^2(F_2^{\text{fit}} - \sigma_r^{\text{measured}})$.

Figure 6 shows the result of such an extraction of F_L by the H1 collaboration. A measurement of F_L allows to make another consistency check of the DGLAP formalism, by comparing the measured F_L to that predicted from the parton densities in the fit. The method is experimentally challenging, because the region of high sensitivity to F_L coincides with a region where backgrounds are high and the quality of the positron detection degrades. In addition, it relies on assumptions made in the extrapolation.

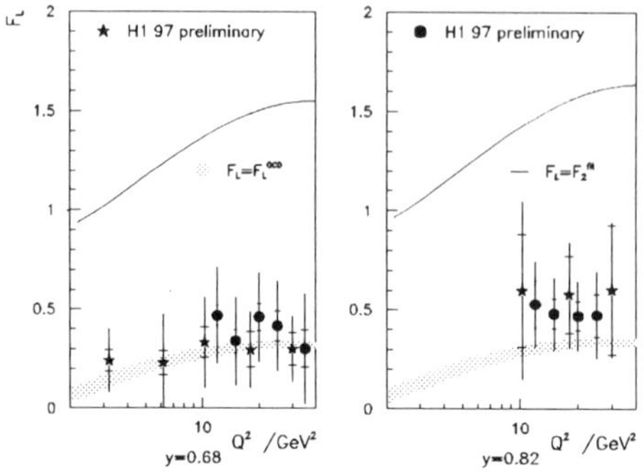

Figure 6: Extracted F_L from a comparison of the measured reduced cross section to an extrapolation of F_2, displayed as F_L vs. Q^2 in two bins of y. Circles and stars denote results from two different extrapolation methods. Systematic errors for points in the same y bin are correlated. The shaded band indicates F_L as calculated from the result of a DGLAP fit to F_2, the solid line the upper limit $F_L = F_2$ (the lower limit is $F_L = 0$).

3 Total Photon-Proton Cross Sections at Low Q^2

In order to study experimentally the region of very low Q^2, i.e. significantly below $1\,\mathrm{GeV}^2$, special detectors for the scattered positron close to the beam line are necessary. The ZEUS detector was upgraded by the installation of the Beam Pipe Calorimeter (BPC) in 1995 and the Beam Pipe Tracker (BPT) in 1997, a small electromagnetic sampling calorimeter and a tracking detector of two silicon microstrip planes, detecting scattered positrons at scattering angles as low as 1.1°–1.9°, thus making the region $0.045\,\mathrm{GeV}^2 < Q^2 < 0.65\,\mathrm{GeV}^2$ accessible. H1 installed the VLQ (Very Low Q^2) detector, a device very similar to BPC and BPT, in 1998, and is working on the analysis of the data.

In Fig. 7 the ZEUS results of the F_2 measurement with the BPC and BPT, converted to total virtual photon-proton cross sections, are shown. The data are compared to two parameterizations by Donnachie and Landshoff (DL [1], DL98 [3]), to ALLM97 [4] and to a ZEUS Regge fit [5] of the measured data which resembles DL. The rise of $\sigma_{\mathrm{tot}}^{\gamma^* p}$ with W^2 is compatible with a soft behavior for Q^2 up to $0.5\,\mathrm{GeV}^2$, and becomes steeper at higher Q^2. At the low Q^2 values of this analysis, $\sigma_{\mathrm{tot}}^{\gamma^* p}$ is nearly independent of Q^2 (in contrast, it falls like $1/Q^2$ at higher Q^2). This can be exploited to constrain the extrapolation $\sigma_{\mathrm{tot}}^{\gamma^* p} \rightarrow \sigma_{\mathrm{tot}}^{\gamma p}$.

In order to extrapolate the measured values of $\sigma_{\mathrm{tot}}^{\gamma^* p}$ to $Q^2 = 0$, an assumption about their Q^2 dependence is taken from the continuum part of the GVDM [6] prediction on σ_T: $\sigma_{\mathrm{tot}}^{\gamma^* p}(W^2, Q^2) = m_0^2/(m_0^2 + Q^2)\sigma_{\mathrm{tot}}^{\gamma p}(W^2)$ (fitting instead both the σ_T and σ_L terms changes the extrapolated values only within their statistical errors).

The extrapolations are compared to the two direct measurements from ZEUS and H1 and to data at lower energies in Fig. 8, together with two Regge-type fits of the form $\sigma_{\mathrm{tot}}^{\gamma p}(W^2) = A_{\mathrm{I\!R}} W^{2(\alpha_{\mathrm{I\!R}}-1)} + A_{\mathrm{I\!P}} W^{2(\alpha_{\mathrm{I\!P}}-1)}$, and with the DL98 and ALLM97 parameterizations. The extrapolated cross sections have

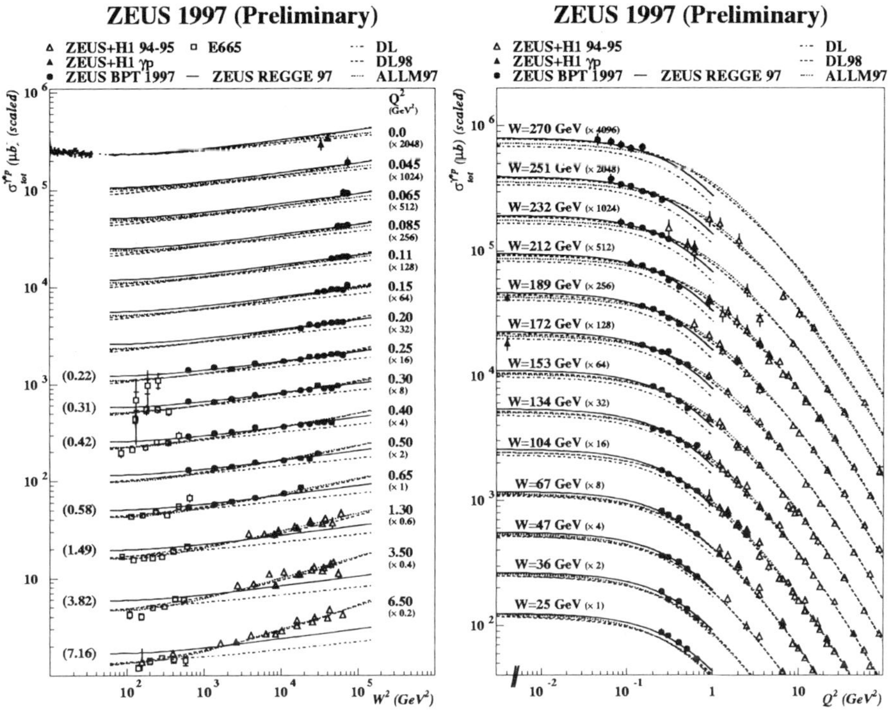

Figure 7: Measured values of $\sigma_{\rm tot}^{\gamma^* p}$, plotted vs. W^2 (left) respectively Q^2 (right). Filled circles denote results from the ZEUS analysis of BPC/BPT data, filled triangles are HERA measurements at $Q^2 = 0$, open triangles represent measurements at higher Q^2, and squares denote results from E665. The data are compared to the parameterizations DL, DL98, and ALLM97, as well as to a ZEUS Regge fit of the measured data.

a tendency to be slightly higher than the directly measured ones. The model-dependence is estimated from extrapolating with the Q^2 dependence of DL, DL98 and ALLM97 (instead of GVDM), and found to be quite large at small values of W^2.

4 Conclusions

The HERA collaborations ZEUS and H1 have delivered a wealth of precise data on the proton structure function F_2. At $Q^2 > 1\,\mathrm{GeV}^2$, the data can be described in a self-consistent way by DGLAP fits. At Q^2 values below $1\,\mathrm{GeV}^2$, the transition region from the perturbative to the non-perturbative regime has been mapped. Regge-inspired models are used to describe the data here.

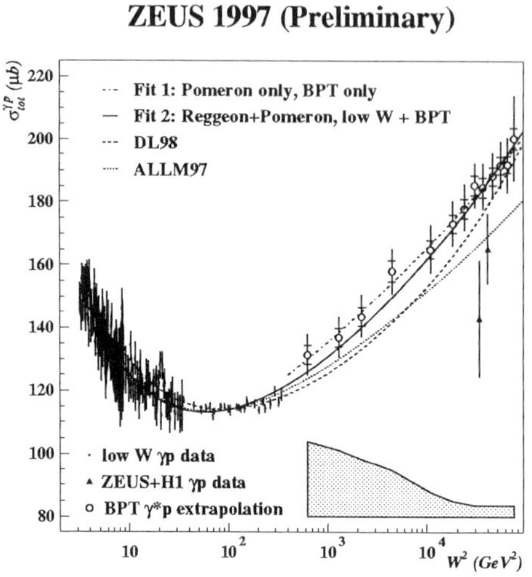

Figure 8: Extrapolated $\sigma_{\rm tot}^{\gamma p}$ versus W^2 from the ZEUS analysis of BPC/BPT data, compared to the two direct measurements from ZEUS and H1 and to data at lower energies. The lines denote two Regge-type fits to the data, and the DL98 and ALLM97 parameterizations. The shaded band is an estimate of the magnitude of the model-dependence due to the assumed Q^2 dependence.

Acknowledgments

I would like to thank the organizers for a conference full of interesting experiences. It is a pleasure to acknowledge the outstanding effort of many colleagues in the H1 and ZEUS collaborations who contributed to the results presented here. I thank E. Hilger and B. Foster for careful reading of the manuscript. This work was supported by a grant from the Bundesministerium für Wissenschaft und Forschung in Germany.

References

1. A. Donnachie and P.V. Landshoff, Z. Phys. **C 61** (1994) 139.
2. V.N. Gribov and L.N. Lipatov, Sov. J. Nucl. Phys. **15** (1972) 438;
   L.N. Lipatov, Sov. J. Nucl. Phys. **20** (1975) 96;
   G. Altarelli and G. Parisi, Nucl. Phys. **B 126** (1977) 298;
   Y.L. Dokshitzer, Sov. Phys. JETP **46** (1977) 641.
3. A. Donnachie and P.V. Landshoff, Phys. Lett. **B 437** (1998) 408.
4. H. Abramowicz and A. Levy, DESY Report 97-251.
5. ZEUS Collaboration, J. Breitweg et al., Eur. Phys. J. **C 7** (1999) 609.
6. J.J. Sakurai and D. Schildknecht, Phys. Lett. **B 40** (1972) 121.

M. N. Kienzle-Focacci

TWO-PHOTON COLLISIONS AT LEP

M. N. KIENZLE-FOCACCI
Département de Physique Nucléaire et Corpusculaire, 24 quai Ernest-Ansermet, CH-1211 GENEVE 4, Switzerland
E-mail: maria.kienzle@cern.ch

Our knowledge of two-photon processes largely increased during the last years thanks to the large samples of $e^+e^- \to e^+e^-$hadrons events obtained at LEP. The two-photon total cross-sections are studied for real and for virtual photons by tagging the scattered electrons. Inclusive c-quark and b-quark production and the differential cross sections of jets and single particle production are a test of NLO perturbative QCD calculations. Two-photon resonance formation gives an insight on the internal structure of the mesons and provides a test for gluonium candidates.

1 Introduction

In recent years hadron production has been studied in a variety of channels in two-photon collisions at LEP (LEP1, $\sqrt{s} \simeq 91$ GeV, and LEP2, $\sqrt{s}>$ 160 GeV). The integrated luminosity available to each of the four LEP experiments at the various energies is indicated in Fig.1. As illustrated in the figure, at LEP2, the two-photon interaction is the main source of hadronic events. It is therefore possible to increase our knowledge on the two-photon processes. This is highly desirable in order to test perturbative and non-perturbative QCD calculations and to better understand the background in new particles searches and at future e^+e^- colliders.

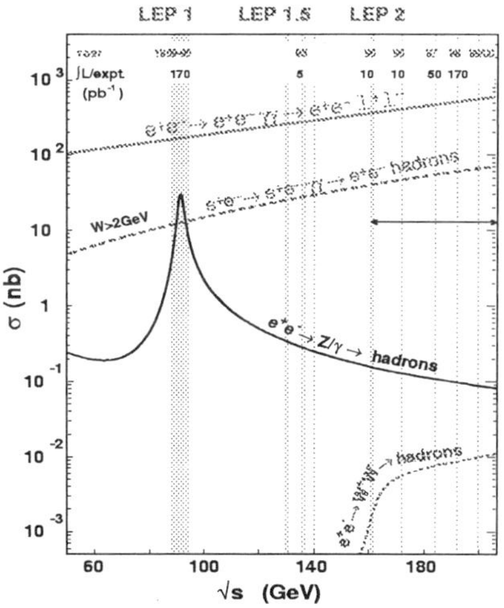

Figure 1: Cross-sections of the main hadronic channels as a function of the LEP energy. The vertical bands indicate the beam energies and the integrated luminosities for each year of data taking.

The Monte Carlo models play a central role in this study. Two generators, PHOJET [1] and PYTHIA [2], aim to a good description of the two-photon interactions. At the present level of theoretical understanding, a perturbative QCD contribution is added to a non perturbative diffractive phenomenology. The interacting photon is considered as a superposition of a *bare* photon and an *hadronic* photon

$$\mid \gamma >= \sqrt{Z_3} \mid \gamma_{bare} > + \frac{e}{f_{q\bar{q}}} \mid q\bar{q} > + \frac{e}{f_{q\bar{q}^*}} \mid q\bar{q}^* > \tag{1}$$

whith $Z_3 = 1 - (e/f_{q\bar{q}})^2 - (e/f_{q\bar{q}^*})^2$. The *bare* photon interacts in a point-like way. For the *hadronic* component it is assumed that the photon fluctuates into a $q\bar{q}$ pair which interacts hadronically. Due to the quantum numbers of the photon, it is expected that the $q\bar{q}$ pairs predominantly form particles with small masses like ρ, ω, ϕ vector mesons (VDM). The hadronic interactions are separated into hard ($q\bar{q}^*$) and soft interactions by a transverse momentum cutoff p_t^{cut}. The subprocesses involving transverse momenta greater than this scale can be calculated by perturbative QCD. The $\gamma^*\gamma^* \rightarrow$ *hadrons* cross section is therefore composed by a "soft" interaction, which can be described by hadronic phenomenology, like the Regge poles, a point-like two-photon reaction, $\gamma\gamma \rightarrow q\bar{q}$ (*direct* process), which is calculable in QED and an "hard component" which implies the knowledge of the quark and gluon distributions (pdf) inside the photon (*resolved* processes). In order to describe the data the models have to include *i)* a consistent transition between low p_t and high p_t events, *ii)* a description of elastic and diffractive reactions like $\gamma\gamma \rightarrow$ VV, *iii)* a description of the structure of the photon (pdf) and of the primordial k_T distribution of the partons inside the photon, *iiii)* a unitarization scheme for the cross sections. In the PHOJET model γp and pp data are used to fix the couplings of the Pomeron and of the Reggeon to the $q\bar{q}$ fluctuation of the photon. The cross sections are then calculated in the framework of a Dual Parton Model, with the unitarization constraint. The model [3] of Schuler and Sjöstrand, implemented in PYTHIA, aims to a smooth superposition of hadron-like and point-like photon interactions. In the following the predictions of the two models will be compared to the LEP results. The data will be also compared to QCD theoretical calculations.

2 Two-photon cross-sections

2.1 Total $\gamma\gamma$ cross-section

At LEP2 the dominant two-photon process $e^+e^- \rightarrow e^+e^-$ *hadrons* is easily separated from the annihilation channels (Fig.2). The analysis can be performed with a low background in a wide range of the two-photon center of mass energy, $W_{\gamma\gamma}$. The total cross-section of two real photons has been measured by L3 [4] and OPAL [5] in the range 5 GeV $< W_{\gamma\gamma} <$ 145 GeV. The value of $W_{\gamma\gamma}$ is

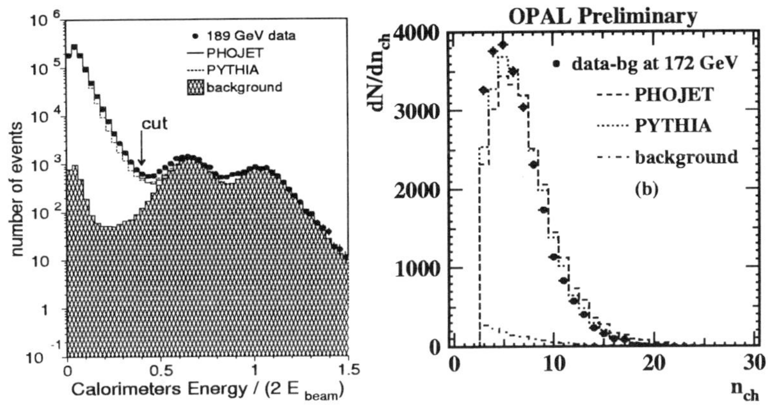

Figure 2: (left) Total energy in the electromagnetic and hadronic calorimeters of the L3 detector, normalised to the centre-of-mass energy. The two-photon events, below the cut, are selected with a low background from $e^+e^- \to e^+e^- \tau^+\tau^-$ and annihilation events. (right) Distribution of charged tracks compared to Monte Carlo predictions, for two-photon events selected by OPAL.

obtained from the effective mass of the measured hadrons, $W_{\rm vis}$, by unfolding the detector acceptance, which is rather poor in the diffractive region, near to the beam direction. The differential cross section $d\sigma/dW_{\gamma\gamma}$, as measured by L3 at different collision energies, is shown in Fig.3. The fast decrease of the cross section as a function of $W_{\gamma\gamma}$ is due to the two-photon luminosity function, $\mathcal{L}_{\gamma\gamma}$, which depends on $\tau = W^2_{\gamma\gamma}/s$. To extract the total cross section of two real photons the photon flux $\mathcal{L}_{\gamma\gamma}$ is calculated [6] and the cross section is extrapolated to $Q^2 = 0$ (Fig.3 right). The unfolding procedure introduces a strong bin-to-bin correlation which must be taken into account when comparing the data to different models. The two generators, employed to correct the data, PYTHIA and PHOJET, give a difference of ~20% in the absolute scale of $\sigma_{\gamma\gamma}$. The measurements of L3 and OPAL agree if the same Monte Carlo is used (Fig.3 right). The two-photon cross-section shows the typical behaviour [7] of hadron-hadron total cross-sections: a fast decrease for $W_{\gamma\gamma} < 20$ GeV, due to reggeon exchange, and a slow increase at higher energies, described by a "soft" Pomeron. The rise of $\sigma_{\gamma\gamma}$ with $W_{\gamma\gamma}$ is faster than the one observed in hadron-hadron or γp cross-sections [8]; a simple factorisation ansatz [9] $\sigma_{\gamma\gamma} = \sigma^2_{\gamma p}/\sigma_{pp}$ is excluded (see the curve from Schuler and Sjöstrand [3] in Fig.3). A fit to the L3 data with the parametrisation $\sigma_{\gamma\gamma} = A \times W_{\gamma\gamma}^{2(\alpha_P(0)-1)}$, for $W_{\gamma\gamma} >$

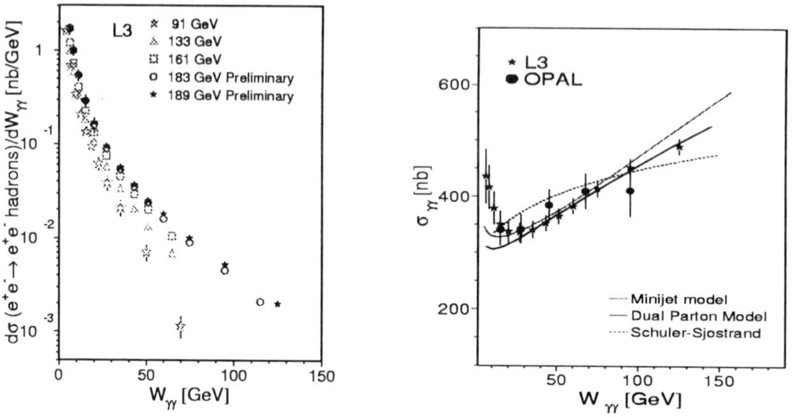

Figure 3: (left) The differential cross section $d\sigma/dW_{\gamma\gamma}$ measured at various beam energies. (right) Two-photon total cross-section as measured by L3 [4] and by OPAL [5], both unfolded with PHOJET. The minijet model [11] and the Dual Parton model [1] reproduce well the data, while the $W_{\gamma\gamma}$dependence predicted by Schuler and Sjöstrand [3] does not agree with the data.

40 GeV, gives a value $(\alpha_P(0) - 1) = 0.14 \pm 0.02$ [a] to be compared with the universal value[8] of 0.095 ± 0.002. Many theoretical predictions [10] exist , the rise of $\sigma_{\gamma\gamma}$ is rather well described by most of the models which take into account the direct coupling of the photon to $q\bar{q}$ as, for example, the minijet model [11].

2.2 Diffractive processes

The different values of $\sigma_{\gamma\gamma}$ obtained by correcting the data with PYTHIA and PHOJET may be related to the description of the diffractive component (Fig.4 left) for which the detectors acceptance is small. L3 has investigated the diffractive contribution by studying exclusive and inclusive ρ^0 production [4] (Fig.5). In the limited kinematical range accessible to the experiment, 3 GeV $< W_{\gamma\gamma} <$ 8 GeV, both generators fail to reproduce the data (Fig.4 right): PHOJET predicts too few ρ^0 events while the almost constant production in $W_{\gamma\gamma}$, predicted by PYTHIA, is not observed. More data are necessary to investigate these processes.

---

[a] The numerical value of $\alpha_P(0)$ depends on the parametrisation used for the fit and on the chosen $W_{\gamma\gamma}$ interval [4].

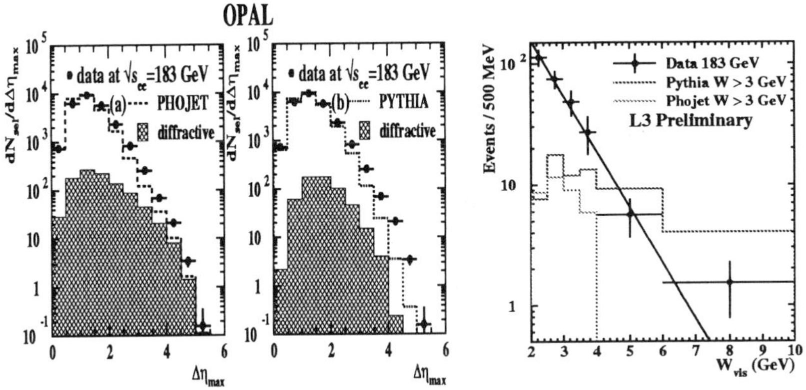

Figure 4: (left) The diffractive component, shown for the distribution of the maximum rapidity gap between any two hadrons, has a different shape in PYTHIA and PHOJET. (right) Number of $\gamma\gamma \to \rho^0 X$ events as a function of $W_{\rm vis}$ compared to PYTHIA and to PHOJET predictions, normalised to the luminosity of the data.

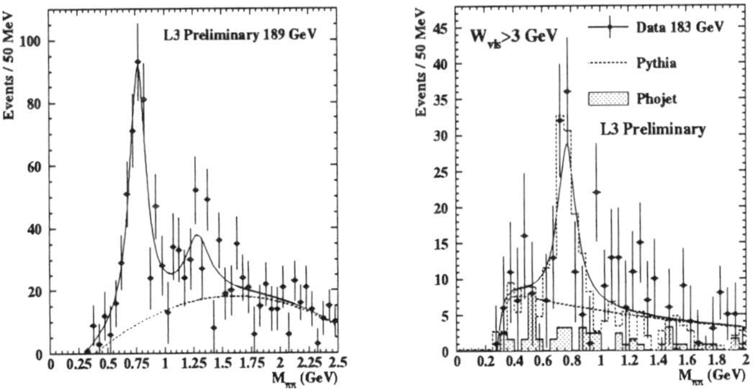

Figure 5: Effective mass of two oppositely charged pions, after subtraction of background estimated from the wrong sign combinations, for two-photon events with $W_{\rm vis} > 3$ GeV (left) in the exclusive channel $e^+e^- \to e^+e^-\pi^+\pi^-\pi^+\pi^-$ and (right) in inclusive $e^+e^- \to e^+e^-\rho^0$ production.

2.3 Double tag measurements

When both scattered electrons are detected, the $W_{\gamma\gamma}$ mass can be measured from the four-momenta of the two scattered electrons, thus strongly reducing the model dependence of the cross-section measurement. The cross-section of $\gamma^*\gamma^*$ collisions has been measured at LEP by L3 [12] and OPAL [13] in the range of $3\ \mathrm{GeV}^2 < Q^2_{1,2} < 37\ \mathrm{GeV}^2$. Since the two photons are highly virtual and, unlike the proton, they do not contain constituent quarks with an unknown density distribution, one may hope to have a complete perturbative QCD calculation under particular kinematical conditions. A recent QCD approach is based on

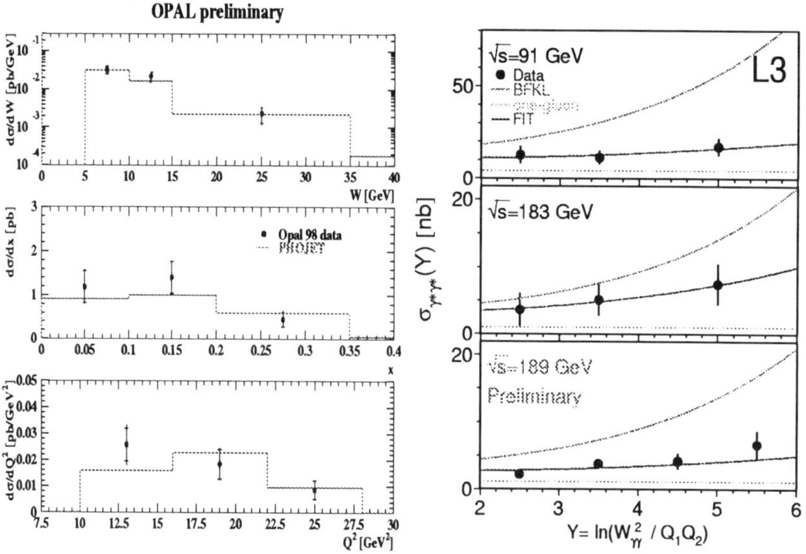

Figure 6: (left) The differential cross-section of double tag events compared to PHOJET Monte Carlo predictions [13] and (right) the two photon cross-sections at LEP1 and LEP2 compared to LO-BFKL calculations after subtraction of the direct contribution [12].

the BFKL equation [14]. The $\gamma^*\gamma^*$ interaction is described as the interaction of two $q\bar{q}$ pairs scattering off each other via multiple gluon exchange. In this scheme the cross-section for the collision of two virtual photons [15,16] depends upon the "hard Pomeron" intercept $\alpha_P - 1 = 0.53$, at the LO, whereas in NLO the BFKL contribution is calculated to be smaller, $\alpha_P - 1 \simeq 0.17$ [17]. The results from L3 and OPAL (Fig.6) are well described by PHOJET, which uses the GRV-LO parton density [18] in the photon and leading order perturbative QCD. The LO BFKL calculations, shown in Fig.6 (right) with dotted lines, are too high. By leaving α_P as a free parameter in the LO formula [15], a combined fit to the L3 data gives a value of $\alpha_P - 1 = 0.29 \pm 0.025$ with χ^2/d.o.f =7/9.

2.4 Inclusive c and b production

Inclusive heavy quark production is well suited to study hard interactions, the high mass of the quark providing a natural hard scale for perturbative QCD calculations [19].

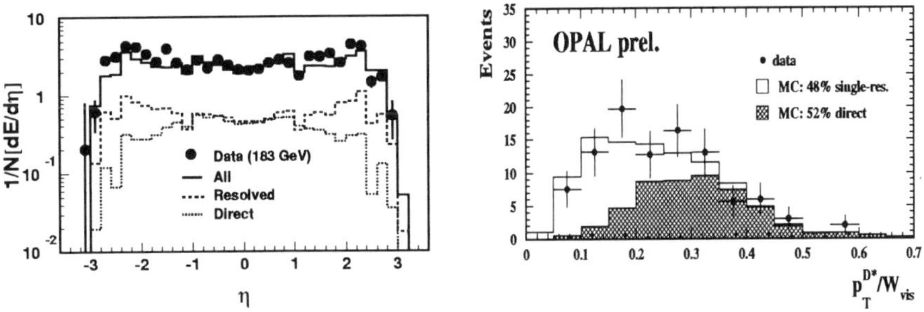

Figure 7: (left) The rapidity distribution of hadrons in events with an electron identifying the heavy-quark semileptonic decay. The data are compared to PYTHIA predictions. (right) The p_T distribution of the D^* normalised to the visible mass of the event. Both direct and resolved contributions are necessary to reproduce the shape of the data.

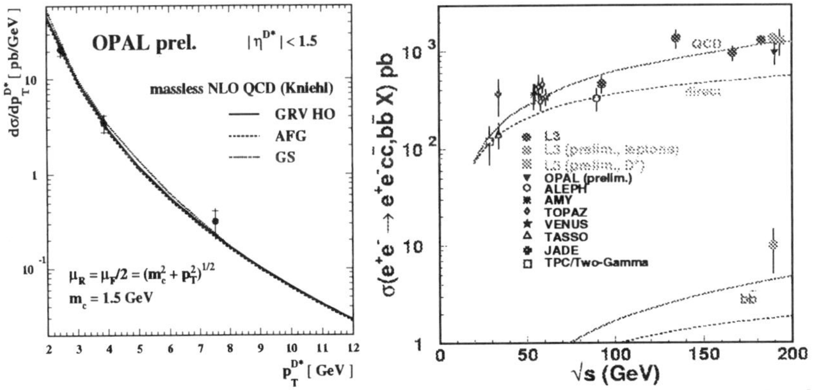

Figure 8: (left) The differential cross-section for inclusive D^* production, compared to NLO QCD calculations [22]. (right) The cross-section for heavy quarks production as measured at LEP and at previous e^+e^- colliders.

At LEP2 energies the direct and resolved processes contribute equally to the cross-section. The resolved contribution is dominated by quark-gluon fusion, $\gamma g \rightarrow q\bar{q}$, thus this measurement provides a strong constraint for the determination of the gluon content in the photon. L3 [20] tags the presence of a charm-quark by identifying a lepton (electron or muon) in the event or by recostructing a $D^* \rightarrow D^0\pi$ in the decay chain $D^0 \rightarrow K^-\pi^+$ and $D^0 \rightarrow K^-\pi^+\pi^0$. OPAL [21] tags the charm-quark with $D^* \rightarrow D^0\pi$, $D^0 \rightarrow K^-\pi^+$ and $D^0 \rightarrow K^-\pi^+\pi^-\pi^+$. The rapidity distribution of the hadrons in events selected with an electron [20], to identify the charm semileptonic decay, is well reproduced by PYTHIA (Fig.7 left) including direct and resolved processes. A good separation of the two contributions is obtained by OPAL by associating the D^* to a dijet analysis or by inspection of the p_t distribution of the D^* (Fig.7 right). The differential D^* cross section (Fig.8 left), measured inside the kinematical range of the detector acceptance, is almost independent on the Monte Carlo models used for efficiency corrections. It agrees well with NLO calculations [22]. The dependence on the pdf's, on the factorisation and renormalisation scales is small. With a stronger model dependence the data are extrapolated to the total charm production cross-section (Fig.8 right). The rise of the cross-section above the charm threshold and the importance of a full QCD calculation [19] are well demonstrated by the LEP data. The calculations depend on the charm mass, which can be determined once more data is available. In Fig.8 (right) also the b production cross-section is shown, as it has been measured for the first time by L3 [20], by tagging the b-quark with high p_t leptons.

3 Jets and single particle production

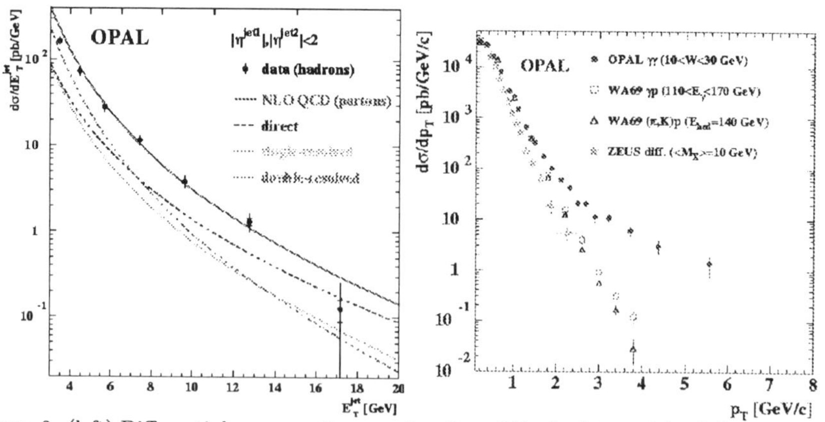

Figure 9: (left) Differential cross-section as a function of E_T for jets and (right) as a function of p_t for charged hadron production [23]. The p_t spectrum is harder in $\gamma\gamma$ than in γp and $(\pi,K)p$ interactions.

Until now only OPAL[23] published measurements of high p_t jets and inclusive charged hadron and K^0_S production. The jet and inclusive single-particle differential cross sections agree well with NLO calculations[24], using GRV[18] pdf (Fig.9 left). All processes are necessary: the direct contribution is dominant at high p_t, the double-resolved processes (ex.: $gg \to gg$) are necessary at low p_t values.

It is also interesting to compare the p_t distribution of charged hadron produced by different interactions ($\gamma\gamma$, γp and πp) (Fig.9 right). While the p_t spectrum of hadrons in photoproduction is similar to the one observed in pure hadronic interactions, the spectrum measured in two-photon interactions is harder due to the point-like photon coupling. OPAL notices also that PHOJET and PYTHIA both underestimate the K^0_S production by $\simeq$ 30 %. For dijet events the x_γ [b] distribution separates the direct from resolved processes. The angular distribution in the dijet center-of-mass shows a clear difference between the direct process, where a quark is exchanged in the t-channel and the double-resolved processes, dominated by gluon exchange (Fig.10).

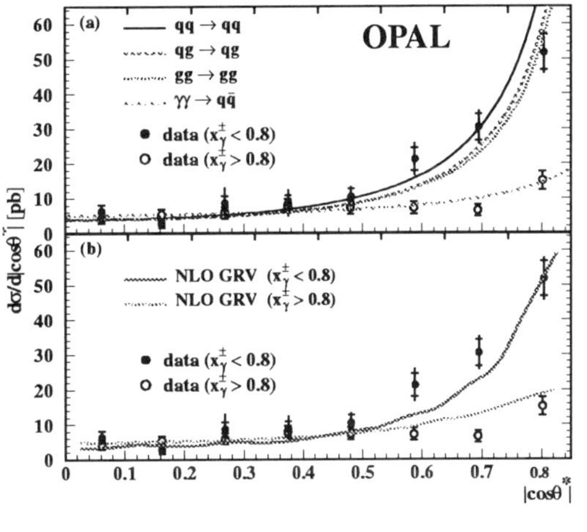

Figure 10: The angular distribution in the dijet center-of-mass system for *direct* ($x_\gamma > 0.8$) and *double-resolved* ($x_\gamma < 0.8$) events[23].

4 Resonance formation

Resonance formation in $\gamma\gamma$ interactions offers a clean experimental environment to study the spectrum of $q\bar{q}$ bound states. Only C=+1 states can be formed and the spin-parity quantum numbers are limited by the two-photon initial state. The measurement of the two-photon width, $\Gamma_{\gamma\gamma}$, and of the meson form factor, $F(Q^2)$, give an insight on the charge and the wave function of the valence quarks. The presence of a gluon component can be tested and pure glueball states are expected to have a two-photon width much smaller than

[b]The variable x_γ measures the momentum of the interacting particle, relative to the photon momentum, $x_\gamma = 1$ for a direct process, $x_\gamma < 1$ for a resolved process.

$q\bar{q}$ states. For heavy quarks, $c\bar{c}$ states, a measurement of α_s is possible by comparing two-photon to two-gluon partial widths. At LEP many exclusive channels are studied as shown in Table 1.

4.1 The η'(958) *form factor and the recurrence of* a_2(1320)

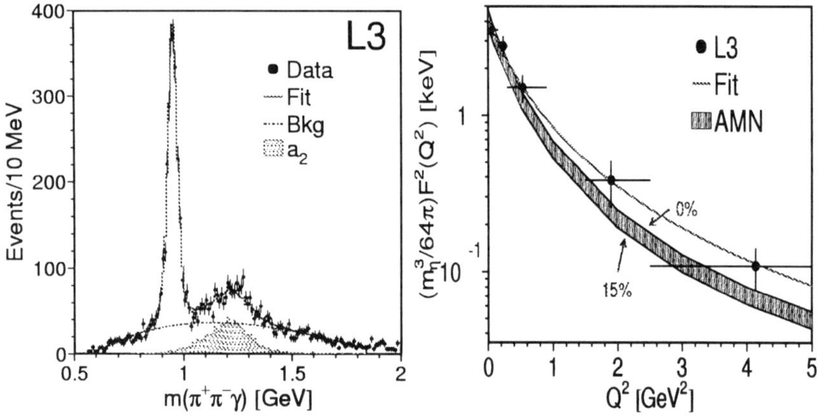

Figure 11: (left) The $\pi^+\pi^-\gamma$ mass spectrum for data [25] at $\sqrt{s}$=91 GeV. (right) Measured form factor fitted by a VDM pole form and compared to QCD predictions [34].

The $\pi^+\pi^-\gamma$ spectrum (Fig.11) is dominated by the peak of the $\eta'(958)$ resonance, the nearby enhancement is due to the a_2 (1320) $\rightarrow \pi^+\pi^-\pi^0$, when a photon of the π^0 decay goes undetected. For this spectrum the events with small transverse momentum, $|\vec{p_t}(\pi^+\pi^-\gamma)|^2 \leq 0.01$ GeV^2, are selected. These events allow an accurate measurement of the two-photon width (Table 1). The $e^+e^- \rightarrow e^+e^-\eta'$ cross-section is also measured as a function of Q^2, by using the relation $|\vec{p_t}(\pi^+\pi^-\gamma)|^2 = Q^2$ and the tagged events. The η' form factor thus obtained is well reproduced by the vector dominance form $F(Q^2)= [1/(1 + Q^2/\Lambda^2)]$. A fit to the data (Fig.11 right) gives a pole mass value $\Lambda = 900 \pm 46$ (stat.) $\pm\, 22$ (sys.) MeV, consistent with the VDM prediction [33] of $\Lambda = 830$ MeV. Anisovich, Melikhov and Nikonov [34] describe the perturbative and non-perturbative QCD regions with few parameters. Their predictions are indicated as a shaded area in Fig.11, ranging from no gluonium content (upper line) to 15% gluonium content (lower line). No definite conclusion can be derived from the comparison to the data, since the theoretical Q^2 dependence is somewhat different from the one observed. The perturbative QCD calculations of Jacob, Kroll and Raulfs [35] agree with the experimental measurement for $Q^2 \geq 2$ GeV^2.

The $\pi^+\pi^-\pi^0$ spectrum is dominated by the a_2 (1320) resonance [26]. The spin-parity analysis of the $\rho\pi$ and $f_2\pi$ channels shows evidence of a wide, 2^{++} state with M=1752±21 MeV and Γ=150±110 MeV. This observation confirms the result of a partial wave analysis of the $\pi^+\pi^-\pi^-$ system performed on data obtained at the Serpukhov accelerator [36]. This state can be interpreted as a radial excitation of the a_2 (1320) [37].

4.2 The $K^0_s K\pi$ *channel*

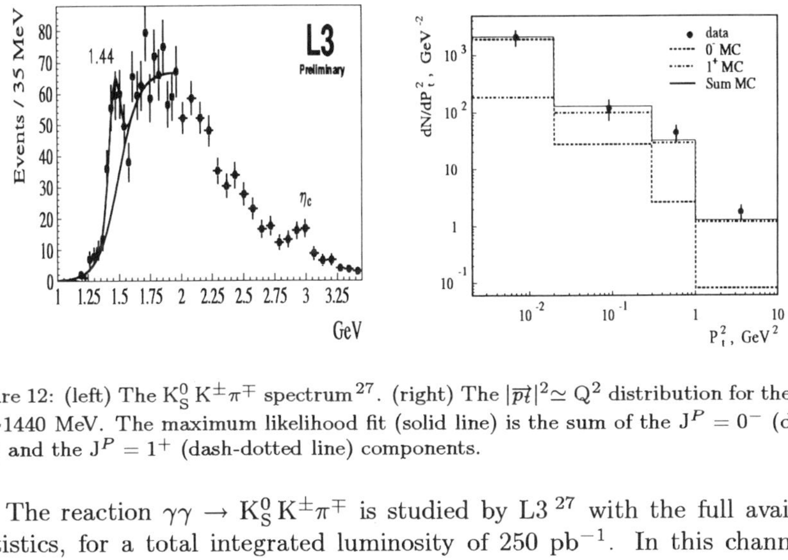

Figure 12: (left) The $K^0_S\, K^{\pm}\pi^{\mp}$ spectrum [27]. (right) The $|\vec{p_t}|^2 \simeq Q^2$ distribution for the peak at ~1440 MeV. The maximum likelihood fit (solid line) is the sum of the $J^P = 0^-$ (dotted line) and the $J^P = 1^+$ (dash-dotted line) components.

The reaction $\gamma\gamma \to K^0_S\, K^{\pm}\pi^{\mp}$ is studied by L3 [27] with the full available statistics, for a total integrated luminosity of 250 pb^{-1}. In this channel, a part from the η_c (2980) peak, there is a peak at threshold containing 119 ± 22 events (Fig.12). A gaussian fit gives the values of M=1465±31 MeV and σ=45±5 MeV, the experimental resolution being 24 MeV. The evolution of this signal with Q^2 (Fig.12 right) is studied with the same method used for the η' form factor. The data cannot be explained by a pure spin-parity state. A confidence level smaller than 1% is obtained for both pure $J^P = 0^-$ and pure $J^P = 1^+$ hypotheses, simulated by Monte Carlo. A fit using a combination of the two waves gives a confidence level of 50% , 49±1 events for the 0^- state, identified with the η (1440), and 63±6 events for the 1^+ state, called f_1 (1450). The two photon width of the η (1440) is determined using events with $|\vec{p_t}(K^0_S\, K^{\pm}\pi^{\mp})|^2 \leq 0.05$ GeV^2, where the 0^- state clearly dominates. Comparing the two-photon width (table 1) to the total width of η (1440) [8], a value of 9.5 ± 3.0 is obtained for the stickiness [38]. This value is similar to the ones obtained for ordinary mesons showing no evidence for a glueball nature of the η (1440).

4.3 $K^0_s K^0_s$ *resonances and glueball search*

The $K^0_s K^0_s$ mass spectrum [27], Fig.13, shows clear evidence for the formation of the $f_2'(1525)$ tensor meson. A Breit-Wigner fit gives 184± 27 events, M=1523±8 MeV and Γ =89±37 MeV. Around 1300 MeV, $f_2(1270) - a_2(1320)$ destructive interference is observed consistent with theoretical predictions [39]. In addition, there is an enhancement of 171±26 events at M=1765±22 MeV and Γ =200±50 MeV which is possibly due to the formation of a radially excited state of the f_2' [37]. A study of the angular distribution of the K^0_S in the two-photon centre-of-mass is in progress.

A search for the glueball candidate $\xi(2230)$ has been performed, motivated by the observation of a narrow peak [40] produced in radiative J/ψ decay. The absence of a signal in the $K^0_s K^0_s$ spectrum gives an upper limit for the two-photon width of 1.5 eV at 95% CL, under the hypothesis of a pure spin 2, helicity 2 state. This low value and a stickiness [38] greater than 68 are inconsistent with a $q\bar{q}$ content, a glueball hypothesis for the $\xi(2230)$ is strongly favoured, if this state exists.

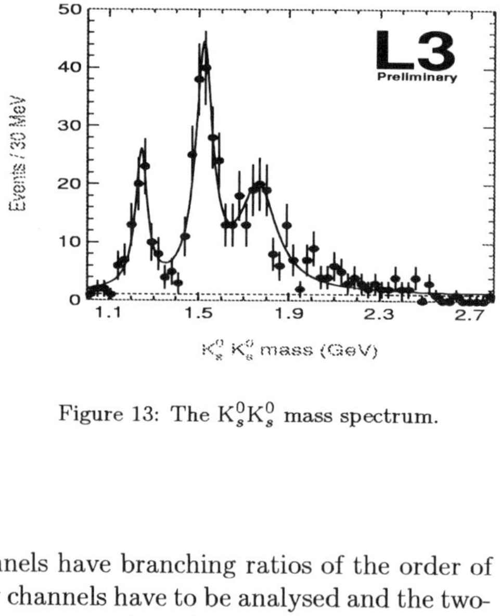

Figure 13: The $K^0_s K^0_s$ mass spectrum.

4.4 Charmonium Production

The individual η_c (2980) decay channels have branching ratios of the order of 1-2 %. As a consequence many decay channels have to be analysed and the two-photon width accuracy is limited by the poor knowledge of the η_c branching ratios[8]. Using LEP I and LEP II data, with a total luminosity of 193 pb^{-1}, L3 [30] measures the η_c (2980) two-photon width (table 1), by combining the events of nine different decay modes. Fig.14 shows the mass distribution of selected events with one of the scattered electron detected in the forward calorimeters. These events allow the measurement of the η_c transition form factor in two

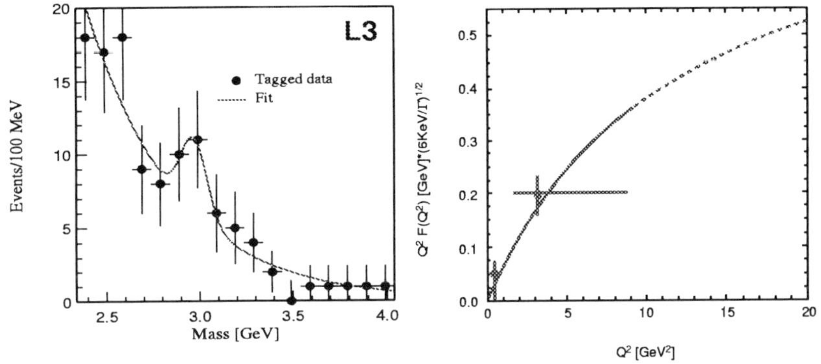

Figure 14: (left) The η_c (2980) signal for single-tag events [30]. (right) The measured η_c form-factor compared to QCD calculations [41].

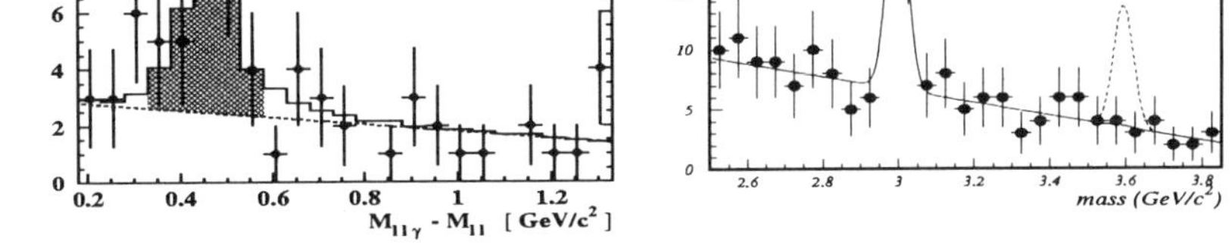

Figure 15: (left) The $\chi_{c2}(3550)$ signal as measured by OPAL [31]. (right) The η_c (2980) signal and the search for η_c'.

Q^2 bins, ($0.2\ \mathrm{GeV}^2 < Q^2 < 9\ \mathrm{GeV}^2$). The η_c form-factor measurement is consistent with a J/ψ pole in the VDM parametrisation and is in agreement with theoretical calculations[41].

The $\chi_{c2}(3550)$ is detected[31,32] through its dominant radiative decay to J/ψ: $\chi_{c2} \to \gamma\ J/\psi \to \gamma e^+e^-$, $\gamma\mu^+\mu^-$ (Fig.15 left). While the two-photon width of the η_c is well in agreement with the value measured in $p\bar{p} \to \gamma\gamma$[42] annihilation , for the χ_{c2} a discrepancy of ~ 3 standard deviations is observed. The combined two-photon results is $\Gamma_{\gamma\gamma}(\chi_{c2}) = 1.23 \pm 0.23$ keV and the new $p\bar{p}$ measurement of E835[44], $\Gamma_{\gamma\gamma} = 0.31 \pm 0.05 \pm 0.04$ keV confirms the previous E760 measurement[43].

No η_c' signal is observed neither at LEP, where DELPHI[45] (Figure 14) and L3[30] searched for it, nor in $p\bar{p}$ annihilation[42].

Table 1: The resonances studied at LEP

Resonance	Final state	J^{PC}	$\Gamma_{\gamma\gamma}$ (keV)
η' (958)	$\pi^+\pi^-\gamma$	0^{-+}	$4.17 \pm 0.10 \pm 0.27$ [25]
a_2 (1320)	$\pi^+\pi^-\pi^0$	2^{++}	$0.98 \pm 0.05 \pm 0.09$ [26]
a_2' (1750)	$\pi^+\pi^-\pi^0$	2^{++}	$0.29 \pm 0.04 \pm 0.02/(BR)$ [26]
η (1440)	$K^0_sK\pi$	0^{-+}	$0.17 \pm 0.05/(BR)$ [27]
f_1 (1450)	$K^0_sK\pi$	1^{++}	Ref. [27]
f_2' (1525)	$K^0_sK^0_s$	2^{++}	$0.09 \pm 0.02 \pm 0.02/(BR)$ [28]
X (1750)	$K^0_sK^0_s$	0 or 2	Ref. [27]
η_c (2980)	12 Channels	0^{-+}	$8.0 \pm 2.3 \pm 2.4$ [29]
η_c (2980)	9 Channels	0^{-+}	$6.9 \pm 1.9 \pm 2.0$ [30]
χ_{c2}(3555)	$J/\psi\ \gamma$	2^{++}	$1.76 \pm 0.47 \pm 0.40$ [31]
χ_{c2}(3555)	$J/\psi\ \gamma$	2^{++}	$1.02 \pm 0.40 \pm 0.17$ [32]

Conclusions and outlook

Our knowledge of two-photon physics has significantly increased with the analysis of LEP data. With a total integrated luminosity of $\geq 500\ \mathrm{pb}^{-1}$ at the end of LEP data taking, we expect a good sensitivity also to low cross-section channels. An improvement on our understanding of perturbative and non-perturbative QCD phenomena is possible, if the Monte Carlo models are carefully studied in order to better reproduce the experimental distributions of the two-photon events.

References

1. R. Engel, *Z. Phys.* C **66**, 203 (1995) ;
   R. Engel and J. Ranft, *Phys. Rev.* D **54**, 4246 (1996).
2. T. Sjöstrand, *Comput. Phys. Commun.* **82**, 74 (1994).
3. G.A. Schuler and T. Sjöstrand, *Z. Phys.* C **73**, 677 (1997).
4. L3 Coll., M. Acciari et al.,*Phys. Lett.* B **408**, 450 (1997);
   L3 Coll., M. Acciari et al., Submitted to EPS, Tampere, 1999.
5. OPAL Coll., G. Abbiendi et al., hep-ex/9906041 to appear in *Eur. Phys. J.*
6. V.M. Budnev, I.F. Ginzburg, G.V. Meledin and V.G. Serbo, *Phys. Rep.* C **15**, 181 (1974);
   G.A. Schuler, *Comput. Phys. Commun.* **108**, 279 (1998).
7. A. Donnachie and P. V. Landshoff, *Phys. Lett.* B **296**, 227 (1992).
8. Review of Particle Physics, *Eur. Phys. J* **3**, 205 (1998).
9. V.N.Gribov and I.Ya.Pomeranchuk, *Phys. Rev. Lett.* **8**, 343 (1962).
   S.J.Brodsky, *J.Phys.Suppl. (Paris)* 35 1974 C2-69.
   K.V.L.Sarma and V. Singh , *Phys. Lett.* B **101**, 201 (1981)
   C. Bourrely, J. Soffer, T.T. Wu, hep-ph/9903438.
10. A. Donnachie, H.G. Dosch and M. Rueter hep-ph/9810206 ;
    P. Desgrolard, M. Giffon, E. Martynov and E. Predazzi hep-ph/9811393
    B. Badelek, J. Kwiecinski and A.M. Stasto hep-ph/9903248 .
11. A. Corsetti, R.M.Godbole and G.Pancheri, *Phys. Lett.* B **435**, 441 (1998);
    R.M.Godbole and G.Pancheri, hep-ph/9903331.
12. L3 Coll., M. Acciarri et al., *Phys. Lett.* B **453**, 333 (1999);
    L3 Coll., M. Acciarri et al., Submitted to EPS, Tampere, 1999.
13. OPAL Coll., G. Abbiendi et al. , Submitted to EPS, Tampere, 1999.
14. E.A. Kuraev, L.N. Lipatov and V.S. Fadin, Sov. Phys. JETP **45** (1977) 199;
    Ya.Ya. Balitski and L.N. Lipatov, Sov. J. Nucl. Phys. **28** (1978)822.
15. S.J. Brodsky, F. Hautmann and D.E. Soper, *Phys. Rev.* D **56**, 6957 (1997).
16. J. Bartels, A. De Roeck and H. Lotter, *Phys. Lett.* B **389**, 742 (1996);
    J. Bartels, A. De Roeck, C. Ewerz and H. Lotter, hep-ph/9710500.
17. S.J. Brodsky, V.S. Fadin, V.T. Kim, L.N. Lipatov and G.B. Pivovarov, hep-ph/99101229.
18. M. Glück, E. Reya and A. Vogt,*Phys. Rev.* D **46**, 1973 (1992)
19. M. Drees, M. Kramer, J. Zunft and P.M. Zerwas, *Phys. Lett.* B **306**, 371 (1993)

20. L3 Coll., M. Acciarri et al.,*Phys. Lett.* B **453**, 83 (1999);
L3 Coll., M. Acciarri et al., Submitted to EPS, Tampere, 1999 ;
L3 Coll., M. Acciarri et al., hep-ex/9909005 to appear in *Phys. Lett.* B.
21. OPAL Coll., G. Abbiendi et al. , Submitted to EPS, Tampere, 1999
22. J. Binnewies, B.A. Kniehl and G. Kramer, *Phys. Rev.* D **53**, 6110 (1996);
J. Binnewies, B.A. Kniehl and G. Kramer, *Phys. Rev.* D **58**, 014014 (1998).
23. OPAL Coll., K. Ackerstaff et al. , *Z. Phys.* C **73**, 433 (1997);
OPAL Coll., K. Ackerstaff et al. , *Eur. Phys. J* **C 6**, 253 (1999)
24. T.Kleinwort and G. Kramer, *Phys. Lett.* B **370**, 141 (1996).
25. L3 Coll., M. Acciarri et al., *Phys. Lett.* B **418**, 399 (1998).
26. L3 Coll., M. Acciarri et al., *Phys. Lett.* B **413**, 147 (1997).
27. L3 Coll., O. Adriani et al., Submitted to EPS, Tampere, 1999.
28. L3 Coll., M. Acciarri et al., *Phys. Lett.* B **363**, 118 (1995).
29. L3 Coll., O. Adriani et al., *Phys. Lett.* B **286**, 403 (1992).
30. L3 Coll., O. Adriani et al., hep-ex/9909008 to appear in *Phys. Lett.* B.
31. OPAL Coll.,K. Ackerstaff et al., *Phys. Lett.* B **439**, 197 (1998).
32. L3 Coll., M. Acciarri et al., *Phys. Lett.* B **453**, 73 (1999).
33. Ll. Ametler, J.Bijnens, A.Bramon, F.Cornet, *Phys. Rev.* D **45**, 986 (1992).
34. V.V.Anisovich, D.I.Melikhov and V.A.Nikonov, *Phys. Rev.* D **55**, 2918 (1997).
35. R.Jacob, P.Kroll and M.Raulfs, *J. Phys. G* **22**, 45 (1996).
36. Yu.M. Antipov et al., *Nucl. Phys.* B **63**, 153 (1973).
37. C. R. Münz, *Nucl. Phys.* A **609**, 364 (1996).
38. M. Chanowitz in Proceedings of the VIth International Workshop on Photon-Photon Collisions, World Scientific, 1984.
39. H. J. Lipkin, *Nucl. Phys.* B **7**, 321 (1968).
40. BES Collab., J. Z. Bai et al., *Phys. Rev. Lett.* **76**, 3502 (1996) ;
Mark III Collab., R. M. Baltrusaitis et al., *Phys. Rev. Lett.* **56**, 107 (1986).
41. T. Feldmann and P. Kroll, *Phys. Lett.* B **413**, 410 (1997).
42. E760 Coll., T.A. Armstrong et al., *Phys. Rev.* D **52**, 4839 (1995).
43. E760 Coll., T.A. Armstrong et al., *Phys. Rev. Lett.* **70**, 2988 (1993).
44. E835 Coll. represented by M. Stancari in PHOTON99, Proceedings to appear in *Nucl. Phys.* B.
45. DELPHI Coll., P. Abreu et al., *Phys. Lett.* B **441**, 479 (1998).

Aleph results in $\gamma\gamma$ physics

M. H. Lehto

Department of Physics and Astronomy,
University of Sheffield,
Hicks Building,
Hounsfield Road,
Sheffield S3 7RH
United Kingdom
E-mail: M.Lehto@cern.ch

$F_2^\gamma(x, Q^2)$ has been measured in two energy regimes and in five Q^2 bins. The results are in agreement with models and previous measurements. Upper limits on the product of two-photon width and $\pi^+\pi^-$ branching ratio have been calculated as $\Gamma_{\gamma\gamma} \cdot \mathcal{BR}(\pi^+\pi^-) < 0.31$ keV for the $f_0(1500)$ and $\Gamma_{\gamma\gamma} \cdot \mathcal{BR}(\pi^+\pi^-) < 0.55$ keV for the $f_J(1710)$, both at 95% C.L. .

1 Introduction

Aleph has completed two measurements of the hadronic photon structure function $F_2^\gamma(x, Q^2)$, one from the LEP I [1] data and one from LEP II [2]. In addition a search has been made on untagged data for evidence of two glueball candidates [3].

The ALEPH detector has been described in detail elsewhere [4,5]. The key components for observing the final state products are the time projection chamber (TPC), the electromagnetic calorimeter (ECAL), and the hadron calorimeter (HCAL). The two luminosity calorimeters LCAL and SiCAL were used to measure the hadronic energy at small angles and to detect the tags used the structure function analyses. The inner tracking chamber (ITC) while primarily designed to supply fast track triggers, provides additional information on charged tracks. Trigger efficiences have been estimated from the data using different trigger combinations and are very high for the structure function measurements ($\simeq 98\%$). The muon chambers, the outermost part of the detector, were used for muon identification.

Structure function measurements rely on finding a "tagged" electron or tag which defines the Q^2 of the event. x_{vis} was reconstructed from $x_{vis} = Q^2/(Q^2 + {W_{vis}}^2)$ by calculating W_{vis} from the detected hadronic tracks and clusters.

2 LEP I Structure Function

ALEPH data taken from 1991-1995 was used with a total integrated luminosity of 162 (155) pb^{-1} for the ECAL (LCAL) tagged samples. The number of $\gamma^*\gamma$ collisions expected in this sample is a few thousand, compared with the 4.9 million Z decays in the same period, so the data selection is important.

In the ECAL case a tag was defined to be an identified electron with energy greater than 12 GeV, with scattering angle in the region $-0.6 < cos\theta_{TAG} < 0.95$. For the LCAL events a cluster of energy greater than 30 GeV, and scattering angle $60 < \theta_{TAG} < 150$ mr was taken as a tag. The event was constrained to have only one tag, and in addition no other cluster of more than 20 GeV in the LCAL or SiCAL.

The following cuts were then applied to the sample (mainly to eliminate Z decays): the number of energy flow objects of all types must be less than 26; the W reconstructed from the observed hadronic state, W_{vis}, must be in $2 < W_{vis} < 40$ GeV; the normalised longitudinal momentum balance (NLMB) must be greater than 0.2 (0.3) in ECAL (LCAL) events, where NLMB $= cos\theta_{TAG}\Sigma p_z/(|cos\theta_{TAG}|\Sigma E)$ (ΣE and Σp_z are respectively the sums of the energies and z components of the momentum of all the energy flow objects in the event, including the tag); there must be no energy flow object within 11.5 degrees of the tag; the total energy in a cone, with opening angle 18.2 degrees, opposite in direction to the tag must be less than 20 GeV; the total electromagnetic energy in identified electrons and photons (excluding the tag) must be less than 20 GeV; apart from the tag, there must be no identified electron with energy greater then 2 GeV; the number of charged tracks must be greater then 2; events with fewer than six charged tracks must have a thrust value less than 0.985; the interaction point must be reconstructed within 6 cm in the z direction and 1cm radially.

The total number of selected events was 163 (3190) for the ECAL (LCAL) analysis. The LCAL sample was split roughly in two: one with $6 < Q^2 < 13$ (1543 events)and one with $13 < Q^2 < 44$ GeV^2 (1647) with respective means of 9.93 ± 0.04 and 20.67 ± 0.16 GeV^2. The estimate of the remaining backgrounds, coming from Monte Carlo simulations, is of order 10%

The generator PHOT02 [6] was used to simulate Vector Meson Dominance (VMD) events and Berends, Darveveldt and Kleiss (BDK) [7] was used to simulate Quark Parton Model events. These contributions were fitted to the data (using variables not closely related to x) and provided the QPM+VMD model.

An alternative simulation was produced, using the HERWIG program with the GRV LO photon parton density function [9]. All parameters of the program were set at their defaults for version 5.9, apart from the intrinsic transverse momentum k_t of the partons in the target photon which was modified according to the scheme described in Ref. [10].

In general both models give an acceptable description of the data. The HERWIG sample was found to have too low a cross section by 20%.

Extraction of Structure Function results

As shown in Fig. 2, the measured value x_{vis} is smeared asymmetrically from the true value so that an unfolding procedure is first necessary to extract the true x distributions from the data. The unfolding was performed using the procedure proposed by Blobel [11]. The results of the unfolding are shown in figure 1. The total error is shown and the cross bar represents the statistical component. Generally the dominant systematic error comes from differences in the hadronization schemes between the two models.

To study the dependence of F_2^γ with Q^2 it is convenient to analyse the data in a fixed x bin. The region $0.1 < x < 0.6$ is chosen, so as to remove the regions at low and high x where the greatest theoretical uncertainty arises, but at the same time to ensure as precise a measurement as possible in all the Q^2 bins used. The results obtained in this way are most suitable for testing the $\log(Q^2)$ dependence predicted by QCD. The results are shown in figure 1. The data are in agreement both with previous data and most models.

3 LEP II Structure Function

In the LEP II analysis events were selected with tags in the LCAL or SiCAL calorimeters. The background is much lower at higher energy so the selection cuts were much simpler: the tag energy had to be greater than 60 (50) GeV and within the angular range $35 \leq \theta_{TAG} \leq 54$ ($60 \leq \theta_{TAG} \leq 155$) mr for SiCAL (LCAL) tags; the SiCAL and LCAL opposite to the tag had to contain no cluster of energy greater than 20 GeV; W_{vis} had to be greater than 2 GeV; there had to be at least 3 charged tracks; there should be no identified lepton with energy greater than 2 GeV in the region $|\cos\theta| < 0.8$.

In this analysis the unfolding was carried out using the maximum entropy method and done separately for each tagging region.

Unfolding procedure

The parameters to be estimated are the cross sections $d^2\sigma/dxdQ^2$ averaged over each of M bins ($\bar{\mu} = \mu_1, ..., \mu_M$). From the data one obtains a histogram ($\bar{n} = n_1, ..., n_N$) which is distorted with respect to $\bar{\mu}$ by acceptance, resolution and statistical fluctuations. The n_i can be treated as a Poisson variable with

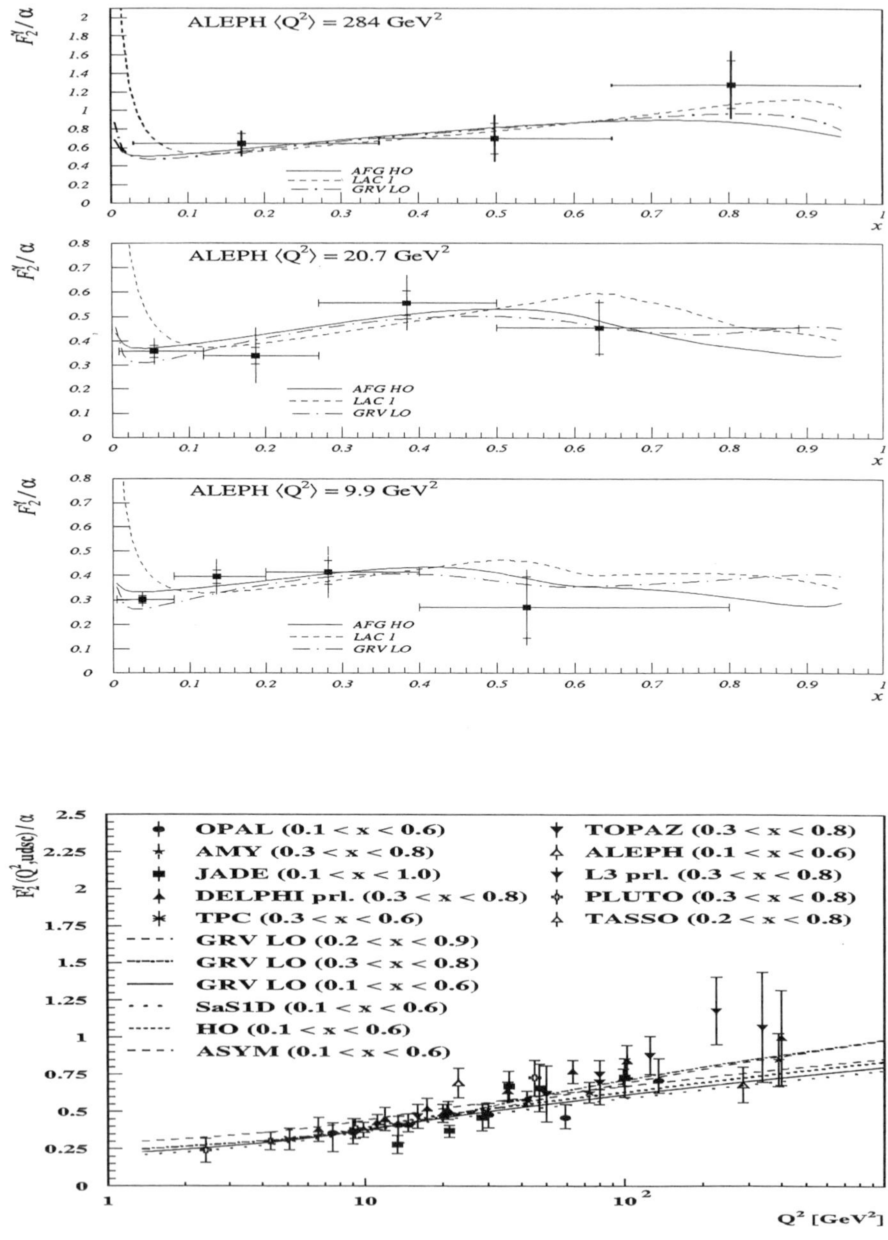

Figure 1: The values of $F_2^\gamma(x)/\alpha$ compared to three parametrisations[9,12,13]. The bottom plot shows the evolution of $F_2^\gamma(x, Q^2)$ with Q^2 compared to various models and other experiments.

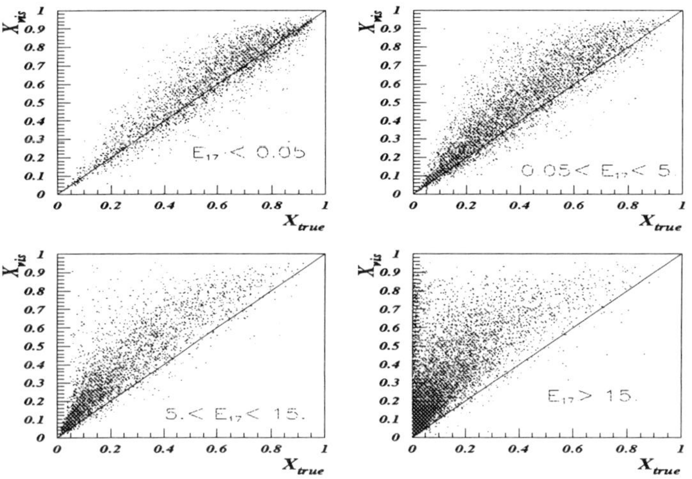

Figure 2: The resolution in x for various E_{17} regions.

expectation value ν_i. The vectors $\bar{\nu}$ and $\bar{\mu}$ are related by $\bar{\nu} = R\bar{\mu} + \bar{\beta}$ where R_{ij} represents the probability that an event with true x in bin j was observed in bin i and the vector $\bar{\beta}$ is the expected background. The log liklihood function is $\log L(\bar{\mu}) = \sum \log(n_i \log \nu_i - \nu_i)$

The maximum liklihood estimators have a very large variances (because of the large smearing in x). These variances can be reduced by maximizing a combination of $\log L$ and a regularization function $S(\bar{\mu})$,

$$\phi(\bar{\mu}, \lambda) = \alpha \log L(\bar{\mu}) + S(\bar{\mu}) + \lambda \left(n_{tot} - \sum \nu_i \right),$$

with respect to the parameters $\bar{\mu}$ and the Lagrange multiplier λ. The function S (in this case the Shannon entropy) is a measure of the smoothness of the unfolded result and α determines the tradeoff between smoothness and liklihood. The entropy based approach makes no reference to the relative location of any of the bins which means there is no special treatment near the edge of a distribution and multidimensional unfolding is a simple extension.

The value of x_{true} is not the only influence on the probability to measure a certain x_{vis}. For example, events where there are an appreciable number of hadrons at low angle to the beam are much more badly measured than ones where this is not so. A variable E_{17}, defined as the energy in a 17° cone around the beam pipe, was used to characterise this forward hadron activity. Figure 2 shows how the x resolution varies with increasing E_{17}, and it can be seen that events with high E_{17} have a much worse resolution. The two dimensional distribution in E_{17} and x is unfolded as above and the result integrated over E_{17} to gove an unfolded x_{true}. This reduces model dependance and tends to give smaller statistical errors by giving a higher weight to those events with a low E_{17} (and thus a better measurement of x).

The two Monte Carlo generators used in this study were HERWIG [8] annd PHOJET [14]. HERWIG in general agrees well with the data and PHOJET is not so good, specially at higher Q^2.

$F_2^\gamma(x, Q^2)$ was extracted from the dN/dx extracted as above using

$$F_2^\gamma(x, Q^2)_{meas} = F_2^\gamma(x, Q^2)_{ref} \cdot \left(\frac{dN}{dx}\right)_{meas} / \left(\frac{dN}{dx}\right)_{ref}$$

The final $F_2^\gamma(x, Q^2)$ results are shown in figure 3. The bin to bin correlations lie in the range -0.7 to -0.28. The results slightly prefer GRV among the models shown. The errors are comparable or less than those from the previous analysis.

4 Searches for Glueball candidates

Quantum chromodynamics (QCD) predicts the existence of bound states of gluons, called glueballs. Lattice calculations have predicted the lightest scalar glueball to be a scalar resonance with mass $1600 \pm 150\text{MeV}^2$ [16,17]. Experimentally two principal candidates for the scalar glueball have been observed, the $f_0(1500)$ and the $f_J(1710)$ [18].

In $\gamma\gamma$ interactions, production of a pure gluon state is suppressed. Measuring the two-photon width $\Gamma_{\gamma\gamma}$ of the $f_0(1500)$ and the $f_J(1710)$, or setting an upper limit on $\Gamma_{\gamma\gamma}$, should indicate whether either is likely to be a pure glueball or has quark content.

Samples of the signal processes $\gamma\gamma \rightarrow f_0(1500) \rightarrow \pi^+\pi^-$ and $\gamma\gamma \rightarrow f_J(1710) \rightarrow \pi^+\pi^-$ were generated using PHOT02 [6]. Each sample was generated as a Breit-Wigner resonance of appropriate mass, width and spin, with the $f_J(1710)$ here assumed to be spin zero. Expected background processes $\gamma\gamma \rightarrow e^+e^-$, $\gamma\gamma \rightarrow \mu^+\mu^-$, and $\gamma\gamma \rightarrow \tau^+\tau^-$ were also generated using PHOT02.

This analysis uses 160.9 $\text{pb}^{-1}$of data taken around $\sqrt{s} = 91$ GeV from 1990 to 1995. Candidate events for $\gamma\gamma \rightarrow \pi^+\pi^-$ are selected according to the following criteria: the event contains only two good tracks, of equal and

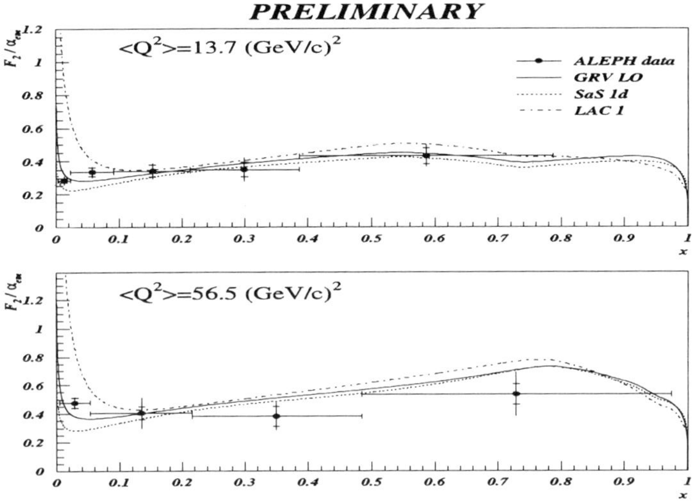

Figure 3: The structure function results for the LEP II data compared to various models.

opposite charge; the primary vertex is reconstructed within ± 2 cm in z; the total energy observed in the event is less than 30 GeV and visible mass less than 10 GeV; the transverse momentum of the final state with respect to the beam axis does not exceed 0.1GeV; the absolute value of the cosine of the angle θ^* between the two tracks in their centre-of-mass system and the boost direction is required to be less than 0.9; $\mathrm{d}E/\mathrm{d}x$ is more than 3σ from an electron and less than 3σ from a pion; no track must be identified as a muon [5];

A sample of 294141 $\pi^+\pi^-$ candidate events is selected. The dominant background is still muon production.

Fitting the mass spectrum

The $\pi^+\pi^-$ invariant mass spectrum obtained is shown in Fig. 4. The spectrum falls off steeply at around 0.5GeV^2 because of the trigger efficiency and is not shown. The clear peak in the spectrum above 1GeV^2 can be identified as the known tensor resonance $f_2(1270)$. No other structure is observed.

A fit to the invariant mass spectrum is performed. For the resonances

$f_2(1270)$, $f_0(1500)$ and $f_J(1710)$, a Breit-Wigner function of the form

$$\frac{mm_0\Gamma(m)}{(m_0^2 - m^2)^2 + m^2\Gamma^2(m)} \tag{1}$$

is used. For the $f_2(1270)$ the mass m_0, total width Γ_0 and overall normalisation are free parameters. For the $f_0(1500)$ and the $f_J(1710)$ the normalisation is a free parameter, while the mass and width are fixed The background spectrum, due to $\gamma\gamma \to \mu\mu$ events and the $\pi\pi$ continuum process, is fitted on the data with a 5th order Chebyshev polynomial.

The fit to the data is shown in Fig. 4 where only the $f_2(1270)$ and the background are fitted. The parameters of all fits are summarised in Table 1. For the fit including the $f_2(1270)$ alone, the χ^2 of 75 for 76 degrees of freedom is already very good. The fitted signals for the resonances are both consistent with 0 events.

The fitted numbers of signal events from the processes $\gamma\gamma \to f_0(1500) \to \pi^+\pi^-$ and $\gamma\gamma \to f_J(1710) \to \pi^+\pi^-$ are used to calculate upper limits on $\Gamma_{\gamma\gamma} \cdot \mathcal{BR}(\pi^+\pi^-)$ for both resonances. In the 1500MeV^2 mass region the average trigger efficiency is $(66 \pm 7)\%$, while for the 1700MeV^2 mass region it is $(77 \pm 7)\%$.

From the Monte Carlo the acceptance and selection efficiency for $\gamma\gamma \to f_0(1500) \to \pi^+\pi^-$ and for $\gamma\gamma \to f_J(1710) \to \pi^+\pi^-$ are determined from Monte Carlo to be $(17.5 \pm 0.4)\%$ and $(16.3 \pm 0.4)\%$. The product branching ratio, $\Gamma_{\gamma\gamma} \cdot \mathcal{BR}(\pi^+\pi^-)$, of each resonance is calculated directly from the fitted number of signal events. The fitted value of $\Gamma_{\gamma\gamma} \cdot \mathcal{BR}(\pi^+\pi^-)$ (which may be negative) and its error are used to define the mean and width of a Gaussian distribution: the area under the Gaussian is integrated above zero to obtain the 95% C.L. limit on $\Gamma_{\gamma\gamma} \cdot \mathcal{BR}(\pi^+\pi^-)$.

The upper limits on the product branching ratios at 95% C.L. are $\Gamma_{\gamma\gamma} \cdot \mathcal{BR}(\pi^+\pi^-) < 0.31$ keV and $\Gamma_{\gamma\gamma} \cdot \mathcal{BR}(\pi^+\pi^-) < 0.55$ keV for the $f_0(1500)$ and $f_J(1710)$ respectively.

5 Conclusion

Inclusive production of hadrons in $\gamma^*\gamma$ interactions have been studied using Aleph LEP I and LEP II data. The results, in terms of measured $F_2^\gamma(x, Q^2)$, are in agreement with both theoretical models and previously published data. Production of the glueball candidates $f_0(1500)$ and $f_J(1710)$ in $\gamma\gamma$ collisions at LEP1 has been studied via decay to $\pi^+\pi^-$. No signal from either resonance is seen.

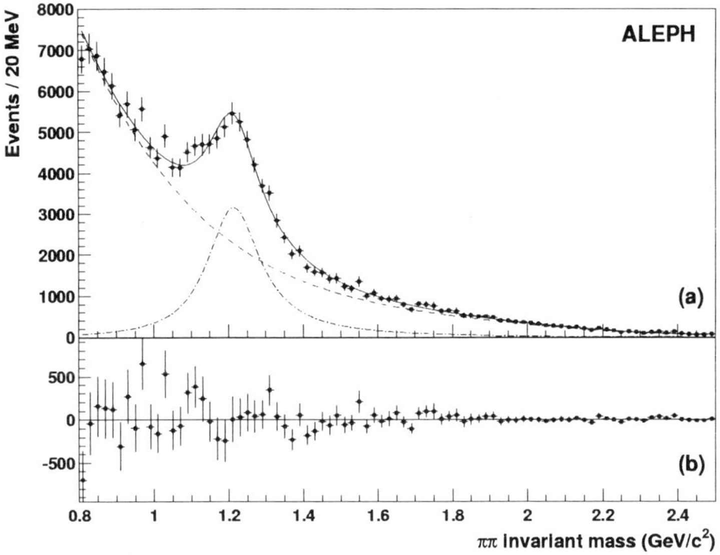

Figure 4: (a) The fit to data with a Breit-Wigner function for the $f_2(1270)$ (dot-dash line), a polynomial for the background (dashed line) and the combination of these functions (solid line); and (b) the data after subtraction of the fitted curve. Error bars indicate statistical errors only.

Table 1: Table of parameters for fits using the Breit-Wigner form of Eqn.1 for resonances.

	Fit for $f_2(1270)$ only
χ^2	74.96
Degrees of freedom	76
Fitted mass of $f_2(1270)$ (MeV^2)	1213.5 ± 3.7
Fitted width of $f_2(1270)$ (MeV^2)	178.3 ± 12.8
	Fit for $f_2(1270) + f_0(1500) + f_J(1710)$
χ^2	73.21
Degrees of freedom	74
Fitted mass of $f_2(1270)$ (MeV^2)	1214.2 ± 3.8
Fitted width of $f_2(1270)$ (MeV^2)	175.5 ± 14.2
No. of $f_0(1500)$ signal events	-671.0 ± 690.0
No. of $f_J(1710)$ signal events	198.6 ± 541.9

References

1. ALEPH Collaboration, e^-Measurement of the hadronic photon structure function at LEP I for Q^2 values between 9.9 and 284 GeV^2, e^-Phys. Lett. B 458 (1999) 152.
2. e^-A Study of the Photon Structure Function using two dimensional unfolding, in Proceedings of Photon'99 (1999).
3. ALEPH Collaboration, e^-Search for the glueball candidates $f_0(1500)$ and $f_J(1710)$ in $\gamma\gamma$ collisions, submitted to e^-Phys. Lett. B.
4. ALEPH Collaboration, e^-"ALEPH: a detector for electron-positron annihilation at LEP", e^-Nucl. Instr. Meth. **A294** (1990) 121.
5. ALEPH Collaboration, e^-"Performance of the ALEPH detector at LEP", e^-Nucl. Instr. Meth. **A360** (1995) 481.
6. ALEPH Collaboration, e^-"An experimental study of $\gamma\gamma \to$hadrons at LEP", e^-Phys. Lett. **B313** (1993) 509.
7. F.A. Berends, P.H. Daverveldt and R. Kleiss, e^-Computer Phys. Commun.
   40 (1986) 285.
8. G. Marchesini et al., e^-Computer Phys. Commun. **67** (1992) 465.
9. M. Glück, E. Reya and A. Vogt, e^-Phys. Rev. **D45** (1992) 3986; **D46** (1992) 1974.
10. The UK phenomenology workshop on LEP2 physics, S. Cartwright et al., e^-J. Phys. G **24** (1998) 457.
11. V. Blobel, in Proceedings of the CERN School of Computing, Aiguablava, Spain (1984), CERN 85-09.
12. P. Aurenche, M. Fontannaz, J.Ph. Guillet, e^-Z. Phys. **C64** (1994) 621.
13. H. Abramowicz, K. Charchula and A. Levy, e^-Phys. Lett. **B269** (1991) 458.
14. R. Engel and J. Ranft, e^-Phys. Rev. D54 (1996) 4244.
15. G.A. Schuler and T. Sjöstrand, e^-Z. Phys. **C68** (1995) 607.
16. UKQCD Collaboration, e^-A Comprehensive Lattice Study of SU(3) Glueballs, Phys. Lett. **B309** (1993) 378.
17. J. Sexton, A. Vaccarino and D. Weingarten, Phys. Rev. Lett **75** (1995) 4563.
18. Particle Data Group, Eur. Phys. J. **C3** (1998) 1.

N. E. Tyurin

UNITARITY AND ANTISHADOW SCATTERING AT HIGH AND SUPERHIGH ENERGIES

S. M. TROSHIN, N. E. TYURIN
Institute for High Energy Physics, Protvino, Moscow Region, 142284 RUSSIA

Theoretical treatment of soft hadronic reactions involves substantial part of phenomenology and uses various model approaches. The basic role in our consideration belongs to unitarity of the scattering matrix.

Since discovery of the total cross-section rise [1] the importance of unitarity became particularly obvious. The amplitude

$$F(s,t) = \Phi[V(s,t)]$$

results in application of unitarization procedure Φ to the input amplitude $V(s,t)$ usually considered as a power function of energy. The two most often used schemes to restore unitarity are based on eikonal and generalized reaction matrix (U-matrix). In the impact parameter representation for the elastic scattering amplitude $f(s,b)$ at high energies:

$$Im f(s,b) = |f(s,b)|^2 + \eta(s,b), \tag{1}$$

where inelastic overlap function $\eta(s,b)$ can be expressed as a sum over n-particle production cross-sections:

$$\eta(s,b) = \sum_n \sigma_n(s,b). \tag{2}$$

Eq. (1) implies the constraint

$$|f(s,b)| \leq 1 \tag{3}$$

Elastic unitarity is often resolved with the use of eikonal

$$f(s,b) = \frac{i}{2}(1 - \exp[i\omega(s,b)]) \tag{4}$$

Assumption on a pure imaginary amplitude is a rather common approximation at high energies and one expands the eikonal unitarization Eq. (4) by choosing pure imaginary eikonal $\omega(s,b) = i\Omega(s,b)$. This leads to the amplitude that a priori presumes inequality

$$|f(s,b)| \leq 1/2 \tag{5}$$

instead of general constraint of Eq. (3). Pure imaginary amplitude is determined by the inelastic processes:

$$f(s,b) = \frac{i}{2}[1 \pm \sqrt{1-4\eta(s,b)}], \tag{6}$$

and Eq. (4) with imaginary eikonal function corresponds to the choice of the particular solution in Eq. (6) with sign minus. Different reasonable parameterizations for opacity lead to the rising total and elastic cross–sections and slope parameter:

$$\sigma_{tot}(s) \sim \sigma_{el}(s) \sim B(s) \sim \ln^2 s \tag{7}$$

and the ratio

$$\frac{\sigma_{el}(s)}{\sigma_{tot}(s)} \to \frac{1}{2}. \tag{8}$$

Eq. (8) has been a priori assumed when black disk limit Eq. (5) was adopted through the parameterization Eq. (4).

Alternative resolution of the unitarity is achieved with the form of the amplitude

$$f(s,b) = \frac{U(s,b)}{1-iU(s,b)}. \tag{9}$$

Unitarity is satisfied provided the inequality

$$ImU(s,b) \geq 0 \tag{10}$$

takes place and inelastic overlap function is

$$\eta(s,b) = \frac{ImU(s,b)}{|1-iU(s,b)|^2}. \tag{11}$$

In Eqs. (9) and (11) function $U(s,b)$ – generalized reaction matrix – is considered as an input dynamical quantity similar to the eikonal function. This form for the amplitude follows from the relativistic generalization of the Heitler equation for the elastic scattering amplitude [2]. The form for scattering amplitude similar to Eq. (9) was obtained by Feynman in his parton model of diffractive scattering [3].

Construction of particular models in the framework of the U-matrix approach proceeds with the same steps as it does for the eikonal function. The basic dynamics as well as the notions on hadron structure are used to obtain a particular form for $U(s,b)$[4,5]. Without going into details of the particular parameterizations we present the general conclusions. The total and elastic cross–sections and slope parameter show the $\ln^2 s$ rise:

$$\sigma_{tot}(s) \sim \sigma_{el}(s) \sim B(s) \sim \ln^2 s \tag{12}$$

similar to conclusions of the eikonal unitarization.

However, these two unitarization schemes lead to different predictions for the inelastic cross-sections and for the ratio of elastic to total cross-section when $s \to \infty$

$$\frac{\sigma_{el}(s)}{\sigma_{tot}(s)} \to 1, \tag{13}$$

i.e. this ratio reaches its maximal possible value. Eq. (13) reflects the fact that the bound for the partial wave amplitude in the U-matrix approach is $|f| \leq 1$ in conformity with the unitarity while the respective bound for the case of imaginary eikonal is $|f| \leq 1/2$ (black disk limit).

When the amplitude exceeds the "black disk limit" (in the central collisions at high energies) then the scattering at such impact parameters turns out to be of ANTISHADOW nature. It corresponds to the particular solution of the unitarity Eq. (6) with plus sign. In this ANTISHADOW scattering mode the elastic amplitude increases with decrease of the inelastic channels contribution.

Usually the shadow mode is considered as the only possible one. But the two solutions of unitarity have an equal meaning. The both modes are realized in a natural way in the U-matrix approach [6] despite the two modes are described by the two solutions of unitarity.

With conventional parameterization of the U-matrix the inelastic overlap function $\eta(s,b)$ increases with energy at modest values of s. It reaches its maximum value $\eta(s,b=0) = 1/4$ at some $s = s_0$ and beyond this energy the antishadow scattering mode appears at small values of b. The region of energies and impact parameters corresponding to the antishadow scattering mode is determined by the conditions $Imf(s,b) > 1/2$ and $\eta(s,b) < 1/4$. The quantitative analysis of the experimental data [7] gives the threshold value: $\sqrt{s_0} \simeq 2$ TeV.

Thus, the function $\eta(s,b)$ becomes peripheral when energy is increasing. At such energies the inelastic overlap function reaches its maximum value at $b = R(s)$ where $R(s)$ is the interaction radius. So, beyond the transition threshold there are two regions in impact parameter space: the central region of antishadow scattering at $b < R(s)$ and the peripheral region of shadow scattering at $b > R(s)$. At $b = R(s)$ the maximal absorption (black ring) takes place (Fig. 1).

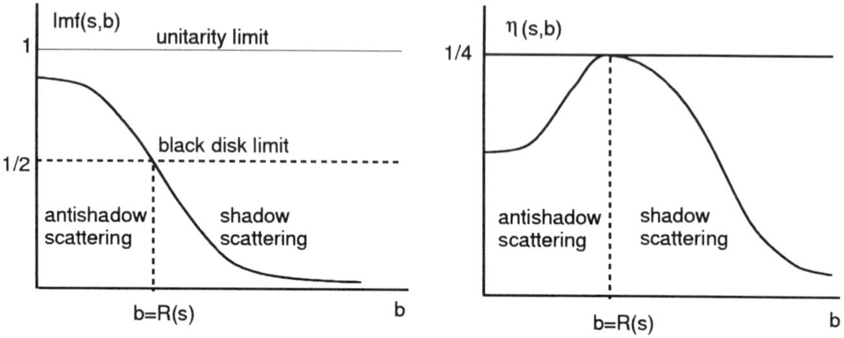

Figure 1: Shadow and antishadow scattering regions

The transition to the antishadow scattering results also in a relatively slow rise of inelastic cross–section:

$$\sigma_{inel}(s) = 8\pi \int_0^\infty \frac{ImU(s,b)}{|1 - iU(s,b)|^2} \sim \ln s. \tag{14}$$

at $s \to \infty$.

It should be noted that appearance of the antishadow scattering mode does not contradict to the basic idea that the particle production is the driving force for elastic scattering. Indeed, the imaginary part of the generalized reaction matrix is the sum of inelastic channel contributions:

$$ImU(s,b) = \sum_n \bar{U}_n(s,b), \tag{15}$$

where n runs over all inelastic states and

$$\bar{U}_n(s,b) = \int d\Gamma_n |U_n(s,b,\{\xi_n\}|^2 \tag{16}$$

and $d\Gamma_n$ is the n–particle element of the phase space volume. The functions $U_n(s,b,\{\xi_n\})$ are determined by the dynamics of $2 \to n$ processes. However, unitarity leads to self–damping of the inelastic channels [8] and increase of the function $ImU(s,b)$ results in decrease of the inelastic overlap function $\eta(s,b)$ when $ImU(s,b)$ exceeds unity.

In what follows we give analysis of the appearances of the antishadow scattering mode. We consider hadron scattering in the framework of the effective chiral quark model where hadron dynamics is due to effective Lagrangian approach [9]. The principal features of the model are as follows:

1. A hadron is represented as consisting of inner region where valence constituent quarks are located and the outer region filled with quark condensate. Such picture implies that overlapping and collision of the peripheral condensates in hadrons occurs at the first stage of interaction.

2. As a result virtual massive quarks appear. Being released a part of hadron energy goes to the generation of massive quarks. Besides the mass, quark acquires an internal structure and a finite size. Quark radii are determined by the radii of the respective clouds. Strong interaction radius of quark Q is determined by its Compton wavelength

$$r_Q = \frac{\xi}{m_Q} \tag{17}$$

with ξ universal for the different flavors.

3. Valence quarks are supposed to scatter in a quasi-independent way in the field V_{eff} generated by the produced massive quarks (at given impact parameter) and by the other valence quarks.

The function $U(s, b)$ is chosen as a product of the averaged quark amplitudes

$$U(s, b) = \prod_{Q=1}^{N} \langle f_Q(s, b) \rangle. \tag{18}$$

The b–dependence of the function $\langle f_Q \rangle$ related to the quark formfactor $F_Q(q)$ has a simple form

$$\langle f_Q \rangle \propto \exp(-m_Q b/\xi).$$

Thus, the basic dynamical quantity in this approach gets the following form(in a pure imaginary case)

$$U(s, b) = ig\left[1 + \alpha\frac{\sqrt{s}}{m_Q}\right]^N \exp(-Mb/\xi), \tag{19}$$

where $M = \sum_{q=1}^{N} m_Q$.

In the model the total cross–section

$$\sigma_{tot}(s) = \frac{\pi\xi^2}{m_Q^2}\Phi(s, N). \tag{20}$$

The function Φ has the following behavior:

$$\Phi(s, N) = \begin{cases} (8g/N^2)\left[1 + N\alpha\sqrt{s}/m_Q\right], & s \ll s_0, \\ \ln^2 s, & s \gg s_0. \end{cases} \tag{21}$$

Here $\sqrt{s_0} \equiv m_Q/\alpha$ determines the range of "moderate energies" $(s \ll s_0)$. One should note that linear with $\sqrt{s}$ preasymptotic rise of the total cross-sections is in agreement with the data up to $\sqrt{s} \sim 0.5$ TeV [7]. In the model such behavior reflects the energy dependence of a number of virtual quarks generated under "collisions of condensates".

Asymptotically $(s \gg s_0)$

$$\sigma_{inel}(s) = \frac{4\pi\xi^2}{Nm_Q^2}\ln s \tag{22}$$

$$\frac{\sigma_{el}(s)}{\sigma_{tot}(s)} = 1 - \frac{4}{N\ln s}. \tag{23}$$

It is important to note that such a behavior of $\sigma_{el}(s)/\sigma_{tot}(s)$ and σ_{inel} results from self-dumping of inelastic channels at small impact distances which starts to develop at $s > s_0$. Numerical estimates show that the ratio Eq. (23) becomes close to its asymptotic value 1 at extremely high energies $\sqrt{s} \simeq 500$ TeV.

Unitarization drastically changes the scattering picture: at lower energies inelastic channels provide dominant contribution and amplitude has a shadow origin while at high energies elastic scattering dominates over inelastic contribution and the scattering corresponds to the antishadow mode. s–dependence of total cross-section at $s \ll s_0$ is described by a simple linear function of $\sqrt{s}$. This is a preasymptotic dependence and it has nothing to do with the true asymptotics of the total cross-sections.

As well as elastic scattering the diffractive production is the basic process which would lead to understanding of large distance dynamics and hadron structure.

To obtain the cross-section of the diffractive process $h_1 + h_2 \to h_1 + h_2^*$ we should single out the respective final states in Eq. (15). Considering for simplicity the pure imaginary case we can represent the differential cross-section in the form

$$\frac{d\sigma_{diff}}{dM^2} = 8\pi\int_0^\infty bdb\frac{U_{diff}(s,b,M)}{[1+U(s,b)]^2} \tag{24}$$

where the function $U_{diff}(s,b,M)$ includes contributions from all the final states $|n\rangle_{diff}$ which result from the decay of the excited hadron h_2^* of mass M: $h_2^* \to |n\rangle_{diff}$.

In the framework of the picture of hadron interaction discussed in connection with the elastic scattering it is natural to assume [9] that the origin of diffractive production process is the excitation of one of the constituent valence quarks in colliding hadron: $Q \to Q^*$, its subsequent scattering and decay into the final state. The excited constituent quark is scattered similar to other valence quarks in a quasi-independent way. The function $U_{diff}(s,b,M)$ can be represented then as a product

$$U_{diff}(s,b,M) = \langle f_{Q^*}(s,b,M_{Q^*})\rangle \prod_{Q=1}^{N-1} \langle f_Q(s,b)\rangle, \tag{25}$$

where M_{Q^*} is the mass of excited constituent quark, which is proportional to the mass M of excited hadron h_2^* for large values of M. The b–dependence of the amplitude $\langle f_{Q^*}\rangle$ is related to the formfactor of excited quark whose radius is detemined by its mass M_{Q^*}.

The cross-section of diffractive dissociation process is given by expression (24) and has the following s and M^2 dependence

$$\frac{d\sigma_{diff}}{dM^2} \simeq \frac{8\pi g^* \xi^2}{M^2}\eta(s,\xi/M) \tag{26}$$

We obtained familiar $1/M^2$ dependence of the diffraction cross-section which is related in this model to the geometrical size of excited constituent quark.

The energy dependence of single diffractive cross-section has the following form ($\xi/M \simeq 0$)

$$\sigma_{diff}(s) = 8\pi g^* \xi^2 \eta(s,0) \int_{M_0^2}^{M_1^2} \frac{dM^2}{M^2} = 8\pi g^* \xi^2 \eta(s,0) \ln \frac{s(1-x_1)}{M_0^2}, \tag{27}$$

where x_1 is the lower limit of the relative momentum of hadron $h_1 (x_1 \simeq 0.8 - 0.9)$ which corresponds to the experimental constraint on diffractive process definition. Eq. (27) shows that the total cross-section of diffractive dissociation has a non-trivial energy dependence which is determined by the contribution of inelastic channels into unitarity equation at close to zero value of impact parameter. From Eqs. (11) and (19) it follows that $\eta(s,0)$ increases with energy at $s \leq s_0$, where s_0 is determined by equation $|U(s_0,0)| = 1$. This increase as it follows from Eq. (19) and from the experimental data is rather slow one. However at $s \geq s_0$, $\eta(s,0)$ reaches its maximum value $\eta(s,0) = 1/4$ and at $s > s_0$, the function $\eta(s,0)$ decreases with energy. At $s \to \infty$:

$$\sigma_{diff}(s) \propto \left(\frac{1}{\sqrt{s}}\right)^N \ln s \tag{28}$$

since $\eta(s,0) \propto (1/\sqrt{s})^N$ in this limit.

Thus at asymptotical energies the inelastic diffraction cross section drops to zero. Decrease is due to the fact that $\eta(s,b)$ becomes peripheral at $s > s_0$, i.e. transition to the antishadow scattering mode in the multi-TeV energy range. The development of the antishadow mode in head-on pp and $p\bar{p}$–collisions could be associated with new phenomena in the central hadronic collisions where the temperatures are high and the energy density can be up to several GeV/fm^3. Of course, there might be different reasons for decrease of $\sigma_{diff}(s)$. The decreasing energy dependence of $\sigma_{diff}(s)$ was also predicted in[10,11].

The recent straitforward interpretations of the data on the deep–inelastic scattering together with the analysis of the data on hadron–hadron scattering could lead to the unexpected conclusion on the existence of the various Pomerons.

Indeed, soft hadronic reactions imply that the Pomeron's intercept $\alpha_{\mathcal{P}} = 1.08$ [12], and small–x dependence of the structure function $F_2(x,Q^2)$ leads to $\alpha_{\mathcal{P}} = 1.4$-$1.5$ [13,14] and the measurements of the diffractive cross–section in the deep inelastic scattering provide $\alpha_{\mathcal{P}} = 1.23$ [15]. The above conclusions are based on the presumed dominance of the Pomeron contribution already in the preasymptotic energy region. What is called a Pomeron is to be interpreted as a true asymptotical contribution of the driving mechanism.

We would like to argue that all the three classes of the processes described above are related to the similar mechanisms and the corresponding energy dependence of the cross-sections can be well described by the universal functional energy dependence of the type $a + b\sqrt{s}$. Beyond the preasymptotic energy region unitarity changes the picture drastically. We have already discussed fit to the total hp cross-sections. The value of $\sqrt{s_0} \simeq 2\,\mathrm{TeV}$ is on the verge of the preasymptotic energy region, i.e. the Tevatron energy is at the beginning of the road to the asymptotics. Evidently the HERA energy range $W(=\sqrt{s_{\gamma p}}) \leq 300$ GeV is in a preasymptotic domain as well.

The chiral quark model gives the linear with $\sqrt{s}$ dependence for the total cross–sections :

$$\sigma_{tot}^{hp,\gamma p} = a + b\sqrt{s}, \tag{29}$$

where parameters a and b are different for different processes and the same is true for the scale s_0. Eq. 29 is in a good agreement with the data[7]. The same dependence

$$\sigma_{tot}^{\gamma^* p}(W) = a(Q^2) + b(Q^2)W \tag{30}$$

assumed by the small–x behavior of the structure function $F_2(x,Q^2)$ [13,14] and obtained in [16]:

$$F_2(x,Q^2) = a(Q^2) + b(Q^2)/\sqrt{x}. \tag{31}$$

From Eqs. (29) and (30) we arrive to conclusion on the universal c.m.s. energy dependence in the framework of the used model.

The third value for the Pomeron intercept $\alpha_{\mathcal{P}} = 1.23$ has been obtained from the analysis of the experimental data on the diffractive cross–section in DIS [15] which demonstrate linear rise of the differential cross–section $d\sigma^{diff}_{\gamma^* p\to XN}/dM_X^2$ with W, i.e. just the same functional dependence on the c.m.s. energy as for $\sigma_{tot}^{hp,\gamma p,\gamma^* p}$.

Having in mind the preasymptotic nature of the interaction mode ($s \ll s_0$) we arrive to the universal energy dependence for the diffractive cross–sections which are again determined by the generic form

$$\frac{d\sigma^{diff}_{hp\to XN}}{dM_X^2} \propto a + b\sqrt{s}. \qquad (32)$$

and

$$\frac{d\sigma^{diff}_{\gamma^* p\to XN}}{dM_X^2} \propto a(Q^2) + b(Q^2)W. \qquad (33)$$

The above linear dependences for the cross–sections of different processes is the generic feature associated with the preasymptotic nature of the interaction dynamics at $s \ll s_0$. As one goes above this energy range the unitarity starts to play the major role and provides the $\ln^2 s$ rise of the total cross–sections at $s \gg s_0$ and also leads to the decreasing dependence of the inelastic diffractive cross–section at $s \to \infty$

$$\frac{d\sigma^{diff}}{dM_X^2} \propto \left(\frac{1}{\sqrt{s}}\right)^N. \qquad (34)$$

for the hp, γp and $\gamma^* p$ processes [17]. Eq. 34 is associated with the antishadow scattering mode which develops at small impact parameters at $s > s_0$.

Conclusions

The genesis of hadron scattering when energy increases can be described as transition from the grey to black disk and, eventually, to the black ring with development of the antishadow scattering mode in the center. Such transition is under control of unitarity. The threshold value is estimated by $\sqrt{s_0} \simeq 2$ TeV.

The appearance of antishadow scattering mode could be revealed through impact parameter analysis of elastic scattering and directly in the measurements of the inelastic diffractive cross section. Unitarity leads to the decreasing inelastic diffractive cross-sections for hp, γp and $\gamma^* p$ processes at $s \to \infty$.

It would be interesting to discover correlations between the antishadow scattering mode and the new phenomena expected at high energies and usually associated with the formation of quark–gluon plasma (and disoriented chiral condensate).

In the framework of effective chiral quark model it was argued that the most of the currently available data on diffractive scattering should be attributed to the preasymptotic energy range. We have demonstrated the possibility of the unified description of the processes of hp, γp and $\gamma^* p$ diffraction scattering with the universal energy dependence of the type $a + b\sqrt{s}$. Beyond the preasymptotic region the unitarity changes the picture drastically. Thus we might expect the different asymptotia and universal preasymptotic regimes for the considered different classes of diffraction processes.

References

1. S. P. Denisov et al., Phys. Lett. **36B**, 415, 528 (1971).
2. A. A. Logunov, V. I. Savrin, N. E. Tyurin and O. A. Khrustalev, Teor. Mat. Fiz. **6** (1971) 157;
3. F. Ravndal, Int. J. Mod. Phys. **A8** 4369, 1993
4. N. E. Tyurin and O. A. Khrustalev, Teor. Mat. Fiz. **24** (1975)291;
5. S. M. Troshin and N. E. Tyurin, Sov. J. Particles and Nuclei **15** (1984) 25;
6. S. M. Troshin and N. E. Tyurin, Phys. Lett. **B 316** (1993) 175;
7. P. M. Nadolsky, S. M. Troshin and N. E. Tyurin, Z. Phys. C **69,** 131 (1995).
8. M. Baker and R. Blankenbecler, Phys. Rev. **128** (1962) 415.
9. S. M. Troshin, N. E. Tyurin and O. P. Yuschenko, Nuovo Cim. **91A** (1986) 23;
10. T. Chou and C.N. Yang, Phys. Rev. **D32** (1985) 1692.
11. L. Durand and H. Pi, Phys. Rev. **D38** (1988) 78.
12. P. V. Landshoff, The Two Pomerons, Talk given at PSI school at Zuos, August 1994
    M. Bertini, M. Giffon and E. Predazzi, Two Pomerons?, INFN 1995
13. H1 Collaboration, Nucl. Phys. B **407,** 515 (1993) ibid. B **439,** 379 (1995).
14. ZEUS Collaboration, Phys. Lett. **B316,** 412 (1993); Z. Phys. C **65,** 379 (1995).
15. ZEUS Collaboration, Z. Phys. C **70,** 391 (1996).
16. S. M. Troshin and N. E. Tyurin, Europhys. Lett., **37**, 239 (1997).
17. S. M. Troshin and N. E. Tyurin, Z. Phys. C **64,** 311 (1994).

E. Predazzi

THE POMERON IN THE IMPACT PARAMETER

B.Z. KOPELIOVICH[1,2], I.K. POTASHNIKOVA[1,2], B. POVH[1], E. PREDAZZI[3]
[1] *Max-Planck Institut für Kernphysik, Postfach 103980, 69029 Heidelberg, Germany*

[2] *Joint Institute for Nuclear Research, Dubna, 141980 Moscow Region, Russia*

[3] *Università di Torino and INFN, Sezione di Torino, I-10125, Torino, Italy*

The analysis of the t-dependent differential elastic $pp \to pp$ and diffractive $pp \to pX$ scattering data shows that the effective Pomeron intercept varies with the impact parameter. In elastic scattering the intercept $\Delta(b) = \alpha_{eff}(b) - 1$ is nearly zero at $b \ll 1\,fm$; this is a consequence of unitarity corrections [1]. With increasing b, however, $\Delta(b)$ keeps rising and reaches values few times larger than the one known from total cross sections. Such a behavior would be impossible if the Pomeron were a true Regge pole; standard Regge phenomenology predicts a constant $\Delta(b)$ in absence of unitarity corrections. On the other hand, the data on diffraction show a very different behavior of $\Delta(b)$ which grows with b only up to some small value $\Delta(b) \approx 0.05$ and does not show any trend to grow further. We interpret high-energy elastic scattering on the basis of the Leading-Log-Approximation in the parton model and arrive to a natural explanation for the above observations. Details of the work will be presented elsewhere [2].

1 Introduction to the problem, outline of the method and results.

The logarithmic increase of the total hadronic cross section at high energies is well known since the ISR experiments in early 70's. This discovery came just on time for the Regge theory conjecture. Furthermore, a simple parametrisation of the hadronic cross section, $\sigma_{tot} \propto s^{\Delta}$, made an interpretation of the high energy interaction in the frame of the Regge theory very appealing. The exponent in the energy term of the cross section is related to the Pomeron pole intercept, $\alpha_P(0) = 1+\Delta$. The power growth with energy violates the Froissart bound [3] at asymptotically high energies, but breaks down the unitarity limit for the elastic partial amplitude at small impact parameters at medium high available energies and Regge cuts are supposed to restore unitarity [4,5]. The assumption that the hadronic elastic amplitude at high energies is governed by a Regge pole called Pomeron has no theoretical justification beyond simplicity and faces problems interpreting data from HERA which demonstrate that Δ substantially (by factor ~ 5) increases with Q^2 (see, however, [6]).

The data on hadronic elastic scattering at small t, *i.e.* total cross sections, slope and ratio of real to imaginary parts of the elastic amplitude can be successfully fitted by many phenomenological more or less *ed hoc* models based on different assumptions. A firm point, however, is that the imaginary part of the forward elastic amplitude is connected by the unitarity relation (optical theorem) to the total inelastic cross section, *i.e.* is the shadow of inelastic processes. This implies that one must cope with the dynamics of inelastic collisions to understand the forward elastic scattering. This can be done in the case of perturbative QCD for DIS where inelastic collisions are associated with gluon bremsstrahlung. Depending on the approximation used, two models for the Pomeron are known, the BFKL Pomeron [7] and the double-leading-log DGLAP (see in [8,9]). The rising total cross section results then from the growth with energy of the phase space for radiated gluons. Although the perturbative methods cannot be applied to soft hadronic interaction, we expect a similar source of energy dependence and develop a QCD motivated phenomenology for inelastic hadronic collisions.

In the standard Regge phenomenology the energy dependence of the total cross section and the interaction radius (related to the slope parameter for the forward elastic differential cross section) are controlled by independent parameters, the intercept and the slope of the Pomeron trajectory, respectively. However, one may expect them to be correlated since in QCD the cross sections depend on the hadronic sizes due to color screening [10,11,12,13,14]. An attempt to incorporate this property was made recently in [15]. What we have done is to develop this approach treating more consistently the phase space for the radiated gluons. For this, the comparison with data is made in the impact parameter representation where turns out to be more effective for various reasons connected with unitarity. We refer the interested reader to the anticipation of our method [1] and to our future extended publication [2].

Here, suffices to say that we extend the procedure used by Amaldi and Schubert [16] to extract information on the partial elastic amplitude in the impact parameter representation from the elastic pp and $\bar{p}p$ scattering data. For this, however, we need to adapt the procedure of Ref.[[16]] in order to reduce its energy model dependence on the normalization of differential elastic cross sections (for details, see [2]). The data selected for the analysis cover the energy range from ISR to $Sp\bar{p}S$ but do not include Tevatron data since they have too narrow range of t to perform Fourier transform. The imaginary and real parts of the elastic scattering amplitude are parametrized in a model independent way as,

$$\mathrm{Im}\, f(t) = \sum_{i=1}^{3} a_i\, e^{b_i\, t}; \tag{1}$$

$$\mathrm{Re}\, f(t) = \sum_{i=1}^{2} c_i\, e^{d_i\, t} \ , \tag{2}$$

where t is the four-momentum transfer squared (which is negative), and a_i, b_i, c_i, d_i are the adjustable parameters of the fit. We use the following normalisation,

$$\frac{d\sigma}{dt} = \left[\mathrm{Re}\, f(t)\right]^2 + \left[\mathrm{Im}\, f(t)\right]^2 \ ; \tag{3}$$

$$\sigma_{tot} = 4\sqrt{\pi}\, F(0) \ . \tag{4}$$

We performed two variants of fit, one *variant I* with $c_2 = 0$ and the second *variant II* where both c_1 and c_2 are used a free parameters in the fit. As soon as the parameters are fixed, one can calculate at each energy the partial amplitude in the impact parameter representation,

$$\Gamma(b) = \frac{1}{2\,\pi^{3/2}} \int d^2b\, e^{i\,\vec{q}\cdot\vec{b}}\, f(-q^2) \ , \tag{5}$$

where $\vec{q}$ is the transverse component of the transferred momentum, $t \approx -q^2$ and where the normalization is

$$\sigma_{tot} = 2 \int d^2b\, \mathrm{Im}\, \Gamma(b) \tag{6}$$

The results for $\Gamma(b)$ show that at $b = 0$ the amplitude nearly saturates the unitarity limit and hardly changes with energies, while at larger impact parameters the amplitude grows considerably with energy.

The fit to the logarithmic derivative of $\Gamma(b)$, *i.e.* to the slope $\Delta(b)$ as found by the fit is shown in Fig.1

As one can see, $\Delta(b)$ is tiny, ~ 0.05 at small b, but increases enormously, up to $\Delta(b) \sim 0.3$ at large $b = 3\, fm$. Thus, the energy dependence of the total cross section originates mainly from peripheral interactions.

The systematic uncertainty of the analysis is related mainly to the choice of parameterization of the elastic amplitude. The results corresponding to variant II of the fit are depicted by open points; the systematic errors increase substantially at $b > 2\, fm$, while at smaller values of b they are rather small.

A QCD based parton model has been developed to derive the elastic amplitude from the dynamics of inelastic collisions and is discussed in details elsewhere ([2]). Assuming the t-dependence of the elastic amplitude $f_n(s,t)$ corresponding to the Fock component with n gluons to be exponential, the full elastic amplitude is found to be

$$f_P(s,t) = N \sum_{n=0}^{\infty} \frac{(\Delta \ln s)^n}{n!} \left(R_0^2 + \frac{9}{16}\lambda^2\, n\right) \exp\left[\frac{1}{4}\, t\, (R_0^2 + \lambda^2\, n)\right] \ . \tag{7}$$

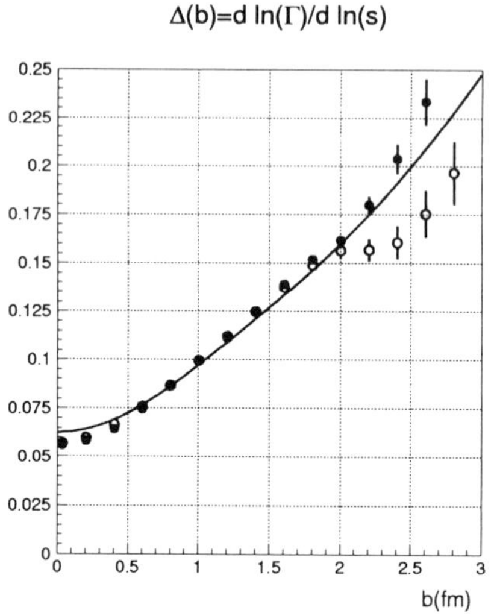

Figure 1: The variation with b of the exponent $\Delta(b)$ as found by the fit with the power energy dependence at each value of b. The black and open points correspond to the fits with parametrizations I and II respectively.

and the imaginary part of the partial amplitude in the impact parameter representation is given by Fourier transform of (7),

$$\tilde{f}_P(s,b) = \frac{N}{\pi} \sum_{n=0}^{\infty} \frac{(\Delta \ln s)^n}{n!} \left(\frac{R_0^2 + (9/16)\,\lambda^2\, n}{R_0^2 + \lambda^2\, n} \right) \exp\left(-\frac{b^2}{R_0^2 + \lambda^2\, n} \right) \qquad (8)$$

This amplitude grows with energy and violates the unitarity bound $f(s,b) \leq 1$, therefore it has to be unitarized. This is quite an uncertain procedure, for which many recipes are known (but all give, qualitatively similar results).

With some work, one calculates the *unitarized* $\Delta_{eff}(b)$ which is shown by the solid curve in Fig.1. We conclude that our model for the Pomeron reproduces the data for $\Delta(b)$ quite well (much better than other models) and that unitarization is very important at small b as one could expect. On the other hand, as we mentioned this is quite an uncertain procedure and one should not expect an excellent agreement, especially when the data are as precise as in our case.

The similar analysis carried out for diffractive data shows a rather different behavior and the analysis is still not complete.

References

1. B. Z. Kopeliovich, B. Povh and E. Predazzi, Phys. Lett. **B 405** (1997) 361.
2. B. Z. Kopeliovich, B. Povh, I.K. Potashnikova and E. Predazzi, to be published.
3. M. Froissart, Phys. Rev. **123** (1961) 1053
4. J.L. Cardy, Nucl. Phys. **B75** (1974) 413
5. M.I. Dubovikov, B.Z. Kopeliovich, L.I. Lapidus and K.A. Ter-Martirosyan, Nucl. Phys. **B123** (1977) 147
6. A. Capella, A. Kaidalov, C. Merino and J. Tran Thanh Van, Phys. Lett. **B337** (1994) 358
7. E. A. Kuraev, L. N. Lipatov and V. S. Fadin, Z. Eksp. Teor. Fiz. **72** (1977), 199; Ya. Ya. Balitsky and L. N. Lipatov, **28** (1978), 1957 (Sov. J. Nucl. Phys. **28** (1978), 822); Sov. Phys. JETP **63** (1986), 904. For a more recent updating, see, V. S. Fadin and L. N. Lipatov, *Proceedings of the LISHEP International Conference*, Rio de Janeiro, February 1998 (to be published by World Scientific).
8. Yu.L. Dokshitzer, V.A. Khoze, A.H. Mueller and S.I. Troyan, Basics of Perturbative QCD, Editions Frontieres, ADAGP, Paris 1991
9. B. Badelek, K. Charchula, M. Kravczyk and J. Kwiecinski, Rev. Mod. Phys. **64** (1992) 927
10. F.E. Low, Phys. Rev. **D12** (1975) 163
11. S. Nussinov, Phys. Rev. Lett. **34** (1975) 1986
12. J.F. Gunion and D.E. Soper, Phys. Rev. **D15** (1977) 2617
13. Al.B. Zamolodchikov, B.Z. Kopeliovich and L.I. Lapidus, Sov. Phys. JETP Lett. **33** (1981) 612
14. J. Hüfner and B. Povh, Phys. Rev. **D46** (1992) 990
15. B. Povh, Hadron Interactions - Hadron Sizes, hep-ph/9806379
16. U. Amaldi, K.R. Schubert, Nucl. Phys. **B166** (1980) 301

HIGH-ENERGY HADRONIC INTERACTIONS AND DIFFRACTION IN DIS

A. KAIDALOV
ITEP, Moscow

A connection between small x deep inelastic scattering (DIS) and high-energy hadronic interactions is discussed. An important role of multi–Pomeron exchanges for supercritical Pomeron is emphasized. It is demonstrated that the multi–Pomeron exchanges are very important for understanding of many characteristic features of multiparticle production in hadronic interactions as well as properties of diffractive dissociation in these processes. It is argued that in DIS these contributions are much less important and energy behavior of the structure function in DIS in the region of not too large Q^2 is determined mostly by the bare Pomeron exchange. An explicit model for contribution of multi–Pomeron exchanges to amplitudes of virtual photon - proton interaction is proposed. It is shown that this model gives a unified description of structure function F_2 in a broad region of Q^2 and of diffraction dissociation of a virtual photon. Predictions of the model in the region of intermediate Q^2 are used as initial conditions for QCD evolution at very large Q^2. This approach allows to reproduce all main features of experimental data at very small x observed at HERA. An important test of the model is provided by the data on small–x behaviour of structure functions of nuclei. It is shown that the model correctly predicts shadowing corrections for nuclear structure functions. It is emphasized that inclusive spectra and multiplicity distributions of final hadrons in DIS give an important information on dynamics of the process. The AGK–cutting rules allow to express the properties of multiparticle production in DIS in terms of the fragmentation functions found from analysis of hadronic interactions. Predictions of this approach are in a good agreement with HERA data.

HELICITY FLIP IN DIFFRACTIVE DIS

N.N. Nikolaev
Institut f. Kernphysik, Forschungszentrum Jülich, D-52450 Jülich, Germany
L.D.Landau Institute for Theoretical Physics, 142432 Chernogolovka, Russia
E-mail: N.Nikolaev@fz-juelich.de

I review new results on s-channel helicity nonconservation (SCHNC) in diffractive DIS. I discuss how by virtue of unitarity diffractive DIS gives rise to spin structure functions which were believed to vanish at $x \ll 1$. These include tensor polarization of sea quarks in the deuteron, strong breaking of the Wandzura-Wilczek relation and demise of the Burkhardt-Cottingham sum rule.

1 Introduction

The unitarity is a recurrent theme at the Blois Woprkshops on diffractive and elastic scattering. The often repeated fundamental argument is that diffractive scattering is driven via unitarity by absorption due to multiparticle production. The common wisdom is that this entails vanishing spin-dependence of diffractive scattering. The related QCD argument has been that quark helicity conservation entail the s-channel helicity conservation (SCHC) at small x, i.e., decoupling of QCD pomeron from helicity flip. Here I review the recent discovery [1,2,3,4,5,6] of substantial SCHNC in diffractive DIS into the both continuum and vector mesons. Furthermore, SCHNC diffractive DIS in conjunction with unitarity changes dramatically the small-x behaviour of spin structure function g_2 and leads to the demise of the Burkhardt-Cottingham sum rule and the departure from the Wandzura-Wilczek relation. Still another diffraction driven spin effect is the tensor polarization of sea quarks in the deuteron.

2 Is SCHNC compatible with quark helicity conservation?

The backbone of DIS is the Compton scattering (CS) $\gamma^*_\mu p \to \gamma^{*'}_\nu p'$. The CS amplitude $A_{\nu\mu}$ can be written as $A_{\nu\mu} = \Psi^*_{\nu,\lambda\bar{\lambda}} \otimes A_{q\bar{q}} \otimes \Psi_{\mu,\lambda\bar{\lambda}}$ where $\lambda, \bar{\lambda}$ stands for $q, \bar{q}$ helicities, $\Psi_{\mu,\lambda\bar{\lambda}}$ is the wave function of the $q\bar{q}$ Fock state of the photon. The QCD pomeron exchange $q\bar{q}$-proton scattering kernel $A_{q\bar{q}}$ does not depend on, and conserves exactly, the $q, \bar{q}$ helicities. For nonrelativistic massive quarks, $m_f^2 \gg Q^2$, one only has transitions $\gamma^*_\mu \to q_\lambda + \bar{q}_{\bar{\lambda}}$ with $\lambda + \bar{\lambda} = \mu$. However, the relativistic P-waves give rise to transitions of transverse photons $\gamma^*_\pm$ into the $q\bar{q}$ state with $\lambda + \bar{\lambda} = 0$ in which the helicity of the photon is transferred to the $q\bar{q}$ orbital momentum. Consequently, the QCD pomeron exchange SCHNC

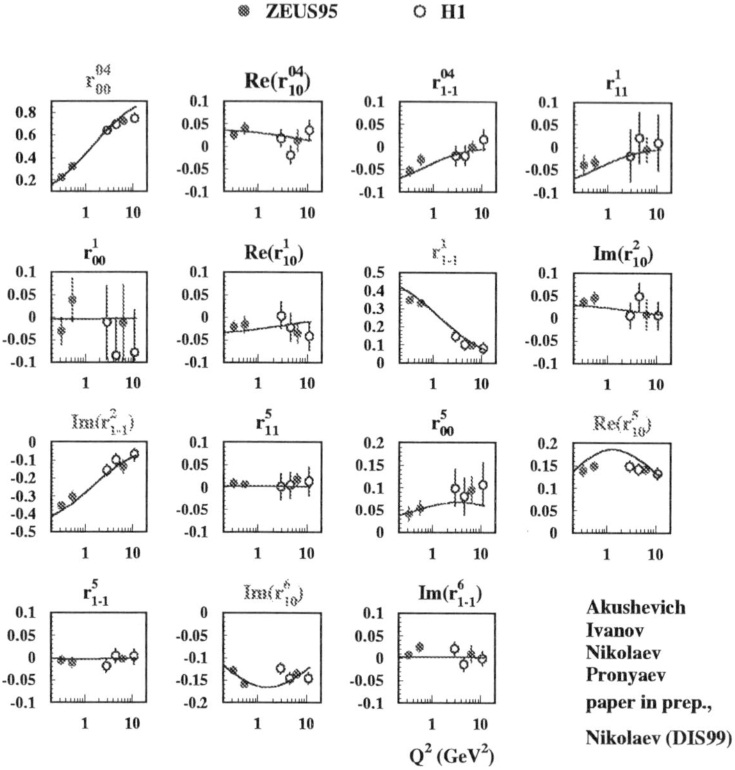

Figure 1: Our prediction [5] for the spin density matrix r^n_{ik} of diffractive ρ^0-meson vs. the experimental data from ZEUS [8] and H1 [9].

transitions $\gamma^*_\pm \to (q\bar{q})_{\lambda+\bar{\lambda}=0} \to \gamma^*_L$ and $\gamma^*_\pm \to (q\bar{q})_{\lambda+\bar{\lambda}=0} \to \gamma^*_\mp$ are allowed [1,2] and SCHNC persists at small x. We emphasize that the above argument for SCHNC does not require applicability of pQCD.

3 LT interference and SCHNC in diffractive DIS

The first ever direct evaluation [1] of SCHNC effect in QCD - the LT-interference of transitions $\gamma^*_L p \to p'X$ and $\gamma^*_\pm p \to p'X$ into the same continuum diffractive states X - has been reported in 1997. Experimentally, it can be measured at HERA by both H1 and ZEUS via azimuthal correlation between the (e, e') and (p, p') scattering planes and can be used the determination of the otherwise elusive $R = \sigma_L/\sigma_T$ for diffractive DIS is found in [2]. The principal issue is that this asymmetry persists, and even rises slowly, at small $x_{\mathbb{P}}$.

The azimuthal correlation of the (e, e') and (p, V) planes with the vector meson decay plane, and the in-decay-plane angular distributions of decay products, allow the experimental determination of all helicity amplitudes $A_{\mu\nu}$ for diffractive $\gamma^* p \to V p'$. One finds that helicity flip only is possible due to the transverse and/or longitudinal Fermi motion of quarks and is extremely sensitive to spin-orbit coupling in the vector meson, I refer for details to [3,4]. The consistent analysis of production of the S-wave and D-wave vector mesons is presented only in [4]. One would readily argue that by exclusive-inclusive duality [7] between diffractive DIS into continuum and vector mesons [1,2] the dominant SCHNC effect in vector meson production is the interference of SCHC $\gamma_L^* \to V_L$ and SCHNC $\gamma_T^* \to V_L$ production, i.e., the element r_{00}^5 of the vector meson spin density matrix. The overall agreement between our theoretical estimates [5] of the spin density matrix r_{ik}^n for diffractive ρ^0 and the ZEUS [8] and H1 [9] experimental data is very good. There is a clear evidence for $r_{00}^5 \neq 0$.

In the D-wave state the total spin of $q\bar{q}$ pair is predominantly opposite to the spin of the D-wave vector meson. As a results, SCHNC in production of D-wave vector mesons is much stronger [4] than for the ground state S-wave mesons, which may facilitate the disputed D-wave vs. $2S$-wave assignment of the $\rho'(1480)$ and $\rho'(1700)$ and of the $\omega'(1420)$ and $\omega'(1600)$. Striking predictions for D-wave meson production include abnormally large higher twist corrections [4] and non-monotonous Q^2 dependence of $R^D = \sigma_L/\sigma_T$.

4 Impact of diffraction upon g_2: breaking of the Wandzura-Wilczek relation and demise of the Burkhardt-Cottingham sum rule

The transverse spin asymmetry A_2 in polarized DIS is proportional to the amplitude of forward CS $\gamma_L^* p\uparrow \to \gamma_T^* p\downarrow$ which is proportional to $g_{LT} = g_1 + g_2$. In the standard two-gluon t-channel tower approximation the cross-talk of the target and beam helicity flip only is possible at the expense of suppression of the small-x behaviour of $x^2 g_{LT}$ by the extra factor $\sim x$ compared to F_1, because not both gluons in the pomeron can have the nonsense polarization simultaneously.

The more familiar argument for the vanishing A_2 has been the parton model Wandzura-Wilczek relation between g_{LT} and g_1 [11]:

$$g_{LT}(x, Q^2) = \int_x^1 \frac{dy}{y} g_1(x, Q^2)\,,$$

which entails $g_{LT} \sim g_1$. Because the diffractive pomeron exchange does not contribute to g_1, the WW relation can be reinterpreted as a vanishing pomeron exchange contribution to g_{LT}. This vanishing of the pomeron

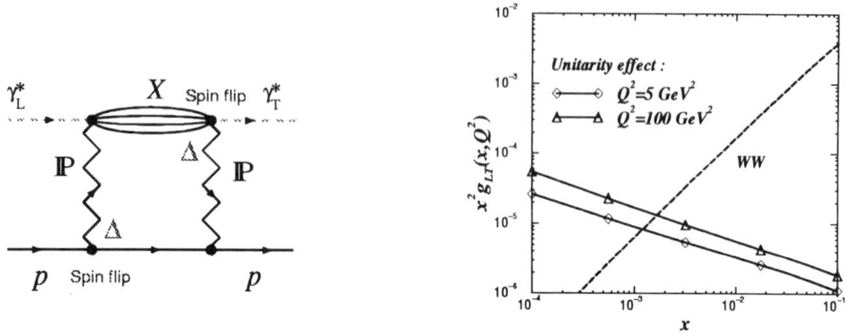

Figure 2: The LHS: The unitarity diagram for diffractive contribution to g_{LT}. The RHS: The unitarity correction to, and WW relation based evaluation of, g_{LT}.

contribution has been the principal motivation behind the much discussed Burkhardt-Cottingham [12] (BC) sum rule $\int_0^1 dx g_2(x, Q^2) = 0$.

Our recent discovery is that diffractive SCHNC destroys via unitarity the both WW relation and BC sum rule [6]. The unitarity diffractive contribution (fig. 2) to CS has as building blocks diffractive amplitudes $\gamma^* p \to p' X$ in which there is a helicity flip sequence, $\gamma_L^* \to X_L \to \gamma_T^*$ in the top blob and helicity flip sequences either $p\uparrow \to p'\uparrow \to p\downarrow$ or $p\uparrow \to p'\downarrow \to p\downarrow$ in the bottom blob. The both blobes are proportional to $\mathbf{\Delta}$ and vanish for forward produced X, but upon the integration over the phase space of $p'X$ one finds the nonvanishing $\int d^2\mathbf{\Delta}\Delta_i\Delta_k$ and unitarity driven transition $\gamma_L^* p\uparrow \to \gamma_T^* p\downarrow$ which does not vanish in the forward direction. The principal point is that the unitarity diagram furnishes the cross-talk of the beam and target helicity flip with pure nonsense polarizations of all the four exchanged gluons in the two pomerons.

Our result [6] for the diffraction-driven g_{LT} reads

$$g_{LT}(x, Q^2) \propto \frac{1}{x^2} r_5 \int_x^1 \frac{d\beta}{\beta} g_{LT}^D(x_{\mathbf{IP}} = \frac{x}{\beta}, Q^2)\,. \tag{1}$$

It rises steeply at small x. It is the scaling function of Q^2 because the diffractive LT structure function $g_{LT}^D(x_{\mathbf{IP}}, Q^2)$ is the scaling one. The resulting asymmetry $A_2 \propto x g_{LT}/F_1$ does not vanish at small x, furthermore, at a moderately small x it even rises because $g_{LT}^D(x_{\mathbf{IP}}, Q^2) \propto G^2(x_{\mathbf{IP}}, \overline{Q}^2)$ where $G(x_{\mathbf{IP}}, \overline{Q}^2)$ is the gluon SF of the proton and $\overline{Q}^2 \sim 0.5$-1 GeV^2.

In fig. 3 we show how the steeply rising unitarity correction overtakes at small x the standard g_{LT} evaluated from the Wandzura-Wilczek (WW) relation starting with fits to the world data on g_1. As such our unitarity effect is the

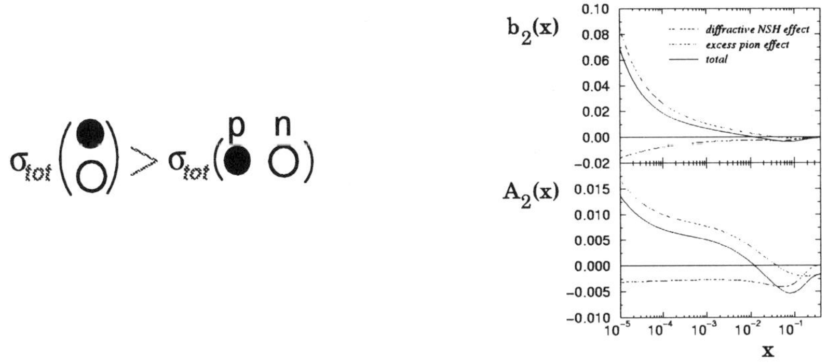

Figure 3: The LHS: The unitarity diagram for diffractive contribution to g_{LT}. The RHS: The unitarity correction to, and WW relation based evaluation of, g_{LT}.

first nontrivial scaling departure from the WW relation

Finally, the above breaking of the WW relation implies $g_{LT} \gg g_1$ and $g_2 = g_{LT}$ at very small x. Consequently, the unitarity-driven rise of g_2 destroys the BC sum rule because the BC integral would diverge severely. Incidentally, the BC sum rule has always been suspect.

5 Diffraction, unitarity and tensor structure function of the deuteron

The deuteron is an unique spin-1 target. Because of the S-D wave interference it is a dumbbell and spatial orientation of nucleons and their Fermi motion in the deuteron depend on its polarization which may give rise to the dependence of parton densities on the tensor polarization of the deuteron. In the impulse approximation the tensor structure function $b_2(x)$ is found to be negligible, per mill, effect [14] and satisfies the Close-Kumano sum rule [15] $\int_0^1 \frac{dx}{x} b_2(x) = 0$. Close and Kumano conjectured this is a generic QCD sum rule.

Fig. 3 makes it obvious that unitarity-driven diffractive eclipse effect in the deuteron depends on its orientation which is controlled by tensor polarization. The results of the calculation [13] of the eclipse effect are shown in fig. 3. The tensor polarization of the sea measured by tensor asymmetry A_2 is quite substantial and persists at small x. Apart from diffractive eclipse effect, it receives a small contribution also from the pion exchange [13].

Conclusions

Drastic revision of our prejudice on spin-independence of diffraction is called upon. The recent fundamental finding is that QCD pomeron exchange does not conserve the s-channel helicity. The mechanism of SCHNC is well understood. SCHNC offers an unique window at the spin-orbit coupling in vector mesons. SCHNC in diffractive DIS drives, via unitarity relation, a dramatic small-x rise of the transverse spin structure function g_2 which breaks the Wandzura-Wilczek relation and invalidates the Burkhardt-Cottingam sum rule. The related unitarity effect is the tensor polarization of sea quarks in the deuteron which persists at small x.

Acknowledgments. Thanks are due to Prof. V. Petrov for the invitation to VIIIth 'Blois Workshop' in Protvino.

References

1. N.N. Nikolaev and B.G. Zakharov, Deep Inelastic Scattering and QCD, Proceedings of DIS'97, Chicago, IL, 14-18 April 1997, AIP Conference Proceedings No.407, editors J. Repond and D. Krakauer, pp. 445-455.
2. N.N. Nikolaev, A.V. Pronyaev and B.G. Zakharov, *Phys. Rev.* **D59** (1999) 091501.
3. E.V. Kuraev, N.N. Nikolaev and B.G. Zakharov, *JETP Lett.* **68** (1998) 667.
4. I.P.Ivanov and N.N.Nikolaev, *JETP Letters* **69** (1999) 268.
5. I. Akushevich, I. Ivanov, N.N. Nikolaev and A. Pronyaev, paper in preparation.
6. I. Ivanov, N.N. Nikolaev, A. Pronyaev and W. Schaefer, *Phys. Lett.* **B457** (1999) 218.
7. M.Genovese, N.N.Nikolaev and B.G.Zakharov, *Phys. Lett.* **B380** (1996) 213..
8. ZEUS Collab: A. Savin, DIS'99 and **DESY-99-102**.
9. H1 Collab: B. Clerbaux, DIS'99 and **DESY-99-010**.
10. W. Schäfer, Deep inelastic scattering and QCD, Proceedings of DIS'98, World Scientific, eds. Gh.Corenmans and R.Roosen, pp. 404-407.
11. S.Wandzura and F.Wilczek, *Phys. Lett.* **B72** (1977) 195.
12. H.Burkhardt and W.N.Cottingham, *Ann. of Phys. (USA)*, **56** (1970) 453.
13. N.N.Nikolaev and W.Schaefer, *Phys. Lett.* **B398** (1997) 245
14. P.Hoodbhoy and R.L.Jaffe, *Nucl. Phys.* **B312** (1989) 571. A.Umnikov, *Phys. Lett.* **B391** (1997) 177.
15. F.E.Close and S.Kumano, *Phys. Rev.* **D42** (1990) 2377.

REGGE INTERCEPTS AND COUPLINGS FROM HIGH ENERGY FORWARD SCATTERING DATA

J.R. Cudell[a], V. Ezhela[b], K. Kang[c], S. Lugovsky[b], and N. Tkachenko[b]

a. Inst. de Physique, Université de Liège, Bât. B-5, Sart Tilman, B4000 Liège, Belgium

b. COMPAS Group, IHEP, Protvino, Russia

c. Department of Physics, Brown University, Providence RI 02912 USA

This is the first part of the two presentations on the results of the analytic models for high energy forward scattering. We present primarily the results of the intercepts and couplings of the soft Pomeron as a Regge pole as well as those of the ρ/ω and f/a effective Reggeons from all accumulated data set available for total cross section and real part of the hadronic amplitudes. Factorization and quark counting rules are satisfied by the Pomeron couplings to within 10%, which allows us to make predictions on γp and $\gamma\gamma$ total cross sections with the help of the generalized vector dominance idea. In addition we check the range of validity in energy for the Regge pole model from the criteria for acceptable fits and establish the bounds on the Odderon and hard Pomeron contributions.

1 Introduction

Universal description of forward scattering by hadrons and of rising total cross section have been with us over two decades [1]. The interest in this topic has been revived by the recent activities at HERA on deep-inelastic and diffractive scattering. While the deep inelastic scattering data from HERA shows a qualitative rise in the off-shell cross sections extrapolated to low Q^2, the hadronic total cross sections measured from forward scattering reveal a very slow rise in energy.

The simplest idea for the slow rise with energy of the hadronic total cross sections is to assume a Regge trajectory with an intercept slightly larger than 1, having quantum members of vacuum, *i.e.* the Pomeron as a simple Reggeon with $\alpha_P(0) = 1+\varepsilon$, ε being a small positive number. This approach can predict the universal rise with energy of the total cross sections, the factorization of the Pomeron coupling and the quark counting based on the additive coupling of the Pomeron to the constituent quarks of hadrons. In fact our experience of several decades has proven that the Pomeron is a simple Regge pole to a good approximation. Regge theory has been a successful framework for describing soft hadronic interactions and offers a useful guideline in searching a fundamental description for the soft processes based on QCD. The soft Pomeron intercept is a crucial element in HERA analyses and provides the starting point at low x and low Q^2, from which QCD evolution can be performed. It also determines the Pomeron flux factor, which enters gap cross section, and has a bearing on the extrapolation of total cross section to higher energies.

The idea to start from perturbative QCD assumes that a higher order effects would unitarize the amplitude and tame the fierce rise observed at large Q^2 to something compatible with the Froissart bound. But no one has reliably unitarized QCD amplitudes. Moreover such unitarisation will involve necessarily multi-gluon exchanges between the quarks and therefore will require detailed quark structure, i.e., the hadronic wave functions. In such scheme, one is likely to lose the factorization property of the Pomeron as well as the quark counting rules and the universality property [2]. In addition, such unitarisation will lead to cut singularity rather then a pole type so that a power behavior in log s instead of s will be likely for the total cross sections.

Thus the simple pole model for the Pomeron remains to be the simplest and perhaps the most attractive option for near forward hadron scattering at high energies, even though this model does not offer a simple and automatic extension to the off-shell particle scattering and in particular to the deep inelastic scattering. A possible suggestion is to assume an extra "hard Pomeron" [3] that decouples at $Q^2 = 0$. Such extra trajectory is to some extent confirmed by the DGLAP evolution but the BFKL re-summation of energy logarithms is generally believed to be relevant to the high energy hadron processes. It has recently been shown[4] that the next-to-leading order corrections can drastically lower the intercept of the BFKL Pomeron, leading to a very weak dependence on Q^2 result that can be consistent with the soft Pomeron behavior exhibited by the hadronic cross section. However this is achieved by a fine-tuned compensations between the leading order and next-to-leading order corrections in BFKL re-summation, which offers no simple clue for the off-shell extension.

Mainly because of the simplicity, the simple Regge pole idea has been pushed by Donnachie and Landshoff (DL) [5,6] to fit all soft data for scattering of on-shell particles even at non-zero t. The DL Regge model contains an exchange-degenerate meson Reggeon term in addition to the Pomeron term. However it was shown by Cudell, Kang and Kim (CKK) [7] that the assumption of an exchange-degenerate meson trajectory is not even supported by the fits to the total cross sections alone and fails to give stable parameters when fitted to the total cross sections and the real parts of the forward scattering amplitudes. The simplest modification of the DL model suggested by CKK [7] that contains exchanges of two non-degenerate ρ/ω and f/a trajectories is called the model

RRP in our work [8] and in the second presentation [9] by V.V. Ezhela.

Then the question is to what extent can one say that the RRP model is a unique analytic possibility to describe all hadronic soft data. Since it is clear that the Froissart bound, which is a consequence of the unitarity and positivity of the imaginary part of the amplitude in the Lehmann ellipse, will be violated eventually by the bare Pomeron term with an intercept $\alpha_P(0) = 1 + \varepsilon$ in the RRP model, the amplitude parameterization of RRP will not be valid at high energies where the unitarisation by multi-Pomeron exchanges is needed. Thus the validity region in energy of RRP model will necessarily have both a minimum cut-off energy below which the lower secondary trajectory and multi-Regge exchange effects are important and a maximum energy above which the unitarisation is needed. It was suggested in [7] that the stability of the values and errors of the parameters would be a good working criteria to decide the minimum cut-off energy of the validity region. This criteria is also adopted in the present work [8]. The bare Pomeron term can lead to either $\ln^2 s$ or $\ln s$ behavior for the cross section upon making an eikonal unitarisation [10]. Combined with the correct analyticity and crossing symmetry of the near forward amplitude, the results of the parameterization are given by either the model RRL2, which saturates the Froissart bound, or the model RRL1, which has a $\ln s$ behavior for the total cross section in our works [8,9]. We note in addition that the analytic parameterizations of RRL2 and RRL1 can easily be derived as solutions of the derivative dispersion relations [11], which approximate the region of analyticity at high energies but with a full unitarity of the scattering amplitude, or from the prescription of Block and Cahn [12]. In particular, RRL2 is often called as the Amaldi model [1] but has been around since 1974 [11] while RRL1 was proposed by Block, Kang and White [13] in 1990.

The results of comparative study of these three analytic models are presented in the second part [9], which shows that the simple pole model RRP is only one possibility of analytic amplitude models along with RRL2 and RRL1. Consideration of the χ^2 of the fits alone, though the largest data set [14] available is used, does not discriminate between the three analytic possibilities, implying that the effect of unitarisation is negligible and the maximum cut-off energy of the validity region is not evident up to the Tevatron energies.

In this presentation we concentrate on the results of the RRP model fits to the total cross sections and the ratio parameter ρ, *i.e.*, the ratio of the real part to the imaginary part of the forward scattering amplitudes. In particular we present the determination of the intercepts and couplings of the Pomeron and meson Reggeons. As we shall see, the only discrimination that the soft data can bring is lies in the confirmation of the properties that follow from that the Pomeron is a simple Regge pole coupled to constituent quarks, *i.e.*, factorization, quark counting and universality, though such property is also exhibited to some extent by RRL2 and RRL1. In the interest of saving the repetition, we refer to Ezhela presentation [9] for the notations of the RRP parameters. There are three Regge terms coming from the Pomeron and $C = \pm 1$ meson trajectories in RRP model for a given $h\ p$ scattering process where $h = p^{\pm}$, $\pi^{\pm}$ and $K^{\pm}$. In addition, $\gamma\ p$ and $\gamma\ \gamma$ cross section are described by two terms representing contributions from the bare Pomeron and $C = +1$ (a/f) Reggeon. Since the intercepts of the Pomeron and meson trajectories are universal, there are altogether 16 parameters for fitting the RRP model to the data set of total cross sections and ρ parameter of $h\ p$ scattering, ($h = p^{\pm}, \pi^{\pm}$, and $K^{\pm}$) and the total cross sections of $\gamma\ p$ and $\gamma\ \gamma$ collisions. The results presented in this and the second talks [9] complement the earlier preliminary results [15,16].

2 Dataset and Statistical Procedure

We have used the largest data set [14] prepared and maintained by the COMPAS group for the cross sections and ρ parameters. They contain 2747 (303) data points for total cross sections (ρ parameters) for $\sqrt{s} \geq 3$ GeV, and 271 (112) data points for total cross sections (ρ parameters) for $\sqrt{s} \geq 9$ GeV. The number of data points used in our fits is shown as a function of the minimum cut-off energy in Fig.1. As the dataset is large enough and contains no substantial inconsistencies, we have used the conventional definition of χ^2. Note however that the most interesting parameter ε is sensitive to the highest energy region where the data are scarce. One may thus consider a scheme giving more weight to the high energy data. Also one may remember that the total cross sections are better measured than the ρ parameters because of lingering doubt on the Coulomb - nuclear interference calculations. One further point is that the RRP model can not describe the resonance region which is also the case for other asymptotic analytic models and they can be trusted only above a certain minimum energy $\sqrt{s}_{min}$, which should be well above the threshold of resonance production. Thus the minimum cut-of energy $\sqrt{s}_{min}$ could be in principle process dependent. In spite of these worries we see no reason to use non-conventional definition of χ^2 and to invent different weighting schemes [6] for different energy data points but using the inverse squares of their total errors as weights. We demand that $\chi^2/d.o.f. \leq 1$ and that the values of the parameters and their errors remain stable with respect to the minimum cut-off energy $\sqrt{s}_{min}$, in order for the fits to be acceptable. This criterion implies that the Pomeron parameters are stable with respect to $\sqrt{s}_{min}$ and the low energy data becomes of no primary importance. The resulting $\chi^2/d.o.f.$ as well as the stability pattern of the parameters of RRP model is also shown in Fig.1.

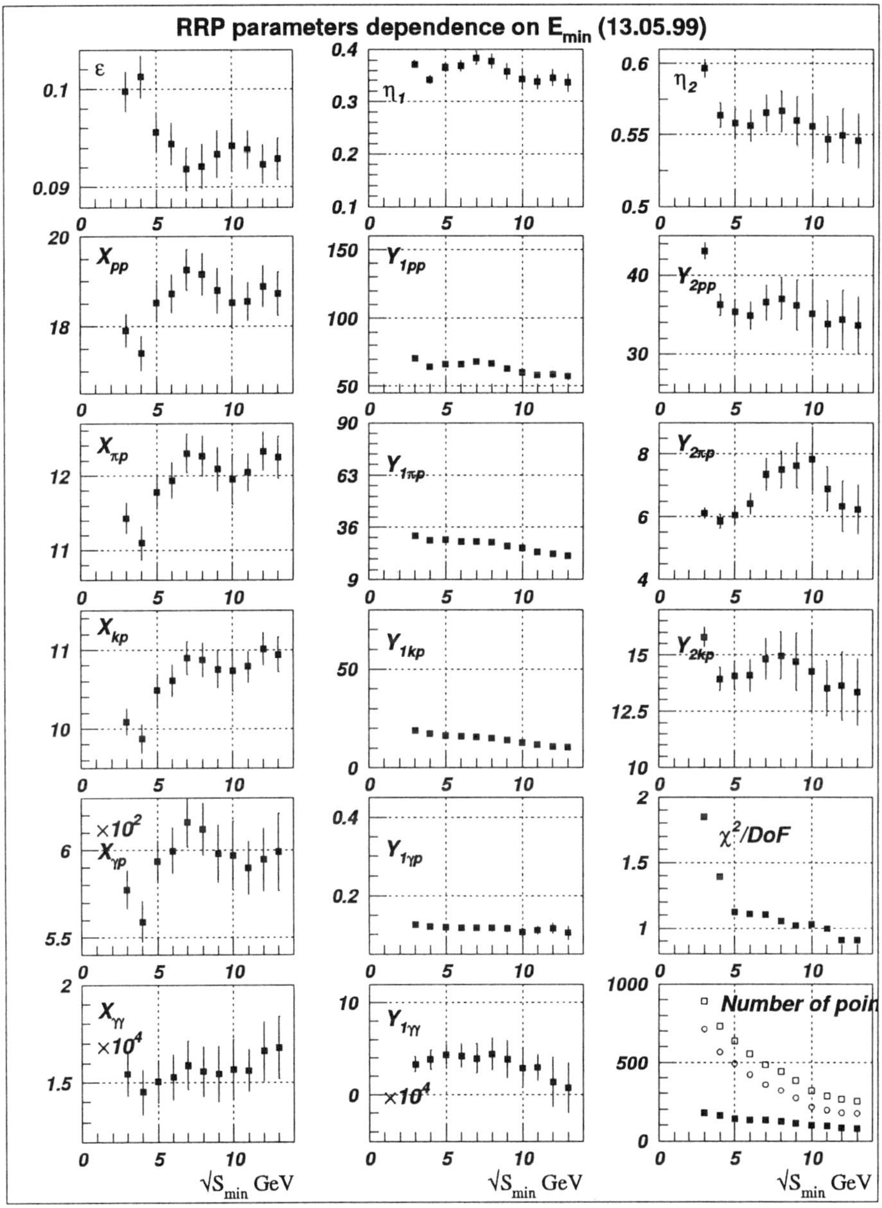

Figure 1: Parameters of the RRP model, as functions of the minimum energy considered in the fit

ε	η_1	η_2	$\chi^2/d.o.f.$	Number of data points	
0.0933(24)	0.357(15)	0.560(17)	1.02	383	
	pp	πp	Kp	$\gamma p \times 10^{-2}$	$\gamma\gamma \times 10^{-4}$
X_1 (mb)	18.79(51)	12.08(29)	10.76(23)	5.98(17)	1.55(14)
Y_1 (mb)	63.0(2.3)	26.20(74)	14.08(57)	11.64(88)	3.9(2.0)
Y_2 (mb)	36.2(3.2)	7.63(72)	14.7(1.3)		

Table 1: Summary of the fit results for $E^{cm}_{min} = 9$ GeV

Clearly, the fit is bad for small energies and in particular the criterion for the acceptable fits can be satisfied for $\sqrt{s}_{min} \geq 9$ GeV. As in the previous work of CKK [7], the proposed criteria fails to meet for the exchange-degenerate meson trajectories because $\chi^2/d.o.f.$ is large and of order 1.3 and the parameter values and their errors are unstable when fitted to the total cross sections and ρ parameters. In fact the results confirm the conclusion reached by CKK [7] that the assumption of exchange-degeneracy for $C = \pm 1$ meson tr ajectories is not supported even by the fits to the total cross sections alone. We need $C = \pm 1$ meson trajectories that are non-degenerate, particularly when we want to fit ρ parameters in addition to the total cross sections. Since the minimum cut-off energy $\sqrt{s}_{min}$ turns out to be as high as 9 GeV, the Pomeron parameters are reasonably well determined while those of the lower trajectories are determined less accurate. As it can be seen from Fig.1, the Pomeron intercept and couplings are stable with respect to $\sqrt{s}_{min}$ beyond 8 GeV. But as for the lower trajectories though $C = \pm 1$ intercepts and the $C = -1$ coupling are stable, $C = -1$ couplings do have $\sqrt{s}_{min}$ dependence within large errors. Note also that the $C = +1$ parameters are highly correlated to the Pomeron parameters so that we can not determinate the a/f couplings reliably. With these in mind, we give in Table 1 the values of the parameters determined for $\sqrt{s}_{min} = 9$ GeV.

We see that the global $\chi^2/d.o.f.$ is good, even though the data is not filtered as in CKK [7], where it was shown that χ^2 per data points for total cross section (ρ parameter) was high for $p\bar{p}$ (pp) scattering due to possible inconsistencies within the data but could drastically be lowered by fitting the data without changing the determination of the parameters. In the present work, we see also high χ^2 per data points for these two processes but a few of these high χ^2 do not affect much for the global $\chi^2/d.o.f$ and our conclusions. The RRP fits for the total cross sections and ρ parameters for $\sqrt{s}_{min} = 9$ GeV, extrapolated down to $\sqrt{s} \leq 5$ GeV, are shown and compared with the other two competing models in Figs.1 and 2 in the second part [9] of our presentations. Although the value of $\chi^2/d.o.f$ goes above 2 as it can be seen from Fig.1, so that statistically unacceptable, the fits look deceptively good. This is why we need a careful statistical analysis of the fits with physically sound criteria imposed on.

3 Discussion on the Results

3.1 Universality

As we see it from the comparative study [8, 9] of the three popular competing models, the χ^2 value alone can not differentiate the RRP model because other fits fare also as well. Thus the support for RRP model which treats the Pomeron as a simple Regge pole must come from some other evidence. In Regge theory, the l-plane singularities are universal, be they poles or cuts. Hence the energy dependence of the data has to be a combination of points which rise or fall with energy in a process-independent manner. Within the context of perturbative QCD, one could *a priori* expect a small deviation from ality [2], because the hadronic wave functions must come into calculation of the various terms. The universality of the Pomeron and meson trajectory intercepts is closely linked to the structure function F_2 at HERA [17]. There the effective Pomeron intercept, extracted from the power behavior in $1/x$ from $F_2(x)/x$, seems to depend on Q^2, the negative mass squared of the virtual photons. In order to check if a target mass M-dependent behavior of the Pomeron intercept is seen on the other side of $M^2 = 0$, we made a partial fit, fixing $C = \pm 1$ meson intercepts, at $\sqrt{s}_{min} = 9$ GeV with the errors in the intercept corresponding to a change of 1 unit in the $\chi^2/d.o.f.$ The result is shown in Fig. 2 which seems to imply that the soft Pomeron intercept may be universal and independent of the target mass within the errors, though not terribly overwhelming.

3.2 Factorization and Quark Counting Rule

The couplings of the Reggeon exchanges are expected to factorize into a product of two couplings, one for each interacting hadron. Also the Pomeron couples to single quarks like a $C = +1$ photon in Regge theory. The Pomeron being an extended object, this will be a viable scenario only for constituent quarks if the Pomeron as a simple pole. The result of Table 1 gives

$$\frac{(X_{pp}/X_{\pi p})}{3/2} = 1.04 \pm 0.11$$

$$X_{Kp}/X_{\pi p} = 0.89 \pm 0.05$$

$$X_{\gamma p} / \left\{ g_{em}^2 \left[\frac{1}{f_p^2} + \frac{1}{f_\omega^2} + \frac{1}{f_\rho^2} \right] (1+\delta) X_{\pi p} \right\} \approx \frac{213.9 X_{\gamma p}}{X_{pip}} = 1.06 \pm 0.04$$

$$X_{pp} X_{\gamma\gamma} / X_{\gamma p}^2 = 0.78 \pm 0.15$$

The first and second relations reflect the quark counting rule, the third is the factorization combined with generalized vector dominance [18] where the contribution from off-diagonal terms δ is expected to be about 15%, and the forth is an example of factorization. From these, we see that the properties of factorization and quark counting, expected from the Pomeron as a simple Regge pole, seem to hold within 10 %. Note however that quark counting fails to be for meson Reggeons, as they have to probe multi-quark configuration unlike the Pomeron which seems to couple to single quarks.

3.3 Other Possible Trajectories

The sharp rise observed at the HERA might be due to the presence of another singularity, hither to undetected, called the hard Pomeron. Assuming that this is a simple pole [3], one can fit the deep inelastic scattering data. To see the possible manifestation of the hard Pomeron in soft interactions, we tested its contribution assuming the hard Pomeron intercept to be 0.4 and obtained the following 2σ bounds on the hard Pomeron

$$X_{hard}^{pp} / X_{soft}^{pp} < 2 \times 10^{-6}$$

$$X_{hard}^{\pi p} / X_{soft}^{\pi p} \approx X_{hard}^{Kp} / X_{soft}^{Kp} < 3 \times 10^{-2}$$

$$X_{hard}^{\gamma p} / X_{soft}^{\gamma p} < 10^{-4}$$

Hence if the hard Pomeron exists as a simple pole, it much decouple at $Q^2 \leq 0$.

The exchange of a $C = -1$ trajectory [19] with intercept close to 1 is needed within the Donnachie- Landshoff model to reproduce the large $-t$ dip in elastic scattering. Assuming its intercept to be as low as 1, we obtain the 2σ bounds

$$\left| X_{odd}^{pp} / X_{soft}^{pp} \right| < 2 \times 10^{-3}$$

$$\left| X_{odd}^{\pi p} / X_{soft}^{\pi p} \right| < 10^{-3}$$

$$\left| X_{odd}^{Kp} / X_{soft}^{Kp} \right| < 2 \times 10^{-3}$$

Thus such object does not seem to be present at $t = 0$.

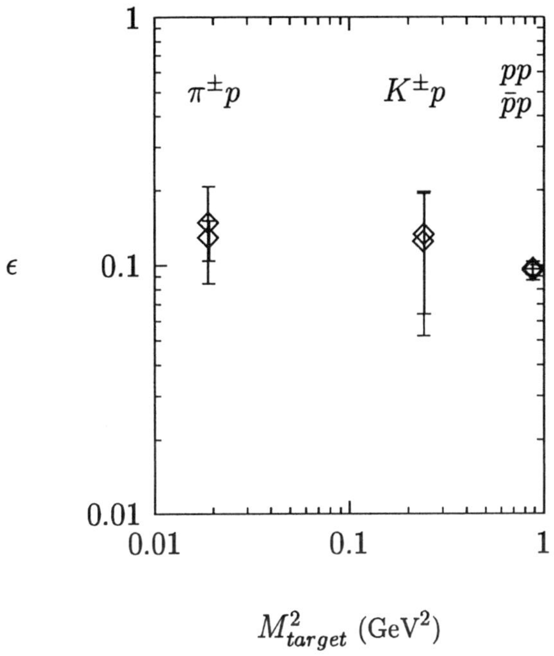

Figure 2: The value of the Pomeron intercept for three different processes.

4 Concluding Remarks

We have shown that RRP model produces very good fits to all data for $\sqrt{s}_{min} \geq 9$ GeV, with $\chi^2/d.o.f \leq 1$ and stable intercepts and Pomeron couplings. The Pomeron intercept is determined to be $\alpha_P(0) = 1.093(3)$ in agreement with the conclusion of CKK [7]. We have shown the $C = \pm 1$ meson trajectories are non-degenerate and have intercepts as given in Table 1. The interplay between the Pomeron and $C = +1$ meson Reggeon contribution makes the determination of the couplings of the latter Reggeon problematic, as there seems to be characteristic $\sqrt{s}_{min}$ dependence. Further stabilization of these couplings is needed. Furthermore we have indicated that the $t = 0$ data, based on the χ^2 consideration alone, can not rule out other analytic models of forward scattering amplitudes, which means that the unitarity effect is in no sight up to the available Tevatron energy. But the factorization and quark counting properties, which a re well respected within 10 %, are difficult to understand outside the context of the Pomeron as a simple Regge pole, which makes the RRP model rather attractive. It will be interesting to be able to enlarge the region of stability in energy, in particular, into the low energy region, while improving the stability of $C = +1$ couplings, either through the use of crossing-symmetric variable $\tilde{s} = (s-u)/2$ or $E_{lab} = (s-u)/4M_{target}$ or by modifying the variable with effective thresholds.

Finally a few comments on an observation form the comparative study of competing analytic models [8,9] are in order: we see from Fig. 3 in the second talk [9] that the stability of the parameters and $\chi^2/d.o.f \leq 1$ set in for RRL1 sooner at $\sqrt{s}_{min} = 5$ GeV than in RRP and RRL2 models. What is more intriguing is that a special case, denoted as RRL2s0, of RRL2 in which the constant term $A = 0$ with varying scale s_0 gives similar $\chi^2/d.o.f$ values as in RRL1 but with a low $\sqrt{s_0}$ about 20 MeV. Thus we seem to have two options, based on this limited study of comparative data fittings, for the Pomeron as a singularity in the l-plane. In view of the phenomenological success of the Pomeron as a simple bare Regge pole with an intercept slightly larger than 1, one may choose to favor the RRL1 model as a unitarized version of such bare Pomeron. On the other hand, the special RRL2s0 model corresponds to the l-plane singularity of a triple pole with the intercept exactly at 1. For such Pomeron, one needs no unitarisation as it respects the Froissart bound already. It will be important to be able to differentiate the two options for the Pomeron by further studies.

5 Acknowledgments

This work is in part supported by the US DOE grant DE-FG02-91ER40688 and also in part by RFBR grants RFBR-96-07-89230 and RFBR-99-07-90356.

References

1. C Bourrely and J.Fisher, Nucl. Phys. B61 (1973) 513; K. Kang and B. Nicolescu, Phys. Rev. D11 (1975) 2461; H. Lipkin, Phys. Rev. D11 (1975) 1827; P. D. B. Collins, F. D. Gault, and A. Martin, Nucl. Phys. B85 (1975) 141; U. Amaldi et al., Phys. Lett. 66B (1977) 390.
2. J.R. Cudell and B.U. Nguyen, Nucl. Phys. **B420** (1994) 669, hep-ph/9310298.
3. A. Donnachie and P.V. Landshoff, Phys. Lett. **B437** (1998) 408, hep-ph/9806344; J.R. Cudell, A. Donnachie and P.V. Landshoff, Phys. Lett. **B448** (1999) 281, hep-ph/9901222.
4. V. Kim, in this Proc., hep-ph/9911228.
5. A. Donnachie and P.V. Landshoff, Nucl. Phys. **B244** (1984) 322; A. Donnachie and P.V. Landshoff, Nucl. Phys. **B267** (1986) 690. For the earlier attempts, see for example, P.D.B. Collins, F.D. Gault and A. Martin (Ref. 1).
6. A. Donnachie and P.V. Landshoff, Phys.Lett. **B296** (1992) 227.
7. J.R. Cudell, K. Kang and S.K. Kim, Phys. Lett. **B395** (1997) 311. See also R. J. M. Covolan, J. Montanha and K. Goulianos, Phys. Lett. **B389** (1996) 176.
8. J. R. Cudell, V. Ezhela, K. Kang, S. Lugovsky and N. Tkachenko, Phys. Rev. **D** (in press), hep-ph/9908218.
9. V. V. Ezhela, in this Proc., BROWN-HET-1194.
10. J. Finkelstein, H. M. Fried, K. Kang and C.-I Tan, Phys. Lett. **B232** (1989) 257.
11. K. Kang and B. Nicolescu (Ref. 1) in which References for the earlier attempts can be found. See also K. Kang, Nucl. Phys. **B** (Proc. Suppl.) **12** (1990) 64.
12. M. M. Block and R. N. Cahn, Rev. Mod. Phys. **57** (1985) 563.
13. M. M. Block, K. Kang and A. R. White, J. Mod. Phys. **A7** (1992) 4449. See also K. Kang, P. Valin, and A.R. White, Nuovo Cim. **107A** (1994) 2103.
14. Data used were extracted from the PPDS accessible at http://wwwppds.ihcp.su:8001/ppds.html.
Computer readable data files are also available at http://pdg.lbl.gov/xsect/contents.html

15. Particle Data Group, C. Caso et al., Eur. Phys. Jour. C3 (1998) 1
An extension of the preliminary results of our work (Ref.8) has been presented by three of the authors (V.E., S.L., and N.T.) in the Review of Particle Physics.
16. K. Kang *et al.*, hep-ph/9812429, presented at 4th Workshop on Quantum Chromodynamics, Paris, France, 1-6 June 1998 and presented at 4th Workshop on Small x and Diffractive Physics, Batavia, IL, 17-20 Sept. 1998, In Batavia 1998, Small-x and diffractive physics, pp. 203-211.
17. C. Adloff *et al.* [H1 Collaboration], Nucl. Phys. **B497** (1997) 3, hep-ex/9703012; J. Breitweg *et al.* [ZEUS Collaboration], Eur. Phys. J. **C7** (1998) 609, hep-ex/9809005.
18. G. Shaw, Phys. Rev. **D47** (1993) 3676 and Phys. Lett. **B228** (1989) 129; P. Ditsas and G. Shaw, Nucl. Phys. **113** (1976) 246; J.J. Sakurai and D. Schildknecht, Phys. Lett. **B40** (1972) 121.
19. L. Lukaszuk and B. Nicolescu, *Lett. Nuovo Cim.* **8** (1973) 405; G. Bialkowski, K. Kang and B. Nicolescu, Lett. Nuovo Cim. **13** (1975) 401; K. Kang and B. Nicolescu (Ref. 1).

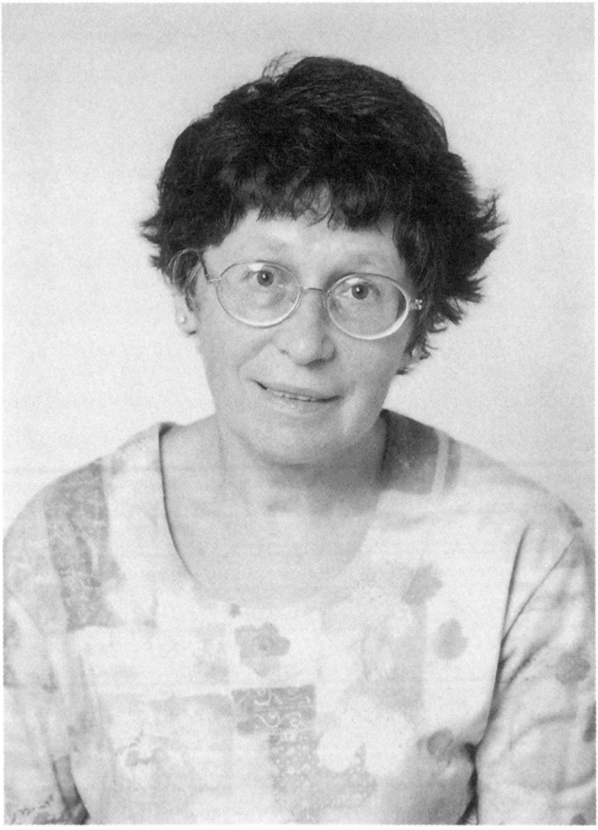

L. G. Dakhno

POMERON IN DIFFRACTIVE PROCESSES $\gamma^*(Q^2)p \to \rho^0 p$ AND $\gamma^*(Q^2)p \to \gamma^*(Q^2)p$ AT LARGE Q^2: THE ONSET OF PQCD

L. G. DAKHNO
Sanct-Petersburg Nuclear Physics Institute, Gatchina,
188300, S.Petersburg district, Russia
E-mail: dakhno@hep486.pnpi.spb.ru

We study the reactions $\gamma^*(Q^2)p \to \rho^0 p$ and $\gamma^*(Q^2)p \to \gamma^*(Q^2)p$ at large Q^2 and W^2/Q^2 and small momentum transfer $\kappa_\perp^2$ to the nucleon where the Pomeron exchange dominates. The amplitudes for both transverse and longitudinal polarizations of the initial photon and outgoing ρ-meson (photon) are calculated in the framework of the BFKL Pomeron exchange. Our calculations show that one cannot expect the early onset of the pure perturbative regime in the discussed diffractive processes: the small interquark distances, less than 0.2 fm, start to dominate not earlier than at $Q^2 \simeq 100$ GeV^2, $W^2/Q^2 \simeq 10^7$ in $\gamma^*(Q^2)p \to \rho^0 p$ and $Q^2 \simeq 50$ GeV^2, $W^2/Q^2 \simeq 10^6$ in $\gamma^*(Q^2)p \to \gamma^*(Q^2)p$.

1 Introduction

We study the onset of the perturbative QCD regime in the processes initiated by virtual photon $\gamma^*(Q^2)$. Two types of hard processes which allow one to study the transition regime from the Strong-QCD physics to the perturbative QCD physics of hard processes at accessible energies: these are (i) hard hadronic form factors and (ii) Pomeron-dominated reactions [1].

An important problem which arises in quantitative description of hard processes is an evaluation of the scale at which the perturbative regime starts to dominate these amplitudes. Some authors[2] advocated an early applicability of pQCD, the others[3] argue in favour of the late onset of the perturbative regime, at very large $Q^2 \simeq 50$ GeV^2. A quantitative study of elastic and transition form factors [4] confirmed that even at Q^2 as large as 20 Gev^2 the form factor is still dominated by soft non-perturbative $q\bar{q}$ kinematical configurations.

The physical picture of such processes at asymptotically large Q^2 consists in the following (see Fig. 1a,b): (*i*) virtual photon produces $q\bar{q}$ pair at small distances, (*ii*) quarks convert into vector meson (or outgoing photon) with the t-channel emission of gluons, (*iii*) gluons at the top of the ladder are at relatively small distances, thus a hard component of the Pomeron (BFKL Pomeron [1]) is selected, (*iiii*) lower part of the gluon ladder is attached to the nucleon at small momentum transfer, so the region of small transverse momenta is selected at the bottom of the ladder. Hence the most realistic is the following scenario: typical virtualities in the gluon ladder change from hard values selected by virtual photon in the hard quark-loop at the top of the

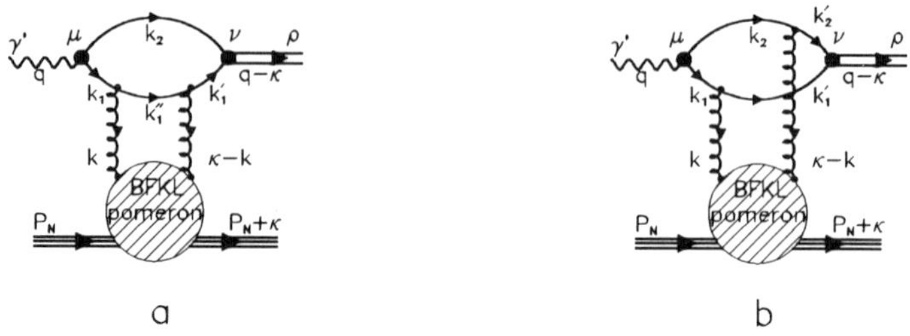

Figure 1: Diagrams for the processes $\gamma^*(Q^2)p \to \rho^0 p$ (a,b).

ladder to the soft values at the bottom, and thus the whole t-channel behaviour is determined by the complicated object which is a superposition of the hard and the soft Pomerons.

The detailed study of the hard upper quark-antiquark loop is based on realistic wave functions of the initial photon and the outgoing vector meson, for it is crucial for understanding the mechanism of the reaction at large but finite Q^2. On the contrary, the details of coupling to the bottom part of gluon ladder, which is related to the BFKL Pomeron attached to the nucleon in the soft regime, weakly influence the onset of the perturbative QCD in the upper block: we have checked this point investigating different variants of Pomeron–nucleon coupling.

So, the problem to be solved may be formulated as a numerical estimation of the scale at which the perturbative regime start to dominate the amplitudes of interest. We found the solution, that is one cannot expect the early onset of a pure perturbative regime in these diffractive processes.

2 Main ingredients of the calculations

These are (*i*) spectral representation of the amplitude, (*ii*) BFKL-pomeron and (*iii*) realistic wave functions of photon and ρ-meson.

2.1 Amplitude representation and loop diagrams for transverse and longitudinal polarizations

The Pomeron is attached to the photon and vector meson through the quark loop diagrams of Fig. 1. We consider separately the cases when both Reggeized gluons are attached to the same and to different quarks in the loop. Corresponding two diagrams depicted in Fig. 1a,b are under detailed consideration, and the following spectral representations are used for the calculation of the invariant (longitudinal and transverse) amplitudes $A_{L,T}$:

$$A^{I}_{L,T} = \int_0^1 \frac{dx}{x(1-x)^3} \int \frac{d^2k_{2\perp}}{(4\pi)^2} \frac{G_\gamma(M^2_{q\bar q})}{M^2_{q\bar q}+Q^2} \frac{G_V(M'^2_{q\bar q})}{M'^2_{q\bar q}-\mu^2_V} \frac{-ig^2}{2q_z} S^{I}_{L,T}, \tag{1}$$

$$A^{II}_{q\bar q} = \int_0^1 \frac{dx}{x^2(1-x)^2} \int \frac{d^2k_{2\perp}}{(4\pi)^2} \frac{G_\gamma(M^2_{q\bar q})}{M^2_{q\bar q}+Q^2} \frac{G_V(M'^2_{q\bar q})}{M'^2_{q\bar q}-\mu^2_V} \frac{-ig^2}{2q_z} S^{II}_{L,T}. \tag{2}$$

The spin factors $S^{I}_{L,T}$ and $S^{II}_{L,T}$ are related to the attachment of gluons of BFKL-pomeron either to one or two quarks and the vertices $G_\gamma(M^2_{q\bar q})$ and $G_V(M'^2_{q\bar q})$ stand for the wave functions of γ-quantum and ρ-meson, correspondingly.

2.2 BFKL-Pomeron couplings to the quark loop and nucleon

To attach the quark loop to target nucleon, several possible scenarios exist, that is due to the ambiguities in the description of the BFKL-Pomeron: in the region of small κ, the coupling is determined by Strong-QCD, thus not obeying the rigid prescriptions of the BFKL machinery. Another ambiguity is related to different possibilities for choosing a scale-factor for the W^2-dependence [1,5].

The BFKL Pomeron is represented by its eigenfunction $E_{0,\nu}(\rho_1,\rho_2) = [(\vec\rho_1-\vec\rho_2)^2\rho_1^{-2}\rho_2^{-2}]^{\frac12+i\nu}$, which is determined in the impact parameter space of gluons $(\vec\rho_1,\ \vec\rho_2)$ (the index-ν integration is performed over the interval $-\infty<\nu<\infty$. The BFKL amplitude is represented as a convolution of eigenfunctions $E_{n,\nu}(\rho_1,\rho_2)$ and $E_{n,\nu}(\rho'_1,\rho'_2)$ where the coordinates $\rho_{1,2}$ and $\rho'_{1,2}$ refer to gluons of the upper and lower blocks, correspondingly.

We consider two possible scenarios for the Pomeron attaching to the nucleon target: *(i)* the BFKL Pomeron is transformed first to the soft Pomeron and then this soft Pomeron is attached to the nucleon, *(ii)* the BFKL Pomeron is directly attached to the nucleon.

In the first case, the ladder is rather long, because of large W; the Pomeron is no longer the perturbative BFKL, but is in a soft regime. So it is attached to the nucleon with a standard soft coupling $F_{PNN}(\kappa^2_\perp)=\tilde g\,\exp(-B\kappa^2_\perp)$, with $B\simeq 2.5GeV^2$.

In the second case the nucleon ladder is rather short, and the BFKL-pomeron is presented in its explicit form (with $\langle r^2\rangle = 0.8$ fm^2):

$$A_{PNN} = \tilde c \int d^2\rho'_1 d^2\rho'_2 \left(\frac{(\vec\rho_1-\vec\rho_2)^2}{\rho'^2_1\rho'^2_2}\right)^{\frac12+i\nu} \exp(i\vec\kappa_\perp\frac{\vec\rho_1+\vec\rho_2}{2})\exp(-\frac{2}{3}\frac{\rho'^2_1+\rho'^2_2}{3\langle r^2\rangle}). \tag{3}$$

2.3 Realistic wave functions of the γ quantum and ρ meson

The light-cone wave function of photon was found in Ref. [4] on the basis of experimental data for the transition form factor $\gamma\gamma^*(Q^2) \to \pi^0$ at $0 \le Q^2 \le 25$ GeV^2; this wave function was successfully applied for the description of data on $\gamma\gamma^*(Q^2) \to \eta$ and $\gamma\gamma^*(Q^2) \to \eta'$. The analysis of the transition form factors has been performed in terms of the double spectral representation over $q\bar{q}$ invariant mass squared, $M^2_{q\bar{q}} = (m^2 + k^2_\perp) / (x(1-x))$ where $k^2_\perp$ and x are the components of the light-cone quark momenta, and m is the constituent quark mass. The photon wave function is represented as a sum of two components which correspond to a direct production of the $q\bar{q}$ pair with a point-like vertex $\gamma^* \to q\bar{q}$ at large $M_{q\bar{q}}$, and to the production in the low-$M_{q\bar{q}}$ region where the vertex has a nontrivial structure due to the soft $q\bar{q}$ interaction. The soft $q\bar{q}$ interaction yields an enhancement of the contribution of low-masses and, respectively, large distances in the quark loop.

The ρ-meson wave function necessary for the description of the reaction $\gamma^*(Q^2)p \to \rho^0 p$ is not known promptly, so we have to use some assumptions concerning the details of its behaviour. The low-$M_{q\bar{q}}$ component of the ρ-meson wave function is supposed to be approximately the same as for a pion. The latter was found in [4] from fitting the data for pion form factor at $0 \le Q^2 \le 10$ GeV^2. Moreover, the ρ-meson low-$M_{q\bar{q}}$ wave function should be close to photon's one due to the vector-meson dominance.

At large $M^2_{q\bar{q}}$ it is reasonable to consider two extreme cases of the behaviour of the $\rho \to q\bar{q}$ vertex: first, $G_\rho(M^2_{q\bar{q}}) \simeq M^{-2}_{q\bar{q}}$ and, second, $G_\rho(M^2_{q\bar{q}}) = const \sim O(\alpha_s)$. In both cases the dominance of the region of small interquark separations $\rho_{q\bar{q}} < 0.2$ fm was not observed at least at $Q^2 \le 50$ GeV^2 and $W^2/Q^2 \ge 10^6$.

3 Calculation strategy and the results

The onset of pQCD regime depends on the selection of small interquark separations in a quark loop with Q^2. So it is important to test which fraction of the cross section is actually gained at small transverse separations. Two cases have been calculated: *(i)* all transverse separations included in the quark loop, *(ii)* a part of the cross section corresponding to small interquark distances $\rho_{q\bar{q}} < 0.2$ fm is calculated. Actually, only this part of the cross section probes the BFKL Pomeron, whereas the cross section corresponding to loop quarks at large distances should be attributed to the soft Pomeron.

Typical behaviour of the cross section is shown in Fig. 2a,b.

The calculations have been performed with different assumptions on the input ingredients. In all considered cases the dominance of the region of small

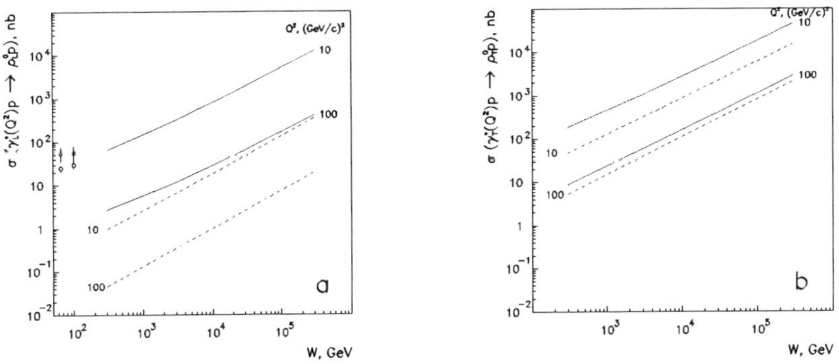

Figure 2: Cross sections in reactions $\gamma^*_{L,T}(Q^2)p \to \rho^0_{L,T}p$. Solid curves correspond to the calculation with all interquark separations in the quark loop; dashed lines stand for the case when the interquark distances $\rho < 0.2$ fm were only taken into account.

interquark separations $\rho_{q\bar{q}} < 0.2$ fm was not observed at least at $Q^2 \leq 50$ GeV^2 and $W^2/Q^2 \geq 10^6$. In the quark loops for the transitions $\gamma^*(Q^2) \to \gamma^*(Q^2)$ and $\gamma^*(Q^2) \to \rho^0$ the soft interquark distances $\rho > 0.2$ fm clearly dominate the amplitude at $Q^2 < 100$ GeV^2. The results of our analysis show that BFKL Pomeron cannot be seen in the diffractive production in the region $Q^2 \leq 100$ GeV^2. Therefore, a theoretical explanation of the change of the W dependence lies in better understanding of properties of the Strong-QCD Pomeron.

Acknowledgments

Thanks are due to V.V. Anisovich, D.I. Melikhov, V.A. Nikonov, M.G. Ryskin, with whom this work has been carried out.

References

1. E.A. Kuraev, L.N. Lipatov, V.S. Fadin, *Sov. Phys. JETP* **44**, 443 (1976); I.I. Balitsky, L.N. Lipatov, *Sov. J. Nucl. Phys.* **28**, 822 (1978).
2. V.L. Chernyak and A.R. Zhitnitsky, *Phys. Rep.* **112**, 173 (1984).
3. N. Isgur and C. Llewellin-Smith, *Phys. Lett.* B **217**, 535 (1989); NPB **317**, 526 (1989).
4. V.V. Anisovich, D.I. Melikhov, and V.A. Nikonov, *Phys. Rev.* D **52**, 5295 (1995); *Phys. Rev.* D **55**, 2918 (1997)).
5. V.S. Fadin, L.N. Lipatov, *Phys. Lett.* B **429**, 127 (1998).

A. V. Prokudin

REGGE-EIKONAL APPROACH AND ITS OFF-SHELL EXTENTION VERSUS EXPERIMENTAL DATA

V. A. PETROV and A. V. PROKUDIN

Institute For High Energy Physics,
142284 Protvino , Russia
E-mail: petrov@mx.ihep.su
prokudin@th1.ihep.su

We develop an off-shell extention of the Regge-eikonal approach which automatically takes into account off-shell unitarity. We argue that exclusive vector-meson production cross-sections measured at HERA can be fairly described with classical universal Regge trajectories. No extra "hard" trajectories of high intercept are needed for that.

1 Introduction

Unitarity condition

$$ImT(s,\vec{b}) = |T(s,\vec{b})|^2 + \eta(s,\vec{b})$$

where $T(s,\vec{b})$ is the scattering amplitude in impact space representation and $\eta(s,\vec{b})$ stands for the contribution of inelastic channels, may be illustrated by the following picture in the momentum space

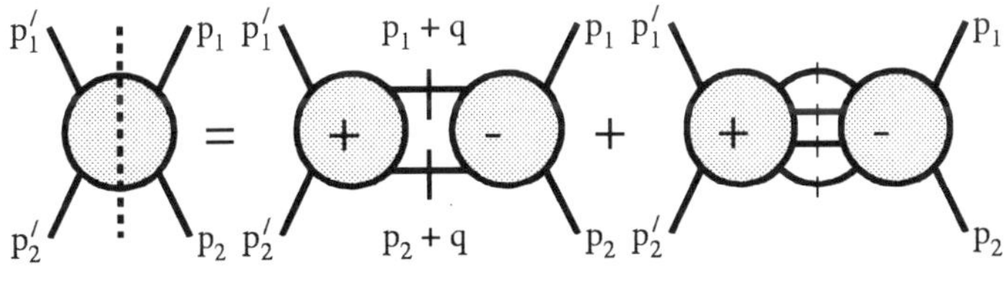

$$ImT(p_1',p_2',p_1,p_2) = \frac{-i}{(2\pi)^4}\int d^4q(-2\pi i)^2\delta((p_1+q)^2-m^2)\delta((p_2-q)^2-m^2)\times$$

$$\times T^{+}(p_1+q,p_2-q,p_1',p_2')T^{-}(p_1,p_2,p_1+q,p_2-q)+... \quad (1)$$

where "..." stands for inelastic channels contribution.

It is worth noting that we can include virtual external particles into consideration and in this case the integration in Eq. (1) still goes over particles on mass shell as it was in the case of real external particles, i.e.

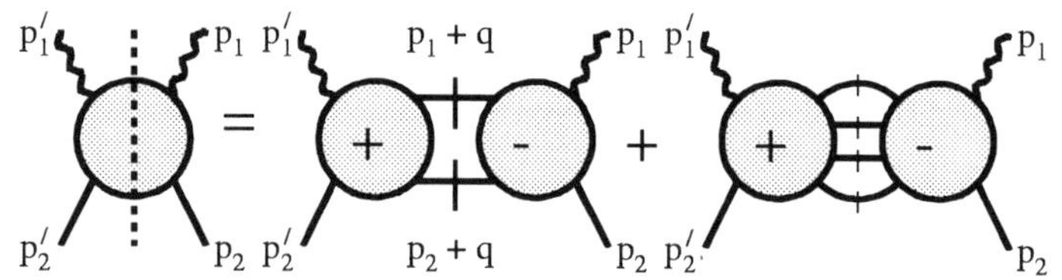

There are many parametrizations that "solve" the s-channel unitarity condition. Examples are U-matrix [1] and Eikonal approach [4].

We choose the eikonal representation as usual

$$T(s,\vec{b})=\frac{e^{2i\delta(s,\vec{b})}-1}{2i}, \quad (2)$$

here $T(s,\vec{b})$ is the scattering amplitude, $\delta(s,\vec{b})$ is the eikonal function. The eikonal function is to be considered as a basic scattering function, which builds up the amplitude. The unitarity condition looks very simple in terms of the eikonal function:

$$Im\delta(s,\vec{b})\geq 0,\ s>s_{inel} \quad (3)$$

In quantum mechanics $\delta(s,\vec{b})$ is related to the (short-range)potential V:

$$\delta(s,\vec{b}) \sim \int_{-\infty}^{+\infty} dz V(z,\vec{b}) \sim$$

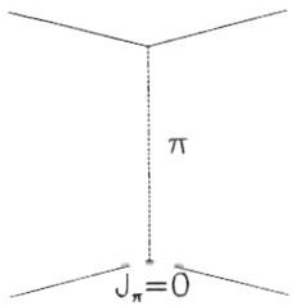

At relativistic energies one can generalize the notion of the potential by introducing a "quasipotential", which is complex and depends on energy [2]. As a concrete realizaton of the quasipotential one can take Van-Hove interpretation of the Regge behaviour as a "sum" of all possible one-particle exchanges in the t-channel [3]. Combination of the eikonal approach and Regge poles was considered for the first time in [4]. It was shown in [5] in the framework of the asymptotic summation of the perturbative series that the eikonal function $\hat{\delta}(s,t)$ behaves like $s^{1+\Delta}$, where $\Delta > 0$.

So one could abstract the idea that there exists some function $\hat{\delta}(s,t)$ with a powerlike growth in s and this is in full conformity with general principles.

We choose the following eikonal function in t space (here t is the transferred momentum)

$$\hat{\delta}(s,t) = c\left(\frac{s}{s_0}\right)^{\alpha(0)} e^{t\frac{\rho^2}{4}} \tag{4}$$

where

$$\rho^2 = 4\alpha'(0)ln\frac{s}{s_0} + r^2 \tag{5}$$

is reffered to as a "reggeon radius".

It means that the eikonal function has a simple pole in J plane and the corresponding Regge trajectory can be written in the following form:

$$\alpha(t) = \alpha(0) + \alpha'(0)t \tag{6}$$

We make use of linear Regge trajectories in this paper, though there exists some evidence of non-linearity of $\alpha(t)$ when t is high enough.

Functions in t- and b-spaces are related by the Fourier-Bessel transformation:

$$\begin{aligned} \hat{f}(t) &= 4\pi s \int_0^\infty db^2 J_0(b\sqrt{-t}) f(b) \\ f(b) &= \frac{1}{16\pi s}\int_{-\infty}^0 dt J_0(b\sqrt{-t})\hat{f}(t) \end{aligned} \tag{7}$$

Making use of Eq. (7) we obtain the following b-representation of the eikonal function:

$$\delta(s,b) = \frac{c}{s_0}\left(\frac{s}{s_0}\right)^{\alpha(0)-1} \frac{e^{-\frac{b^2}{\rho^2}}}{4\pi\rho^2} \tag{8}$$

In what follows the term "pomeron" will mean the leading pole of the eikonal function.

For cross sections we use the following normalizations:

$$\sigma_{tot} = \frac{1}{s} Im T(s, t=0)$$

$$\sigma_{el} = 4\pi \int_0^\infty db^2 |T(s,b)|^2 \qquad (9)$$

$$\frac{d\sigma}{dt} = \frac{|T(s,t)|^2}{16\pi s^2}$$

2 Regge-Eikonal Approach

Let's recall some properties of the Regge-eikonal model.

If $s \to \infty$, then

$$\sigma_{tot} \to 2\pi\rho^2 [c + ln z - Ei(-z)] \qquad (10)$$

where $z = \frac{\frac{c}{s_0}\left(\frac{s}{s_0}\right)^\Delta}{2\pi\rho^2}$, here $(\Delta \equiv \alpha(0) - 1)$ and $Ei(z) = \int_{-\infty}^z \frac{e^x}{x} dx$. At extra high energies we reach the following limit:

$$\sigma_{tot} \to 8\pi\alpha'(0)\Delta\left(ln\frac{s}{s_0}\right)^2 \qquad (11)$$

i.e. the powerlike behaviour of $\hat{\delta}(s,t)$ leads to the asymptotic behaviour of the full scattering ampitude which functionally saturates the Froissart-Martin bound [6]:

$$\sigma_{tot}^{hh} \leq \frac{\pi}{m_\pi^2}\left(ln\frac{s}{s_0}\right)^2, s \to \infty\,. \qquad (12)$$

If $s \to \infty$, then

$$\sigma_{el} \to \pi\rho^2 [c + ln\frac{z}{2} + Ei(-2z) - 2Ei(-z)] \qquad (13)$$

and thus

$$\frac{\sigma_{el}}{\sigma_{tot}} \to \frac{1}{2} \qquad (14)$$

3 The Model

In our model we choose the following contributions for the eikonal function:

$$\delta(s,b) = \delta_{\mathbb{P}}^{+}(s,b) + \delta_{\mathbb{O}}^{-}(s,b) + \delta_{f}^{+}(s,b) + \delta_{\omega}^{-}(s,b) \qquad (15)$$

where $\delta^+_{\mathbb{P}}(s,b)$ is the pomeron contribution, $\delta^-_{\mathbb{O}}(s,b)$ is the odderon[a] contribution, and δ^+_f, $\delta^-_\omega(s,b)$ are contributions of f meson exchange $(C=+1)$ and ω meson exchange $(C=-1)$.

The trajectories of f and ω are[7]:

$$\alpha_f(t) = 0.69 + 0.84t$$
$$\alpha_\omega(t) = 0.47 + 0.93t \tag{16}$$

and

$$\delta^+(s,b) = [i + tg\frac{\pi(\alpha(0)-1)}{2}]\frac{c}{s_0}(\frac{s}{s_0})^{\alpha(0)-1}\frac{e^{-\frac{b^2}{\rho^2}}}{4\pi\rho^2} \quad (C=+1)$$
$$\delta^-(s,b) = [i + ctg\frac{\pi(\alpha(0)-1)}{2}]\frac{c}{s_0}(\frac{s}{s_0})^{\alpha(0)-1}\frac{e^{-\frac{b^2}{\rho^2}}}{4\pi\rho^2} \quad (C=-1) \tag{17}$$

Where $i + tg\frac{\pi(\alpha(0)-1)}{2}$ and $i + ctg\frac{\pi(\alpha(0)-1)}{2}$ are signature coefficients for $C=+1$ and $C=-1$ exchange correspondingly.

4 Fitting procedure

We explore two cases: with and without odderon contribution.

4.1 The Model With the Odderon

The results concerning the trajectories of the pomeron and the odderon are:

$$\alpha_{\mathbb{P}}(0) - 1 = 0.11578 \pm 0.003 \quad , \ \alpha'_{\mathbb{P}}(0) = 0.27691 \pm 0.00434$$
$$\alpha_{\mathbb{O}}(0) - 1 = 0.11578 \pm 0.00711 \ , \ \alpha'_{\mathbb{O}}(0) = 0.27691 \pm 0.00315 \tag{18}$$

where $\alpha_{\mathbb{P}}$ and $\alpha_{\mathbb{O}}$ are the intercepts of the pomeron and the odderon, $\alpha'_{\mathbb{P}}$ and $\alpha'_{\mathbb{O}}$ are the slopes of the pomeron and the odderon respectively. So the fitting procedure lead us practically to the scenario of the weakly degenerate pomeron-odderon trajectory[14].

$\alpha_{\mathbb{P}}$, $\alpha'_{\mathbb{P}}$ and $\alpha_{\mathbb{O}}$, $\alpha'_{\mathbb{O}}$ were chosen so that the following conditions of unitarity[8]

$$\alpha_{\mathbb{P}}(0) \geq \alpha_{\mathbb{O}}(0)$$
$$\alpha'_{\mathbb{P}}(0) \geq \alpha'_{\mathbb{O}}(0) \tag{19}$$

would not be violated. The other parameters are in the Table.(1).

[a]The odderon is the counterpart of the pomeron with $C=-1$

Table 1: Parameters of the model with the odderon.

$c_{\mathbb{P}}$	44.66025	$c_{\mathbb{O}}$	1.92343
c_f	241.69257	c_ω	71.20085
$r^2_{\mathbb{P}}(GeV^{-2})$	14.83200	$r^2_{\mathbb{O}}(GeV^{-2})$	2.18549
$r^2_f(GeV^{-2})$	5.55410	$r^2_\omega(GeV^{-2})$	99.99996

We find out that the odderon contribution is significant when $t \neq 0$ and it allows us to qualitatevely describe the differnce in the structure of $\frac{d\sigma}{dt}$ of $\bar{p}p$ and pp 9, 10 while without this contribution we are able to describe only the slopes of $\frac{d\sigma}{dt}$ when $t = 0$.

It is worth noting that the discrepancy in the description of $\rho = \frac{ReT}{ImT}(s, t = 0)$ can be understood if we remember that the ρ data are not directly mesaured by experimentalists, but rather extracted from $\frac{d\sigma}{dt}$ experimental data in a model-dependent way.

4.2 The Model Without the Odderon

The results concerning the pomeron trajectory is:

$$\alpha_{\mathbb{P}}(0) - 1 = 0.12169 \pm 0.00286 \ , \ \alpha'_{\mathbb{P}}(0) = 0.20933 \pm 0.001379 \qquad (20)$$

The other parameters are in the Table.(2).

Table 2: Parameters of the model without the odderon.

$c_{\mathbb{P}}$	46.599		
c_f	199.630	c_ω	150.530
$r^2_{\mathbb{P}}(GeV^{-2})$	21.204		
$r^2_f(GeV^{-2})$	4.872	$r^2_\omega(GeV^{-2})$	39.967

The intercept is slightly higher then in case of when the odderon contributes to the eikonal. As is shown, the $t = 0$ data can be described without presence of the odderon, while the $t \neq 0$ data can be described, though qualitatively only by means of the odderon. Let's remind our basic assumptions:

- Regge trajectories are linear.
- The t dependence of $\hat{\delta}(s, t)$ is exponential i.e. $\hat{\delta}(s, t) \sim e^{t\rho^2}$.

It is not our aim at present stage to obtain the best fit. What really matters is to fix a set of Regge parameters some of which are considered universal (slopes, intercepts of the trajectories) while others (residues) are not. Universal parameters will be then used for off-shell amplitudes.

5 Off-Shell Extention Of The Regge-Eikonal Approach

As it was shown in [10] the amplitude with one off-mass shell particle both in initial and final states (two asteriks)is related to the off-shell eikonals with one and two off-shell particles and on-shell amplitude in the following way

$$T^{**}(s,b) = \delta^{**}(s,b) - \frac{\delta^{*}(s,b)\delta^{*}(s,b)}{\delta(s,b)} + \frac{\delta^{*}(s,b)\delta^{*}(s,b)}{\delta(s,b)\delta(s,b)} T(s,b) \qquad (21)$$

The expansion Eq. 21 can be illustrated by the following figure:

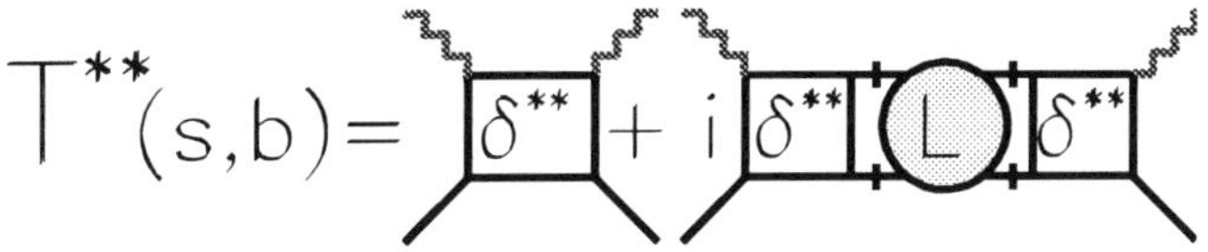

where $L = \sum_{n=2}^{\infty} \frac{(2i\delta)^{n-2}}{n!}$. The case when only one of the particles is off shell can be considered similarly. One has

$$T^{*} = \hat{\delta}^{*} + i\hat{\delta}^{*} \circ L \circ \hat{\delta} \qquad (22)$$

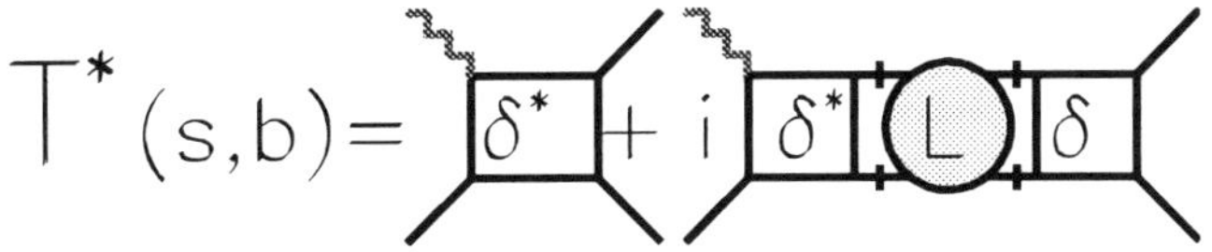

or we rewrite it as follows:

$$T^*(s,b) = \frac{\delta^*(s,b)}{\delta(s,b)} T(s,b) \tag{23}$$

On the basis of Eq.(8) we take the following parametrizations of the off-shell eikonal functions:

$$\delta^*_{\pm}(s,b) = \xi_{\pm} \frac{c_*(Q^2)}{s_0 + Q^2} \left(\frac{s+Q^2}{s_0+Q^2}\right)^{\alpha(0)-1} \frac{e^{-\frac{b^2}{\rho_*^2}}}{4\pi\rho_*^2} \tag{24}$$

where

$$\rho_*^2 = 4\alpha'(0) ln \frac{s+Q^2}{s_0+Q^2} + r_N^2 + r_*^2(Q^2) \tag{25}$$

and

$$\delta^{**}_{\pm}(s,b) = \xi_{\pm} \frac{c_{**}(Q^2)}{s_0 + Q^2} \left(\frac{s+Q^2}{s_0+Q^2}\right)^{\alpha(0)-1} \frac{e^{-\frac{b^2}{\rho_{**}^2}}}{4\pi\rho_{**}^2} \tag{26}$$

where

$$\rho_{**}^2 = 4\alpha'(0) ln \frac{s+Q^2}{s_0+Q^2} + r_N^2 + r_{**}^2(Q^2) \tag{27}$$

Coefficients $c_*(Q^2)$, $c_{**}(Q^2)$ are supposed to weakly (not in a powerlike way)depend on Q^2.

Total cross section

If we consider only the Pomeron exchange (in case of high energies, the main contribution to the scattering amplitude is due to the Pomeron) and set signature coefficient ξ_+ to be equal unity, then we have the following series for the total cross section ($\sigma^{**}_{tot} = \frac{1}{s} Im T^{**}(s, t=0)$):

$$\sigma^{**}_{tot} = \frac{c_{**}(Q^2)}{Q^2 + s_o}\left(\frac{1}{x}\right)^{\Delta} - \frac{c_*^2(\frac{1}{x})^{2\Delta}}{2\pi(Q^2+s_o)^2\rho_*^2} \sum_{n=0}^{\infty} \frac{\left(-\frac{c(\frac{s}{s_o})^{\Delta}}{2\pi s_o \rho^2}\right)^n}{(n+2)!} \cdot \frac{\rho^2}{2\rho^2 + n\rho_*^2} \tag{28}$$

here we used the folowing relation $\frac{s+Q^2}{s_0+Q^2} \simeq \frac{s+Q^2-m_N^2}{Q^2} = \frac{1}{x}$ (if $s \gg Q^2$ and (or)$Q^2 \gg s_0 (Q^2 \gg m_N^2)$) Now we can derive the behaviour of the total cross section in different kinematic limits:

- **Regge Regime**

$$\sigma^{**}_{tot} \to \frac{(s/Q^2)^{\Delta}}{Q^2} \left[c_{**} - \frac{c_*^2}{c}\left(\frac{s_0}{Q^2}\right)^{1+\Delta} \frac{\rho^2}{\rho_*^2}\right] \tag{29}$$

- **Bjorken Regime**

$$\sigma^{**}_{tot} \to \frac{c_{**}(Q^2)}{Q^2}\Big(\frac{1}{x}\Big)^{\Delta} - \frac{c_*^2}{2c}\cdot\frac{1}{Q^2}\cdot\Big(\frac{1}{x}\Big)^{\Delta}\cdot\Big(\frac{s_o}{Q^2}\Big)^{1+\Delta}\cdot\frac{ln\frac{Q^2(1-x)}{s_o x}}{ln\frac{1}{x}} \qquad (30)$$

As we see total cross-section posesses a powerlike behaviour in the Regge limit, but this is not a violation of unitarity, as the Froissart-Martin bound [6] cannot be proven for this case and if we put all particles on the mass shell, then we restore the 'normal' logarithmic asymptotical behaviour $\sigma \sim \ln^2\frac{s}{s_0}$. In the Bjorken limit we have strong (powerlike) violation of scaling in the second term.

Elastic cross-section

For the elastic cross section we have the following expression:

$$\sigma^*_{el} = 4\pi\int_0^\infty db^2\Big|\frac{\delta^*}{\delta}T(s,b)\Big|^2 \qquad (31)$$

As far as $q'^2 = \mu^2$, where μ is the mass of the produced particle, it is natural to set $s_0 = \mu^2$ and now we can derive the following relations:

- **Regge Regime**

$$\sigma^*_{el} \to 16\pi\alpha'(0)\Delta\Big(\frac{c_*}{c}\Big)^2\Big(\frac{\mu^2}{Q^2}\Big)^{2+2\Delta}(ln\frac{s}{\mu^2})^2 \qquad (32)$$

- **Bjorken Regime**

$$\sigma^*_{el} \to 8\pi\alpha'(0)\Big(\frac{c_*}{c}\Big)^2\Big(\frac{\mu^2}{Q^2}\Big)^{2+2\Delta}\frac{(ln(Q^2/x))^2}{ln(1/x)} \qquad (33)$$

As we can easily realize

$$\frac{\sigma^*_{el}}{\sigma^{**}_{tot}} \to 0 \qquad (34)$$

when $s \to \infty$ and $s \gg Q^2$ (compare with 14) As to the Q^2 behaviour, then as far as we have $\Delta = 0.116 \pm 0.003$ (see Eq.(18)), we obtain the following dependence:

$$\sigma^*_{el} \sim \frac{1}{(Q^2)^{2+2\Delta}} = (Q^2)^{-2.232\pm 0.006} \qquad (35)$$

This formula is in close agreement with the results of fitting the Q^2 dependence in [9], where the power of Q^{-2} is measured as 2.24.

6 Vector meson photoproduction

For vector meson photoproduction $\gamma^* p \to Vp$, where $V = \rho_o, \omega, \Phi, J/\Psi, ...$

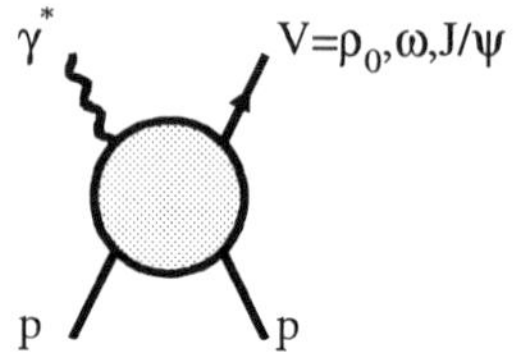

we have the following formula for the amplitude:

$$T^*(s,b) = \frac{\delta^*(s,b)}{\delta(s,b)} T(s,b) \tag{36}$$

here we assume that $T_{\gamma^* p \to Vp} \sim T_{V^* p \to Vp}$ on the basis of vector dominance [13]. As far as in this process reggeons with $C = -1$ do not contribute, we have the following contributions for the eikonal function:

$$\delta(s,b) = \delta^+_{\mathbb{P}}(s,b) + \delta^+_f(s,b) \tag{37}$$

with $(s \equiv W^2)$

$$\delta_\pm(W,b) = \xi_\pm \frac{c}{W^2 - \mu^2 - m_p^2} \left(\frac{W^2 - \mu^2 - m_p^2}{W_0^2 - \mu^2 - m_p^2} \right)^{\alpha(0)} \frac{e^{-\frac{b^2}{\rho^2}}}{4\pi\rho^2} \tag{38}$$

$$\rho^2(W) = 4\alpha'(0) ln \frac{W^2 - \mu^2 - m_p^2}{W_0^2 - \mu^2 - m_p^2} + r_p^2 \tag{39}$$

and

$$\delta^*_\pm(W,b) = \xi_\pm \frac{c_*(Q^2)}{W^2 + Q^2 - m_p^2} \left(\frac{W^2 + Q^2 - m_p^2}{W_0^2 + Q^2 - m_p^2} \right)^{\alpha(0)} \frac{e^{-\frac{b^2}{\rho_*^2}}}{4\pi\rho_*^2} \tag{40}$$

$$\rho_*^2(W) = 4\alpha'(0) ln \frac{W^2 + Q^2 - m_p^2}{W_0^2 + Q^2 - m_p^2} + r_p^2 + r_*^2(Q^2) \tag{41}$$

As far as Regge Eikonal approach does not constrain the Q^2 dependence of residues $c_*(Q^2)$ and radii $r_*^2(Q^2)$ we assume that the main Q^2 dependence is comprised in the other terms of our formulae and residues slowly depend on Q^2 (they posses logarifmic behaviour)and radii depend on Q^2 as $\sim \frac{1}{Q^2}$. Eventually we have the following parametrizations:

$$c_*(Q^2) = c + c_1 ln^3 (\frac{Q_o^2 + Q^2}{Q_o^2}) \tag{42}$$

and

$$r_*^2(Q^2) = r^2 + \frac{r_1^2}{\frac{Q_0^2+Q^2}{Q_0^2}} \tag{43}$$

where $Q_0^2 = 1.0GeV^2$.

The results of fitting the data to these formulae are in Fig.1.

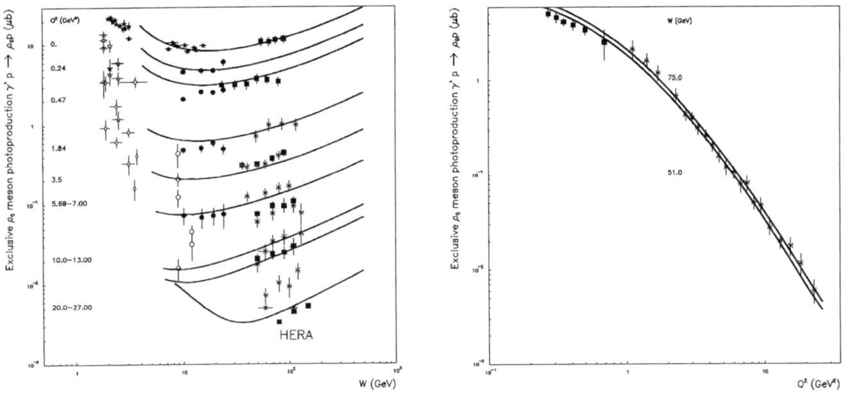

Figure 1: ρ_0 meson exclusive photoproduction cross-section

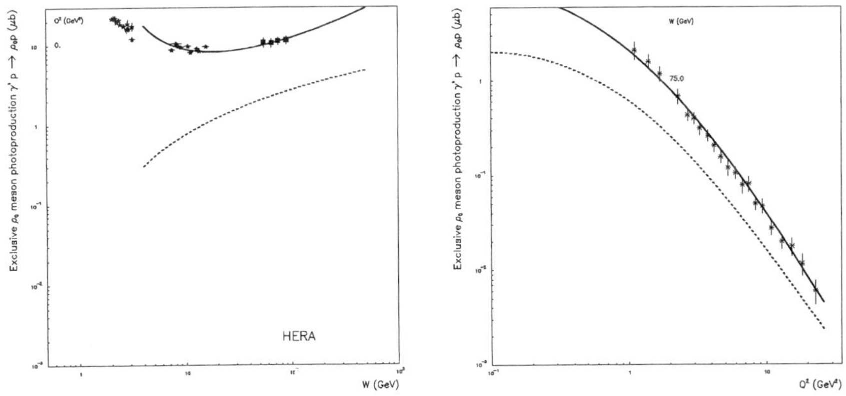

Figure 2: ρ_0 meson exclusive photoproduction cross-section (solid line)Eq. 32 in comparison with its asymptotic behaviour (dashed line) Eq. 33

As one can see from Fig.2, in the region of energies that are available the asymptotic behaviour of exclusive vector meson photoproduction does not take

Results of fitting without the odderon contribution

Hollow dots are pp data and full dots are $\bar{p}p$ data. The dotted (pp) and solid ($\bar{p}p$) curves correspond to the model without the odderon contribution. The shadowed area corresponds to the region available by the uncertainty in the fitting parameters.

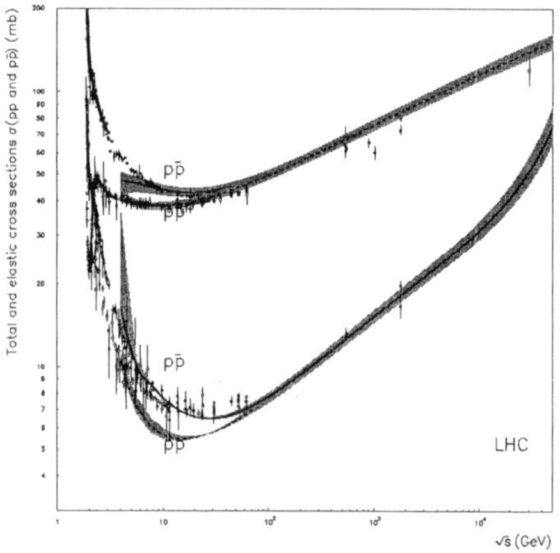

Figure 3: Fitted total and elstic cross sections.

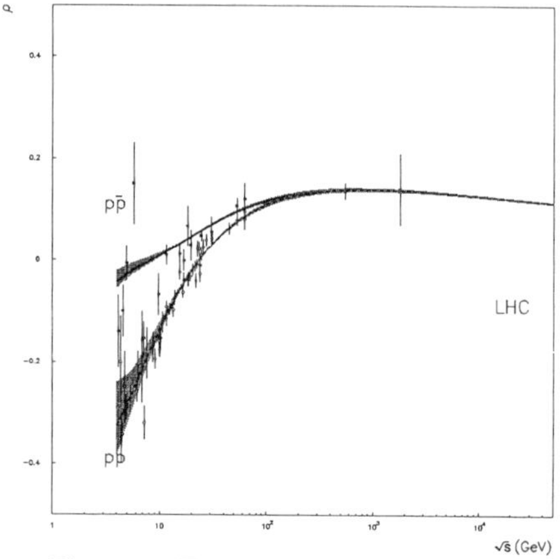

Figure 4: Fitted ratios of real to imaginary forward amplitudes.

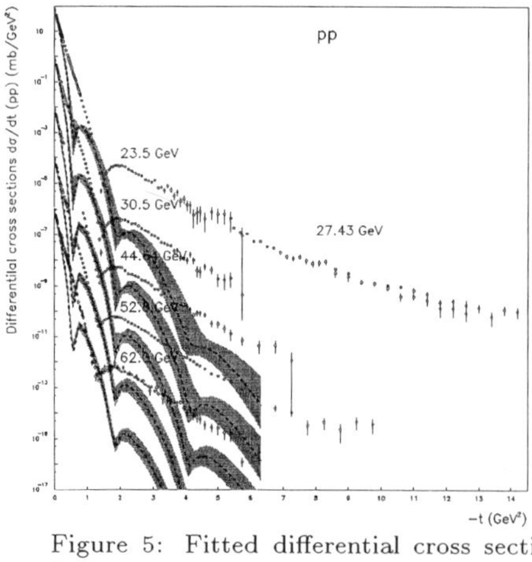

Figure 5: Fitted differential cross sections (pp).

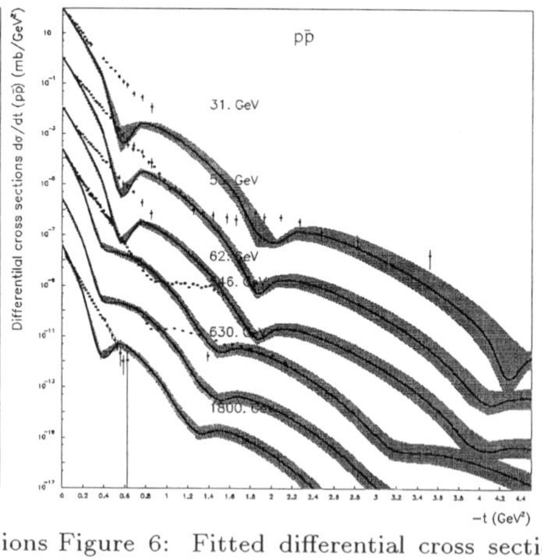

Figure 6: Fitted differential cross sections ($\bar{p}p$).

Results of fitting with the odderon contribution

Hollow dots are pp data and full dots are $\bar{p}p$ data. The dotted (pp) and solid ($\bar{p}p$) curves correspond to the model with the odderon contribution. The shadowed area corresponds to the region available by the uncertainty in the fitting parameters.

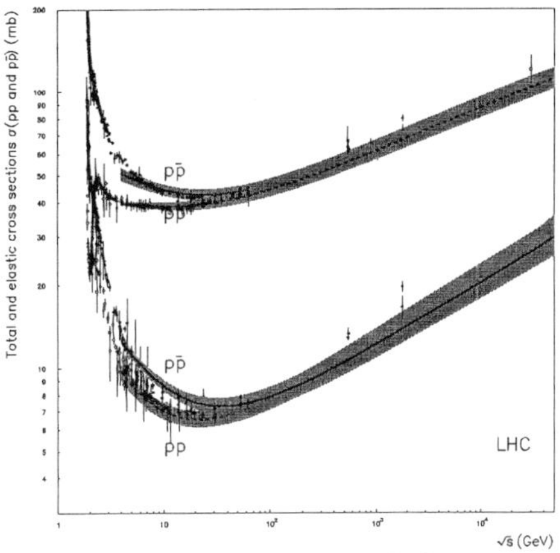

Figure 7: Fitted total and elstic cross sections.

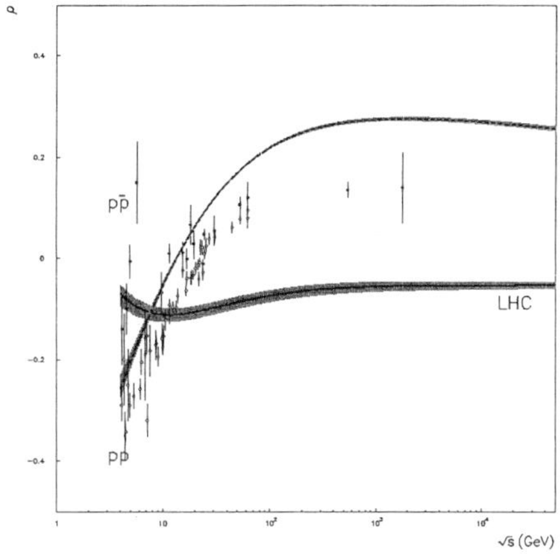

Figure 8: Fitted ratios of real to imaginary forward amplitudes.

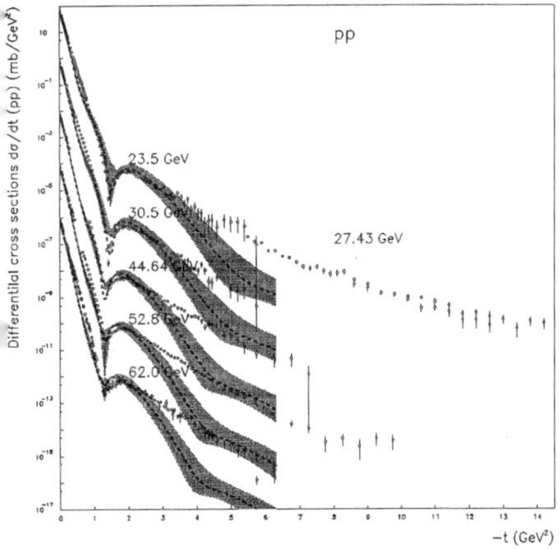

Figure 9: Fitted differential cross sections (pp).

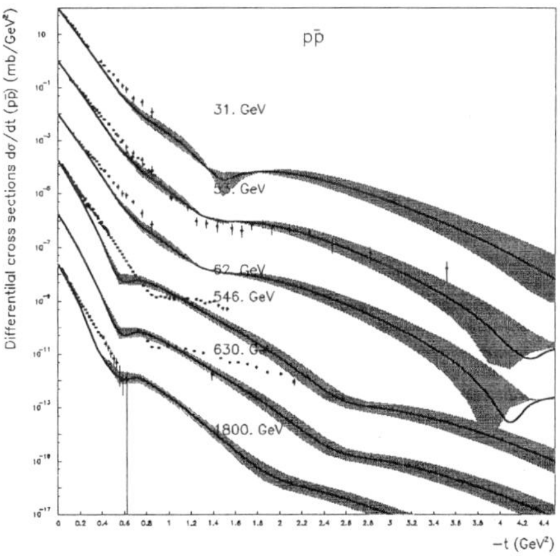

Figure 10: Fitted differential cross sections ($\bar{p}p$).

place and therefore one should take into concideration both Pomeron and f meson exchanges.

For fitting we used the same intercept and slope of the pomeron that have been obtained in the Section 4.1, so that the pomeron in our model is unique and the same for all processes where it takes part.

7 Conclusions

As is seen from Fig.1 and Fig.2, the exclusive vector meson production may be described in the framework of extended Regge-eikonal approach and one does not need 'hard' pomerons with a high intercept in order to do it. The rise of cross sections when Q^2 is high is a transitory phenomenon due to effective delay of asymptotic behaviour which is observed for low Q^2 [11] or [12]. The Q^2-dependence is also described fairly well.

References

1. S. Troshin, N. Tyurin, *Yad. Fiz.* **40**, 1008 (1984).
2. A. A. Logunov A. N. Tavkhelidze , *Nuovo Cimento* **29**, 380 (1963).
3. L. Van-Hove, *Phys. Lett.* B **24**, 183 (1967).
4. R. C. Arnold, *Phys. Rev.* D **153**, 1533 (1967).
5. Cheng H., Wu T. T., *Phys. Rev.* D **186**, 1611 (1969). Cheng H., Wu T. T., *Phys. Rev. Lett.* **24**, 1456 (1970).
6. M. Froissart, *Phys. Rev.* D **123**, 1053 (1961)
   A. Martin, *Phys. Rev.* D **129**, 993 (1963).
7. R. J. M. Covolan, P. Degrolard, M. Giffon, L. L. Jenkovszky, E. Predazzi, *Z. Phys.* C **58**, 109 (1993)
8. J. Finkelshtein, H. M. Fried, K. Kang, C-I Tan, *Phys. Lett.* B **232**, 257 (1989)
   E. S. Martynov, ibid., 367
9. C. Adloff et al., H1 Collaboration submitted to Eur. Phys. J C, DESY 99-010, hep-ex/9902019
10. V. A. Petrov in "Frontiers In Strong Interactions" Edited by P. Chiappetta, M. Haguenauer, J. Trân Thanh Vân. p.139
11. P. Desgrolard, A. Lengyel, E. Martynov, *Eur.Phys.J.* C **7**, 655 (1999)
12. V. Petrov, A. Prokudin, hep-ph/9706257 , *Yad.Fiz* **9** 62 1 (1999)
13. N. M. Kroll, T. D. Lee, B. Zumino *Phys. Rev.* D **157**, 1376 (1967)
14. V. A. Petrov, A. P. Samokhin *Phys. Lett.* B **237**, 500 (1990)

WHAT CAN WE LEARN FROM THE STUDY OF SINGLE DIFFRACTIVE DISSOCIATION AT HIGH ENERGIES?

A. A. ARKHIPOV
Institute for High Energy Physics
Protvino, 142284 Moscow Region, Russia
E-mail: arkhipov@mx.ihep.su

The fundamental relations in the dynamics of single diffraction dissociation and elastic scattering at high energies are discussed.

1 Introduction

I shall start my talk from the end with the short answer to the question in the title. The study of the inclusive reactions in the region of diffraction dissociation at high energies provides a unique possibility to learn on a new type of interactions between elementary particles or a new type of fundamental forces, which the three-body forces are. What was the beginning on?

In 1994 the CDF group at Fermilab published new results on the measurements of $p\bar{p}$ single diffraction dissociation at $\sqrt{s} = 546$ and $1800\ GeV$. They observed that a popular supercritical Pomeron model did not describe new measured values. The statement, made in [1], is as follows: The value of $\sigma_{sd}^{p\bar{p}} = (7.89 \pm 0.33)\ mb$, measured at $\sqrt{s} = 546\ GeV$, is extrapolated by the supercritical Pomeron model to $\sigma_{sd}^{p\bar{p}} = (13.9 \pm 0.9)\ mb$ at $\sqrt{s} = 1800\ GeV$, while the measured value at this energy is equal to $\sigma_{sd}^{p\bar{p}} = (9.45 \pm 0.44)\ mb$. The ratio of the measured $\sigma_{sd}^{p\bar{p}}$ to that obtained by extrapolation is

$$\frac{\sigma_{sd}^{p\bar{p}}(experimental)}{\sigma_{sd}^{p\bar{p}}(extrapolation)}(\sqrt{s} = 1800\ GeV) = 0.68 \pm 0.05. \tag{1}$$

Moreover, at $\sqrt{s} = 20\ GeV$ the experimental $\sigma_{sd}^{p\bar{p}} = (4.9 \pm 0.55)\ mb$ is 4.5 times larger than the value $\sigma_{sd}^{p\bar{p}} = (1.1 \pm 0.17)\ mb$, obtained by the extrapolation of the measured value of $\sigma_{sd}^{p\bar{p}}$ at $\sqrt{s} = 546\ GeV$ down to $\sqrt{s} = 20\ GeV$ with the help of the supercritical Pomeron model. So, the latest experimental measurement of $p\bar{p}$ single diffraction dissociation at c.m.s. energies $\sqrt{s} = 546$ and $1800\ GeV$, carried out by the CDF group at the Fermilab Tevatron collider, has shown that the popular model of supercritical Pomeron does not describe the existing experimental data.

We called the emerged situation as a supercrisis for the supercritical Pomeron model (SCPM).[a] The supercrisis is illustrated on Fig. 1 extracted from

[a] Recent experimental results from HERA [2] lead us to the same conclusion. The soft Pomeron

paper [3].

The attempts undertaken in Refs. [3,4] to save the SCPM are also shown on this figure. Unfortunately GLM paper [4] contains a crude mathematical mistake. The mistake was observed by B.V. Struminsky and E.S. Martynov from Kiev [5]. Besides, in our opinion, an eikonalization procedure cannot be considered as a saving ring for SCPM because this procedure is outside the original Regge ideology. The idea of renormalized Pomeron flux proposed by Goulianos is a good physical idea for an experimentalist, but this idea cannot be a satisfactory one for a theorist because the idea is not grounded by the underlying Regge theory.

Obviously, the foundations of the Pomeron model require a further theoretical study and the construction of newer, more general phenomenological framework, which would enable one to remove the discrepancy between the model predictions and the experiment.

Although nowadays we have in the framework of local quantum field theory a gauge model of strong interactions formulated in terms of the known QCD Lagrangian, its relations to the so called "soft" (interactions at large distances) hadronic physics are far from desired. The understanding of this physics is high interest because it has an intrinsically fundamental nature.

In 1970 the experiments at the Serpukhov accelerator revealed that the K^+p total cross section increased with energy. The increase of the pp total cross section was discovered at the CERN ISR and then the effect of rising total cross sections was confirmed at the Fermilab accelerator.

In spite of more than 25 years after the formulation of QCD we still cannot obtain from the QCD Lagrangian the answer to the question why all the hadronic total cross-sections grow with energy. We cannot predict total cross-sections in an absolute way starting from the fundamental QCD Lagrangian as well mainly because it is not a perturbative problem.

It is well known, e.g., that nonperturbative contributions to the gluon propagator influence the behaviour of "soft" hadronic processes and the knowledge of the infrared behaviour of QCD is certainly needed to describe the "soft" hadronic physics in the framework of QCD. Unfortunately, today we don't know the whole picture of the infrared behaviour of QCD, we have some fragments of this picture though (see e.g. Ref. [6]).

At the same time it is more or less clear now that the rise of the total cross-sections is just the shadow (not antishadow!) of particle production.

Through the optical theorem the total cross-section is related to the imag-

---

phenomenology as currently developed cannot incorporate the HERA data on structure function F_2 at small x and total γ^*p cross section from F_2 measurements as a function of W^2 for different Q^2.

inary part of the elastic scattering amplitude in the forward direction. That is why the theoretical understanding of elastic scattering has the fundamental importance.

From the unitarity relation it follows that the imaginary part of the elastic scattering amplitude contains the contribution of all possible inelastic channels in two-particle interaction. It is clear therefore that we cannot understand the elastic scattering without understanding the inelastic interaction.

Among all the possible inelastic interactions there is a special class of processes which are called a single diffraction dissociation. The single diffraction dissociation is the scattering process where one of two particles in the initial state breaks up during the interaction producing a system of particles in a limited region of (pseudo)rapidity.[b]

Good and Walker have shown [7] that the single diffraction dissociation is predicted by the basic principles of quantum mechanics. However both the elastic scattering and single diffraction dissociation cannot correctly be calculated in QCD due to the non-perturbative nature of the interactions.

The popular Regge phenomenology represents elastic and diffractive scattering by the exchange of the Pomeron, a color singlet Reggeon with quantun numbers of the vacuum. It should be noted that the definition of the Pomeron as Reggeon with the highest Regge trajectory $\alpha_P(t)$, carrying the quantum numbers of the vacuum, is not the only one.[c] There are many other definitions of the Pomeron: Pomeron is a gluon "ladder" [8]; Pomeron is a bound state of two reggezied gluons – BFKL-Pomeron [9]; soft and hard Pomerons [10,11]; etc.[d] This leapfrog is because of the exact nature of the Pomeron and its detailed substructure remains such as that no one knows what it is. The difficulty of establishing the true nature of the Pomeron in QCD is almost obviously related to the calculations of non-perturbative gluon exchange.

Nevertheless in the near past simple formulae of the Regge phenomenology provided good parameterization of experimental data on "soft" hadronic physics and pragmatic application of Pomeron phenomenology had been remarkably successful (see e.g. the latest issue of the Review of Particles Properties).

That was the case before the appearance of the above-mentioned CDF data on single diffractive dissociation and recent results from HERA. Of course, it

---

[b]Pseudorapidity is defined as $\eta = -\ln\tan(\theta/2)$ where θ is the polar angle of the produced particle with respect to the beam direction. Pseudorapidity is frequently used as an approximation to rapidity.

[c]For supercritical Pomeron $\alpha_P(0) - 1 = \Delta \ll 1$, $\Delta > 0$ is responsible for the growth of hadronic cross-sections with energy.

[d]At the Workshop I heard new definition of Pomeron from N.N. Nikolaev: Pomeron is (neither more nor less!) a label of diffraction.

is good that we have a simple and compact form for representing a great variety of data for different hadronic processes, but it is certainly bad that power behaved total cross-sections violate unitarity. Often and often encountered claim, that the model with power behaved total cross-sections is valid in the non-asymptotic domain which has been explored up today, is not correct because the supercritical Pomeron model is an asymptotic one by definition.

We suggested another approach to the dynamical description of one-particle inclusive reactions [12]. The main point of our approach is that new fundamental three-body forces are responsible for the dynamics of particle production processes of inclusive type. Our consideration revealed several fundamental properties of one-particle inclusive cross-sections in the region of diffraction dissociation. In particular, it was shown that the slope of the diffraction cone in $p\bar{p}$ single diffraction dissociation is related to the effective radius of three-nucleon forces in the same way as the slope of the diffraction cone in elastic $p\bar{p}$ scattering is related to the effective radius of two-nucleon forces. It was also demonstrated that the effective radii of two- and three-nucleon forces, which are the characteristics of elastic and inelastic interactions of two nucleons, define the structure of the $p\bar{p}$ total cross-sections in a simple and physically clear form. I have not the possibility to state on paper here some needed details for clarity. I'd like only to stress that the three-body forces appear as a result of consistent consideration of three-body problem in the framework of local quantum field theory. We have used single-time formalism in QFT elaborated in our papers [14,15] as a tool for the study of the three-body systems dynamics. In our approach the three-body forces are an inherent connected part of total three particle interaction which cannot be represented by the sum of pair interactions. It turns out that the global analyticity of the three-body forces results in generalized asymptotic bound [16,17] like Froissart one [18]. The global analyticity may be considered as a direct geometric generalization of the known analytical properties of two-body scattering amplitude strictly proved in the local quantum field theory [19,20,21,22,23]. Unfortunately in the frame of allowed 10 pages I can present here only shortest declaration of the sections content. I refer to Ref. [37] and references therein for the details.

2 Three-body forces in relativistic quantum theory

In this section the definition of three-body forces is given in the framework of single-time formalism in quantum field theory.

3 Global analyticity of the three-body forces

Global analiticity of the three-body forces is formulated and the generalized unitarity bound for three-body forces scattering amplitude is presented here.

4 Scattering from deuteron

Here the structure of the total cross-section in the scattering from deuteron with account of three-body forces has been discussed.

5 Three-body forces in single diffraction dissociation

In this section the formula connecting one-particle inclusive cross-section with the imaginary part of the three-body forces scattering amplitude has been presented and it has been discussed the effect of screening the three-body forces by two-body ones [25]. It is also shown that the effective radius of three-body forces is related to the slope of diffraction cone for inclusive diffraction dissociation processes in the same way as the effective radius of two-body forces is related to the slope of diffraction cone in elastic scattering processes. The property that the slope of diffraction cone for inclusive diffraction dissociation processes at fixed energy decreases with the growth of missing mass is demonstrated too.

6 On the structure of hadronic total cross-sections

Here a formula defining a new nontrivial structure of hadronic total cross-section is presented. Dynamical origin of the new structure is discussed. An attempt to check up the structure on its correspondence to the existing experimental data has been made. The global fit to all the data on proton-antiproton and proton-proton total cross-sections in a whole range of energies available up today was performed with the obtained theoretical formula. The figures 2,3 display a very good correspondence of theoretical formula to the existing experimental data on proton-proton and proton-antiproton total cross-sections.

7 On the slope of diffraction cone in single diffraction dissociation

The fundamental relation between the slopes of diffraction cone in the single diffraction dissociation and elastic scattering is discussed in this section.

8 On total single diffractive dissociation cross-section

Here we have shown (see Fig. 4) that from the generalized asymptotic theorems a là Froissart there follows a simple formula which allows one to match the

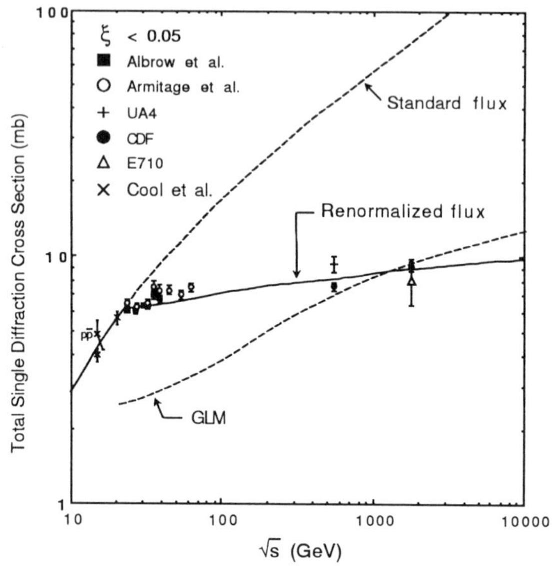

Figure 1: The total single diffraction cross-sections for $p(\bar{p}) + p \to p(\bar{p}) + X$ vs $\sqrt{s}$ compared with the predictions of the renormalized Pomeron flux model of Goulianos [3] (solid line) and of the model Gostman, Levin and Maor [4] (dashed line, labelled GLM).

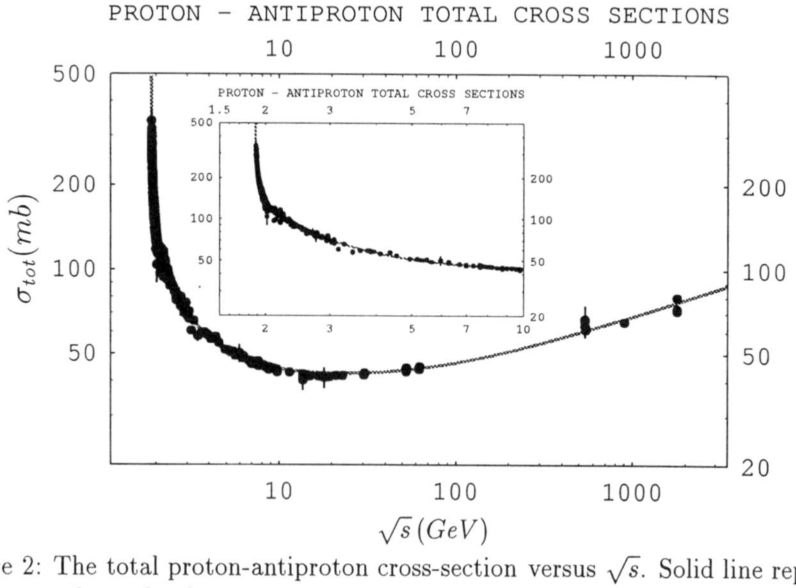

Figure 2: The total proton-antiproton cross-section versus $\sqrt{s}$. Solid line represents our fit to the data.

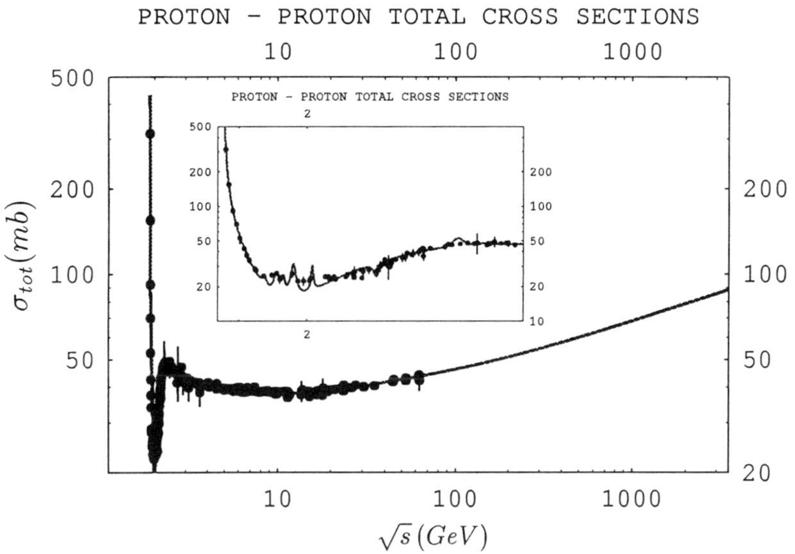

Figure 3: The total proton-proton cross-section versus $\sqrt{s}$. Solid line represents our fit to the data. Statistical and systematic errors added in quadrature.

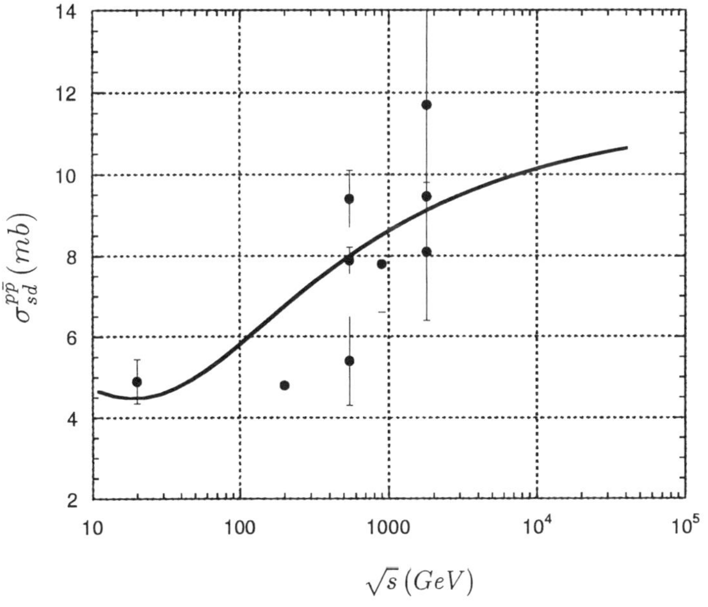

Figure 4: Total single diffraction dissociation cross-section. Solid line represents our fit to the data.

experimental data on $p\bar{p}$ single diffraction dissociation cross-sections at high energies including lower energies as well. We have also found the bound (like Froissart bound!)

$$\boxed{\sigma_{sd}(s) < Const, \quad s \to \infty}. \tag{2}$$

It is pointed out that the understanding of "soft" physics based on general principles of QFT, such as analyticity and unitarity, is so fine confirmed by the experimentally observable picture compared to the models where the general principles have been broken down.

9 On the forms of strong interaction dynamics

A preference of s-channel form of strong interaction dynamics compared to t-channel one is discussed here.

10 Conclusion

In Commemoration of the 200th Anniversary of Alexander S. Pushkin I'd like to conclude my talk with the Ode:

To learn or not to learn?
Of course to learn to be The Forces,
To be The Three-Body Forces as well,
But not Помер-Он *alone.*
That is a sound of Bell!...

Acknowledgements

It is my pleasure to express thanks to the Organizing Committee for the given opportunity to attend the VIIIth Blois Workshop and present the report there. I am indebted to V.V. Ezhela for the access to the computer readable files on total proton-proton and proton-antiproton cross-sections in the IHEP COMPAS database. The friendly encouragement and many pieces of good advice on computer usage from A.V. Razumov are gratefully acknowledged.

References

1. F. Abe *et al*, (CDF) *Phys. Rev.* D **50**, 5535 (1994).
2. A.M. Cooper-Sarkar et al., Preprint DESY 97-226, hep-ph/9712301.
3. K. Goulianos, J. Montanha, *Phys. Rev.* D **59**, 114017 (1999); eprint hep-ph/9805496.
4. E. Gostman, E.M. Levin, U. Maor, *Phys. Rev.* D **49**, R4321 (1994).
5. E.S. Martynov, B.V. Struminsky, in *Proceedings of the XIth Workshop on "soft" physics HADRONS-95, Novy Svet, 1995,* eds. G. Bugrij, L. Jenkovsky, E. Martynov (Kiev, 1995), p.53-60; *Phys. Rev.* D **53**, 1018 (1996); *Sov. J. Yad. Phys.* **59**, 1817 (1996).
6. Proceedings of the Workshop on *QUANTUM INFRARED PHYSICS*, 6-10 June 1994, The American University of Paris, eds. H.M. Freid, B. Muller (World Scientific Singapore, New Jercey, London, Hong Kong, 1995).
7. M.L. Good, W.D Walker, *Phys. Rev.* **120**, 1857 (1960).
8. E.M. Levin, Everything about Reggeons, hep-ph/9710546.
9. V.S. Fadin, E.A. Kuraev, L.N. Lipatov, *Phys. Lett.* B **60**, 50 (1975); E.A. Kuraev, L.N. Lipatov, V.S. Fadin, *Zh. Eksp. Teor. Fiz.* **71**, 840 (1976)[*Sov. J. JETP* **44**, 443 (1976)]; ibid. **72**, 377 (1977)[**45**, 199 (1977)]; Ya.Ya. Balitskii, L.N. Lipatov, *Sov. J. Nucl. Phys.* **28**, 822 (1978).
10. A. Donnachie, P.V. Landshoff, *Z. Phys.* C **2**, 55 (1979); *Phys. Lett.* B **123**, 345 (1983); *Nucl. Phys.* B **231**, 189 (1984); *Nucl. Phys.* B **244**, 322 (1984).
11. J.D. Bjorken, *Nucl. Phys.(Proc. Suppl.)* B**25**, 253 (1992).
12. A.A. Arkhipov, preprint IHEP 96-66, Protvino, 1996; in *Proceedings of the XIIth Workshop on "soft" physics HADRONS-96, Novy Svet, 1996,* eds. G. Bugrij, L. Jenkovsky, E. Martynov (Kiev, 1996), p. 252-263.
13. N.N. Bogoljubov, A.A. Logunov, I.T. Todorov, A.I. Oksak. *General Principles of Quantum Field Theory* (Moscow, Nauka, 1987).
14. A.A. Arkhipov, *Sov. J. Theor. Math. Phys.* **83**, 247 (1990).
15. A.A. Arkhipov, V.I. Savrin, *Sov. J. Theor. Math. Phys.* **16**, 328 (1973); ibid. **19**, 320 (1974); **24**, 78 (1975); **24**, 303 (1975).
16. A.A. Arkhipov, V.I. Savrin, *Sov. J. Theor. Math. Phys.* **49**, 3 (1981).
17. A.A. Arkhipov, *Rep. on Math. Phys.* **20**, 303 (1984).
18. M. Froissart, *Phys. Rev.* **123**, 1053 (1961).
19. H. Lehman, *Nuovo Cim.* **10**, 579 (1958); *Nuovo Cimento Suppl.* **14**, 153 (1959).
20. O.W. Greenberg, F.E. Low, *Phys. Rev.* **124**, 2047 (1961).

21. A. Martin, *Phys. Rev.* **129**, 1432 (1963); *Nuovo Cimento* A **42**, 930 (1966); ibid. **44**, 1219 (1966).
22. G. Sommer, *Nuovo Cimento* A **48**, 92 (1967); ibid. **52**, 373 (1967); **52**, 850 (1967); **52**, 866 (1967).
23. A.A. Logunov, Nguyen Van Hieu, O.A. Khrustalev, in *Problems of Theoretical Physics, Essays dedicated to Nikolai N. Bogoljubov on the occasion of his sixtieth birthday* (Publishing House Nauka, Moscow, 1969) p. 90.
24. A.A. Arkhipov, V.I. Savrin, *Sov. J. Theor. Math. Phys.* **24**, 78 (1975).
25. A.A. Arkhipov, *Sov. J. Theor. Math. Phys.* **49**, 320 (1981).
26. A.A. Arkhipov, P.M. Nadolsky, preprint IHEP 97-6, Protvino, 1997.
27. Yu.A. Troyan, V.N. Pechenov, *Sov. J. Yad. Phys.* **56**, 191 (1993).
28. Yu.A. Troyan, *Sov. J. Physics of Element. Part. and Atomic Nuclei* **24**, 683 (1993).
29. A.W. Chao, C.N. Yang, *Phys. Rev.* D **8**, 2033 (1973).
30. T.T. Chou, C.N. Yang, *Phys. Lett.* B **128**, 457 (1983).
31. A.A. Arkhipov, Single Diffractive Dissociation in $p\bar{p}$ Collisions at High Energy - Talk Presented at "Workshop on Diffractive Physics, LAFEX International School on High Energy Physics (LISHEP-98), Rio de Janeiro, 16 - 20 February, 1998."
32. B.E. Ansorge *et al*, *Z. Phys.* C **33**, 175 (1986).
33. G.J. Alner *et al*, (UA5) *Phys. Rep.* **154**, 247 (1987).
34. D. Bernard *et al*, (UA4) *Phys. Lett.* B **186**, 227 (1987).
35. N.A. Amos *et al*, (E710) *Phys. Lett.* B **243**, 158 (1990); N.A. Amos *et al*, (E710) *Phys. Lett.* B **301**, 313 (1993).
36. R.M. Baltrusaitis *et al* (Fly's Eye Collaboration), *Phys. Rev. Lett.* **52**, 1380 (1984).
37. A.A. Arkhipov, preprint IHEP 99-43, Protvino, 1999; e-print hep-ph/9909531.

A. Martin

THE IMPORTANCE OF REAL PART MEASUREMENTS AT THE LHC

A. MARTIN
Theoretical Physics Division, CERN
CH - 1211 Geneva 23
and
LAPP [a] *- F 74941 Annecy Le Vieux Cedex*
E-mail: martina@mail.cern.ch

After a hitorical review showing how many physical results can be obtained from massive local field theory, including unitarity, we stress the importance of measurements of the total cross sections, to test the saturation of the Froissart bound, and of the real part of the forward amplitude to test locality and get hints of a possible very important underlying structure.

At the present time, almost every physicist believes that the theory of strong interactions is QCD. However, nobody so far has been able to calculate directly from the QCD Lagrangian and from the quark masses the scattering amplitude for – say – proton-proton scattering. So far QCD can only handle "hard processes" or perhaps spectroscopy from the lattice. For soft processes there are of course models but these are only models, even if they are sometimes "QCD inspired". for this reason, use of general principles to obtain information on scattering amplitudes, which was done already back in 1954, is not yet out of fashion as we shall try to show. Before coming to the subject indicated in the title, I would like to present a kind of historical perspective.

In 1954, Gell-Mann, Goldberger and Thirring[1] proved that dispersion relations, previously developed in optics could be established for Compton Scattering: $\gamma P \rightarrow \gamma P$, from the existence of

[a]URA 1436. Associé à l'Université de Savoie.

local fields satisfying the causality property

$$[A(x), A(y)] = 0 \quad \text{for} \quad (x-y)^2 < 0 \; ,$$

i.e., spacelike. This made it possible to express the real part of the forward scattering amplitude as an integral over the imaginary part of the forward scattering amplitude, i.e., by the "optical theorem", an integral over the total cross-section for γP collisions. On this basis, dispersion relations were "proposed" for massive particles in the work of Goldberger on the pion-nucleon scattering amplitude [2]. Soon, his "heuristic proof" was turned into a real proof by various authors [3], using the axioms of local quantum field theory.

Before going on, I would like to explain that these results, even after the discovery that protons and pions are not elementary but made of quarks, are still valid thanks to a fundamental contribution of Zimmermann [4], in which it is proved that to a bound state we can associate a local operator. This constitutes an excellent answer to sceptics like Volodia Gribov [5] or Klaus Hepp [6] (qui brûle ce qu'il a adoré!).

Now I believe that it is necessary to give some technical details, even if many of you know about it.

In 3+1 dimensions (3 space, 1 time) the scattering amplitude depends on two variables energy and angle. For a reaction $A+B \rightarrow A+B$

$$E_{c.m.} = \sqrt{M_A^2 + k^2} + \sqrt{M_B^2 + k^2} \; , \tag{1}$$

k being the centre-of-mass momentum. The angle is designated by θ. There are alternative variables:

$$s = (E_{c.m.})^2 \; , \quad t = 2k^2(\cos\theta - 1) \tag{2}$$

(Notice that physical t is NEGATIVE).

We shall need later an auxiliary variable u, defined by '

$$s + t + u = 2M_A^2 + 2M_B^2 \tag{3}$$

The Scattering amplitude (scalar case) can be written as a partial wave expansion, the convergence of which will be justified in a moment:

$$F(s, \cos\theta) = \frac{\sqrt{s}}{k} \sum (2\ell + 1) f_\ell(s) P_\ell(\cos\theta) \tag{4}$$

$f_\ell(s)$ is a partial wave amplitude.

The Absorptive part, which coincides for $\cos\theta$ real (i.e., physical) with the imaginary part of F, is defined as

$$A_s(s, \cos\theta) = \frac{\sqrt{s}}{k} \sum (2\ell + 1) \ \mathrm{Im}\ f_\ell(s)(\cos\theta) \tag{5}$$

The Unitarity condition, implies, with the normalization we have chosen

$$\mathrm{Im}\ f_\ell(s) \geq |f_\ell(s)|^2 \tag{6}$$

which has, as a consequence

$$\mathrm{Im}\ f_\ell(s) > 0 \ , \quad |f_\ell| < 1 \ . \tag{7}$$

The differential cross-section is given by

$$\frac{d\sigma}{d\Omega} = \frac{1}{s} \ |F|^2 \ ,$$

and the total cross-section is given by the "optical theorem"

$$\sigma_{total} = \frac{4\pi}{k\sqrt{s}} \ A_s(s, \cos\theta = 1) \ . \tag{8}$$

With these definitions, a dispersion relation can be written as:

$$F(s,t,u) = \frac{1}{\pi} \int \frac{A_s(s',t)ds'}{s'-s} + \frac{1}{\pi} \int \frac{A_u(u',t)du'}{u'-u} \tag{9}$$

with possible subtractions, i.e., for instance the replacement of $1/(s' - s)$ by $s^N/s'^N(s'-s)$ and the addition of a polynomial in s, with coefficients depending on t.

The scattering amplitude in the s channel $A + B \to A + B$ is the boundary value of F for $s + i\epsilon$, $\epsilon > 0 \to 0$, $s > (M_A + M_B)^2$. In the same way the amplitude for $A + \bar{B} \to A + \bar{B}$, $\bar{B}$ being the antiparticle of B is given by the boundary value of F for $u + i\epsilon$, $\epsilon \to 0$ $u > (M_A + M_B)^2$. Here we understand the need for the auxiliary variable u.

The dispersion relation implies that, for fixed t the scattering amplitude can be continued in the s complex plane with two cuts. The scattering amplitude possesses the reality property, i.e., for t real it is real between the cuts and takes complex conjugate values above and below the cuts.

In the most favourable cases, dispersion relations have been established for $-T < t \leq 0$, $T > 0$.

In the general case, even if dispersion relations are not proved, the crossing property of Bros, Epstein and Glaser states that the scattering amplitude is analytic in a twice cut plane, minus a finite region, for any negative t [7]. So it is possible to continue the amplitude directly from $A + B \to A + B$ to the complex conjugate of $A + \bar{B} \to A + \bar{B}$. By a more subtle argument, using a path with fixed u and fixed s it is possible to continue directly from $A + B \to A + B$ to $A + \bar{B} \to A + \bar{B}$

Another property derived from local field theory was the existence of the Lehmann ellipse[8], which states that for fixed s, physical, the scattering amplitude is analytic in $\cos\theta$ in an ellipse with foci at $\cos\theta = \pm 1$. $\cos\theta = 1$ corresponds to $t = 0$ the ellipse therefore contains a circle

$$|t| < T_1(s) \tag{10}$$

In the Lehmann derivation $T_1(s) \to 0$ for $s \to (M_A + M_B)^2$ and $s \to \infty$.

The absorptive part is analytic in the larger ellipse, the "large" Lehmann ellipse, containing the circle

$$|t| < T_2(s) \tag{11}$$

with $T_2(s) \to c > 0$ for $s \to (M_A + M_B)^2$, $T_2(s) \to 0$ for $s \to \infty$.

It was recognized very early that the combination of analyticity and unitarity might lead to very interesting consequences and might give some hope to fulfill at least partially the S matrix Heisenberg program. This was very clearly stated already in 1956 by Murray Gell-Mann [9] at the Rochester conference. Later this idea was taken over by many people, in particular by Geff Chew.

It was thought by Mandelstam that the two analyticity properties, dispersion relations and Lehmann ellipses, were insufficient to carry very far the analyticity-unitarity program. he proposed the Mandelstam representation [10] which can be written schematically as

$$\begin{aligned} F = \; & \frac{1}{\pi^2} \int \frac{\rho(s', t') ds' dt'}{(s' - s)\,(t' - t)} \\ & + \text{circular permutations in } s, t, u \\ & + \text{one dimensional dispersion integrals} \\ & + \text{subtractions} \end{aligned} \tag{12}$$

This representation is nice. It gives back the ordinary dispersion relations and the Lehmann ellipse when one variable is fixed, but it was never proved nor disproved for all mass cases, even in perturbation theory,. One contributor, Jean Lascoux, refused to co-sign a "proof", which, in the end, turned out to be imperfect.

One very impressive consequence of Mandelstam representation was the proof, by Marcel Froissart, that the total cross-section cannot increase faster than $(\log s)^2$, the so-called "Froissart Bound" [11].

My own way to obtain the Froissart bound [12] was to use the fact that the Mandelstam representation implies the existence of an ellipse of analyticity in $\cos\theta$ qualitatively larger than the Lehmann ellipse, i.e., such that it contains a circle $|t| < R$, R fixed, independent of the energy. This has a consequence that Im $f_\ell(s)$ decreases with ℓ at a certain exponential rate because of the convergence of the Legendre polynomial expansion and of the polynomial boundedness, but on the other hand the Im $f_\ell(s)$'s are bounded by unity because of unitarity [Eq. (7)]. taking the best bound for each ℓ gives the Froissart bound.

To prove the Froissart bound without using the Mandelstam representation one must find a way to enlarge the "small" and the "large" Lehmann ellipses. In the autumn of 1965, I had very stimulating discussions with Harry Lehmann at the "Institut des Hautes Etudes Scientifiques" about an attempt by Nakanishi which was clever but wrong. Finally, in December 1965 [13], I found the way out. The positivity of Im f_ℓ implies, by using expansion (5),

$$\left|\left(\frac{d}{dt}\right)^n A_S(s,t)\right|_{-4k^2\leq t\leq 0} \leq \left|\left(\frac{d}{dt}\right)^n A_S(s,t)\right|_{t=0} \tag{13}$$

To calculate

$$F(s,t)=\frac{1}{\pi}\int_{s_0}\frac{A_s(s't)ds'}{s'-s}$$

(forget the left-hand cut and subtractions!), for s real $< s_0$ one can expand $F(s,t)$ around $t=0$. From the property (13) one can prove that the successive derivatives can be obtained by differentiating under the integral. When one resums the series one discovers that this can be done not only for s real $< s_0$, but for any s and that the expansion has a domain of convergence in t independent of s. This means that the large Lehmann ellipse must contain a circle $|t|<R$. This is exactly what is needed to get the Froissart bound. In fact, in favourable cases, $R=4m_\pi^2$, m_π being the pion mass. It was already known that for $|t|<4m_\pi^2$ the number of subtractions in the dispersion relations was at most two [14], and it lead to the more accurate bound [15]

$$\sigma_T < \frac{\pi}{m_\pi^2}\ (\log s)^2 \tag{14}$$

Notice that this is only a bound, not an asymptotic estimate.

In spite of many efforts the Froissart bound was never qualitatively improved, and it was shown by Kupsch [16] that if one uses only Im $f_\ell \geq |f_\ell|^2$ and full crossing symmetry one cannot do better than Froissart.

Before 1972, rising cross-sections were a pure curiosity. Almost everybody believed that the proton-proton cross-section was approaching 40 millibarns at infinite energy. Only Cheng and Wu [17] had a QED inspired model in which cross-sections were rising and behaving like $(\log s)^2$ at extremely high energy. Yet, Khuri and Kinoshita [18] took seriously very early the possibility that cross-sections rise and proved, in particular, that if the scattering amplitude is dominantly crossing even, and if $\sigma_t \sim (\log s)^2$ then

$$\rho = \frac{ReF}{ImF} \sim \frac{\pi}{\log s},$$

where ReF and ImF are the real and imaginary part of the forward scattering amplitude.

In 1972, it was discovered at the ISR, at CERN, that the $p-p$ cross-section was rising by 3 millibarns from 30 GeV c.m. energy to 60 GeV c.m. energy [19] (see Fig. 1). I suggested to the experimentalists that they should measure ρ and test the Khuri-Kinoshita predictions. They did it [20] (see Fig. 2) , and, at Fermilab and at CERN it was found very early that ρ, negative at low energies, was becoming positive (If I say that it is because I was the first theoretician speaking after the talk of Ugo Amaldi at CERN, announcing the results of the CERN-Rome group at the ISR!). This kind of combined measurements of σ_T and ReF are still going on. In σ_T we have now more than a 50 % increase with respect to low energy values at the Sp$\bar{\text{p}}$S, and at least 75 % increase at the Tevatron (Fig. 1). The best fit of Augier et al. [21], with

$$\sigma_{tot} = A_1(s/s_0)^{-m} + A_2(s/s_0)^{-n_2} + C_0 + C_2(\ln s/s_0)^\gamma$$

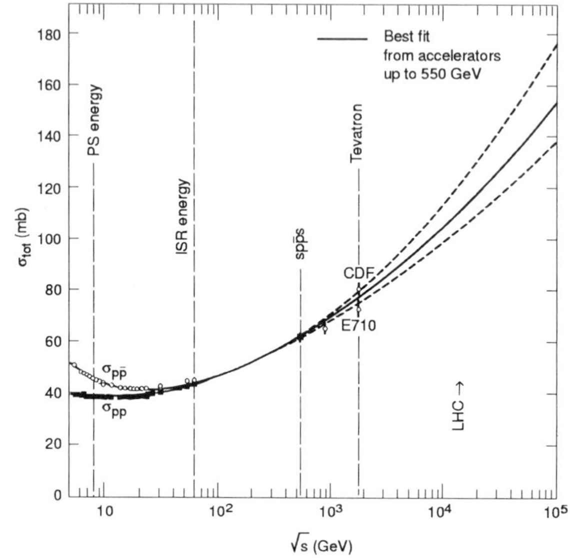

Figure 1: pp and $p\bar{p}$ total cross-sections.

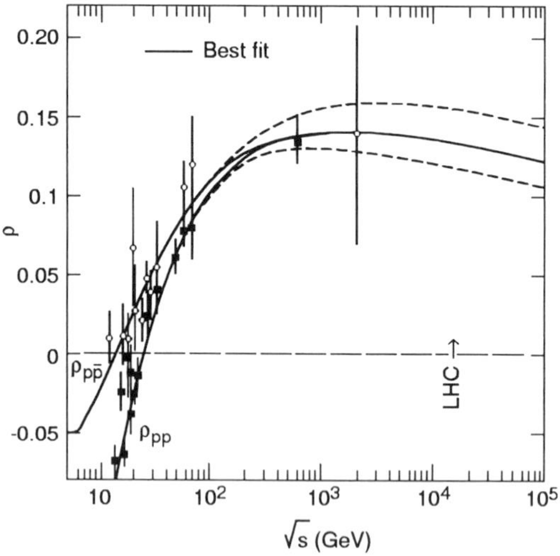

Figure 2: Measurements of the ρ parameter for $\bar{p}p$ and pp scattering are shown together with the result of the dispersion relation fit of Augier et al. [21]

gives $\gamma = 2.2 \pm 0.3$, compatible with a qualitative saturation of the Froissart bound ($\gamma = 2$) and predicts $\sigma_{tot} = 107$ mb at 14 TeV (hopefully the LHC energy). For an up to date review I refer to the article of Matthiae [22]. it is my strong conviction that this activity should be continued with the future LHC. A breakdown of dispersion relation might be a sign of new physics due to the presence of extra compact dimensions of space according to N.N. Khuri [23].

The breakdown of dispersion relations can manifest itself either by the appearance of extra singularities in the complex plane or by the breakdown of polynomial boundedness. In early attempts preserving analyticity but trying to break polynomial boundedness it was thought that the amplitude might grow in complex directions like exp Im k R, i.e., like $\exp C\sqrt{|s|}$. But this is impossible, because if you assume a growth $\exp C|\sqrt{s}|$, you succeed, by using unitarity in the Lehmann ellipse to prove that the physical scattering amplitude is polynomially bounded on the left-hand cut and the right-hand cut, and, using the Phragmén-Lindelöf theorems, in the whole complex plane. The first possible violation is to assume that the scattering amplitude grows like

$$\exp |\mathrm{Im}\ s| R^2$$

in complex directions. Then one must multiply the forward scattering amplitude by a convergence factor, $\exp\left(-2MR^2\sqrt{E_0^2 - E_{\mathrm{lab}}^2}\right)$, where R is the "size" of the extra compact dimensions, E_{lab} the laboratory energy and M the mass of the target. At high energies, $2M\ E_{lab}$ is essentially $(E_{CM})^2$. If $1/R$ is of the order of the supersymmetry breaking scale, which is believed to be at most the energy of the LHC one expects big effects at LHC energy. But, as pointed out by N.N. Khuri, the real part, because of its smallness, is a much more sensitive indicator than the total cross-section and will be seriously affected at appreciably lower energies.

Specifically F will not satisfy a dispersion relation, but instead

$$\hat{F} = \exp -\sqrt{\frac{M^2 - E^2}{E_0^2}}\, F$$

does. The dispersion relations which have been written up to now concern in fact $\hat{F}$ instead of F, but both are very close at low energies. At high energies, however, the real and imaginary parts are very different because of the phase factor

$$\exp\ i\sqrt{\frac{E^2 - M^2}{E_0^2}}$$

if $\sqrt{s_0}$ $(s_0 = 2ME_0)$ is 16 TeV, the phase factor will be $\pi/2$: real and imaginary parts are exchanged! The effect, however, will occur at much smaller energies. For instance, with $\sqrt{s_0} = 16$ TeV, we get at $\sqrt{s} = 5$ TeV a phase factor $\exp i/10$. At this energy, we believe from the dispersion fits that $\rho = 0.13$, but, in fact, it is $\hat{\rho}$, and therefore $\rho \simeq 0.25$, a very drastic change, while the total cross section is almost unaffected. Similarly, if $\sqrt{s_0} = 6$ TeV, large effects on ρ will be seen already at $\sqrt{s} = 2$ TeV.

For that reason accurate measurements of ρ at lower energies (for instance re-doing the Tevatron measurement which is not sufficiently accurate to give any constraint) is extremely desirable. Future experiments, especially for ρ, will be difficult because of the necessity to go to very small angles, but not impossible[24]. We hope very much that these will be done eventally.

References

1. M. Gell-Mann, M.L. Goldberger and W. Thirring, *Phys.Rev.* **95** (1954) 1612.
2. M.L. Goldberger, *Phys.Rev.* **99** (1975) 979.
3. N.N. Bogoliubov, B.V. Medvedev and M.K. Polivanov, Voprossy Teorii Dispersionnyk Sootnoshenii, V. Shirkov et al.

Eds., Moscow 1958;
K. Symanzik, *Phys.Rev.* **105** (1957) 743;
H. Lehmann, *Suppl. Nuovo Cimento* **14** (1959) 153.
4. W. Zimmermann, *Nuovo Cimento* **10** (1958) 597.
5. V. Gribov, private communication (1997).
6. K. Hepp, private communication, Zürich, 1996.
7. J. Bros, H. Epstein and V. Glaser, *Commun.Math.Phys.* **1** (1965) 240.
8. H. Lehmann, *Nuovo Cimento* **10** (1958) 579.
9. M. Gell-Mann, Proceedings of the 6th Annual Rochester Conference, J. Ballam, V.L.Fitch, T. Fulton, K. Huang, R.R. Rau and S.B. Treiman eds., Interscience Publishers, New York 1956, p. 30.
10. S. Mandelstam, *Phys.Rev.* **112** (1958) 1344.
11. M. Froissart, *Phys.Rev.* **123** (1961) 1053.
12. A. Martin, *Phys.Rev.* **129** (1963) 1432, and Proceedings of the 1962 Conference on High energy Physics at CERN, J. Prentki ed., CERN Scientific Information Service, 1962, p. 567.
13. A. Martin, *Nuovo Cimento* **42** (1966) 901.
14. Y.S. Jin and A. Martin, *Phys.Rev.* **B135** (1964) 1375.
15. L. Lukaszuk and A. Martin, *Nuovo Cimento* **52** (1967) 122.
16. J. Kupsch, *Nuovo Cimento* **B70** (1982) 85.
17. H. Cheng and T.T. Wu, *Phys.Rev.Lett.* **24** (1970) 1456.
18. N.N. Khuri and T. Kinoshita, *Phys.Rev.* **B137** (1965) 720.
19. U. Amaldi et al., *Phys.Lett.* **B44** (1973) 112;
S.R. Amendolia et al., *Phys.Lett.* **B44** (1973) 119.
20. V. Bartenev et al., *Phys.Rev.Lett.* **31** (1973) 1367;
U. Amaldi et al., *Phys.Lett.* **66B** (1977) 390.
21. C. Augier et al., *Phys.Lett.* **316B** (1993) 448.
22. G. Matthiae, *Rep.Progr.Phys.* **57** (1994) 743.
23. N.N. Khuri, Rencontres de Physique de la vallée d'Aoste, 1994, M. Greco ed., Editions Frontières 1994, p. 771;

see also: N.N. Khuri and T.T. Wu, *Phys.Rev.* **D56** (1997) 6779 and 6785.

24. Angela Faus-Golfe, private communication.

A. A. Asseev

Extraction of elastically scattered protons at IHEP.

A.A.Asseev
Institute for High Energy Physics
Protvino, Moscow region, 142284 Russia.
E–mail:asseev@mx.ihep.su

Systematic investigations of beam dynamics made it possible to study the alternative possibilities for particle extraction. The data obtained were used for the development of new method of particle extraction which allowed one to extract beams for new experimental facilities where it is not possible to extract charged particles by the classic methods. The new experiments run simultaneously with the existing ones. It improves by 1.3–1.5 times the efficiency of the accelerator use for physics research due to the increase of the number of simultaneously working setups. This has also increased the duration of beam extraction by ~ 2 times. The new method of extraction allowed one to extend the experimental program of the Institute.

1 Non–resonant Slow Extraction.

The idea of such method of a beam extraction from big accelerators was proposed by A.W.Mashke (FNAL) and developed by Ch.Steinbach who evaluated the possibility of its realization for the PS and SPS accelerators of CERN [1]. The evident simplicity and the high expected efficiency of the extraction method on the existing level of the technics development put it in the line with the well known but rather complex and expensive resonant methods of a beam extraction. Not substituting them it has the row of advantages the main of which is the possibility of extraction from the accelerator the proton beam of the intermediate ($10^7 \div 10^{11}$ ppc) intensity, which was not accessible before, at high quality of the time structure.

The calculated efficiency of this method of extraction for the Tevatron (FNAL) is $\sim 98\%$, for the SPS (CERN) is $\sim 95\%$ [1,2]. The method was not realized at FNAL and CERN due to the difficulties connected with the change of the projects of the accelerators which were in the construction stage that time.

The non–resonant slow extraction (NRSE) of protons scattered on internal target under small angles was realized at IHEP for the first time in the practice of particle extraction from the strong–focusing accelerators of such energies [3]. It provided the experimental setups with the intensity by 3–4 orders of magnitude higher than in the case of direct extraction of positive particles from the internal targets.

One have to point out that at this mode of extraction the possibility of other users work with the beam of secondary particles generated by the same internal targets is ensured also. It allows one to gain additionally the rate of the accelerator beam use.

1.1 Principle of Extraction.

There are four different processes (see, for example [4]) which take place when a proton hits a target. They are absorption, elastic scattering, multiple Coulomb scattering and ionization energy loss. Absorption of protons leading to production of secondary particles is the main mechanism of beam intensity decrease during the interaction with an internal target on the flat top of the magnetic cycle. The contribution of elastic nuclear scattering is generally included with inelastic interaction since elastic scattering forces particles to leave the vacuum chamber aperture. As was shown in [3], part of the protons from elastic scattering remain in the region of stable motion for a long time. These particles can also be extracted. Multple Coulomb scattering and ionization energy loss both change the amplitude and phase of betatron oscillations. Ionization energy loss also shifts the closed orbit of the protons. The sum of these processes causes a blow up and a radial shift of the whole proton beam. It throws particles into septum–magnet apertures where they are extracted from the accelerator.

We will show how a target influences the beam dynamics in the accelerator. Only the processes of multiple Coulomb scattering and ionization energy loss are taken into consideration. Elastic nuclear scattering produces an increase of the amplitude of betatron oscillations which deflects particles into the aperture of the first magnetic deflector and concequently kicks them from the accelerator even after a single intersection of a target. Estimates of the significance of such processes are given in [3].

1.2 Effect of multiple Coulomb scattering of particles.

Let us imagine that on a phase plane (see fig.1) normalized for round trajectories (circle 1) a point with coordinates $L_1, \beta x'_1$. Let Θ be the angle of scattering at a certain intersection of a target by a particle and R_0 be its amplitude. Scattering brings the point $(x_1, \beta x'_1)$ to the point $(x_1, \beta x'_2)$ that is away from the first one by $\beta\Theta$ (β is the β–function of the accelerator). As a result, we get a new phase region for the scattered particle (circle 2) which reflects its motion with a new amplitude R.

From the geometric relations and after averaging for all the scattering angles one can get an expression for the new amplitude of the particle:

$$< R^2 >=< R_0^2 > + < (\beta \cdot \Theta)^2 >, \tag{1}$$

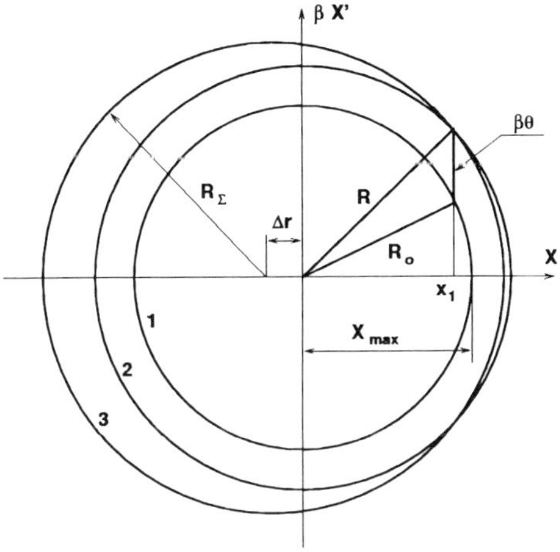

Figure 1: The effect of beam scattering on a phase plane.

It is seen from (1) that the meansquare growth of the betatron amplitude of a particle ΔA_t at the target asimuth after a certain intersection equals:

$$< \Delta A_t >= (< \Delta R_t^2 >)^{1/2} = \beta_t \cdot \Phi, \qquad (2)$$

where β_t is the value of the β–function at the target asimuth. The projection of a meansquare scattering angle of a particle on the appropriate direction is expressed by [5]:

$$\Phi = \sqrt{< \Theta^2 >} = \frac{E_s}{p\beta_v c}\sqrt{\frac{D_t}{L_R}} \cdot (1 + \epsilon). \qquad (3)$$

Here $E_s = 15$ эeV, p is the particle momentum, β_v and c are the relativity factor and speed of light, respectively; D_t, L_R are the thickness of a target along the beam and the radiation length of the material, and ϵ is a correction factor.

One needs to emphasize that if the scattering occurs at the point $u = u_{max}$ (see fig.1) we have growth of the betatron amplitude of the particle that is determined by the second term of the expression (1) independent of the sign of the scattering angle Θ. This effect of betatron amplitude growth (which is stronger than the effect of orbit displacement due to ionization losses) allowed us to do slow extraction of protons from the strong–focusing accelerator.

The dependence for the particle amplitude after interacting with the scattering target is (at the target asimuth):

$$A_{rt}^2 = A_{0rt}^2 + (\beta_t \cdot \Phi)^2, \tag{4}$$

where A_{0rt} is an initial amplitude of the particle. Simple transformations show that the expression for the particle amplitude at the asimuth of the first deflector will be similar:

$$A_{rd}^2 = \left(\frac{\beta_d}{\beta_t}\right) A_{0rt}^2 + \beta_d \cdot \beta_t \cdot \Phi^2, \tag{5}$$

and its growth due to scattering into a target:

$$\Delta A_d = (< \Delta R_d^2 >)^{1/2} = (\beta_d \cdot \beta_t)^{1/2} \cdot \Phi, \tag{6}$$

where β_d is the β–function of the accelerator at the deflector asimuth.

Deflection of a particle into the first deflector aperture occurs when, after some "last" pass through the target the growth of its amplitude satisfies the conditon (provided that the septum is placed at the edge of the undisturbed beam envelope):

$$|(< \Delta R_d^2 >)^{1/2}| > D_d, \tag{7}$$

where D_d is the thickness of the first septum in the deflector system.

1.3 Influence of ionization energy loss.

As we can see both from (1) and fig.1, the change of a particle's angle during its motion in the accelerator is transformed into growth of its amplitude. Analogously, a change of momentum by Δp is transformed into a change of the position of a particle due to displacement of the closed orbit in the horisontal plane. After intersecting a target, the energy loss for a particle is expressed by the formula [5]:

$$\Delta E = -\left(\frac{1}{\rho} \cdot \frac{dE}{dx}\right) \cdot D_t[MeV], \tag{8}$$

where $(1/\rho \cdot dE/dx)$ is the specific ionization loss in the target material. A relative change of the proton momentum for a single intersection of the target $\Delta p/p_0$ leads to a displacement of its orbit to the inside at the target and deflector asimuths of Δr (see fig.1). This is expressed as:

$$\Delta r_{t,d} = -\psi_{t,d} \cdot \frac{\Delta p}{p_0}; \tag{9}$$

where $\psi_{t,d}$ is the dispersion function at the target or a deflector asimuth and p_0 is the initial momentum of the particle.

If $\psi_t' \sim 0$ (which is valid for the case of A–70), the growth of the betatron amplitude is $\Delta A_r = |\Delta r_t|$. If a target is outside of the beam, the maximum of the particle displacement will be after half a period (or any odd number of semiperiods)of horizontal betatron oscillations. At the first deflector asimuth the maximum of the particle displacement will be:

$$\Delta r_{dmax} = -\left[\psi_d \cdot \frac{\Delta p}{p_0} + \left(\frac{\beta_d}{\beta_t}\right)^{1/2} \cdot \psi_t \cdot \frac{\Delta p}{p_0}\right] \approx -2\psi_d \cdot \frac{\Delta p}{p_0}; \qquad (10)$$

The effect can also be used for extraction of particles from the accelerator if the following condition holds:

$$|\Delta r_{dmax}| > D_d, \qquad (11)$$

It is important to note that the scattering target needs to be carefully chosen for this case. That is its mechanical, thermal and nuclear characteristics, as well as the thickness of the target along the beam must be considered. The phase advance of the target placement along the orbit relative to the first deflector is also important. According to calculations in [6] use of the effect allows one to obtain highly efficient ($\varepsilon = 97\%$) slow extraction from the proton storage ring of the Moscow Meson Factory of the Institute of Nuclear Research.

1.4 Joint influence of scattering and energy loss.

Taking account of the above effects, the amplitude of the horisontal oscillation of particles at the deflector aismuth are given by the formula:

$$A_{rd\Sigma} = A_{rd} + \left(\frac{\beta_d}{\beta_t}\right)^{1/2} \cdot \psi_t \cdot \frac{\Delta_p}{p_0}, \qquad (12)$$

where A_{rd} is the amplitude of the betatron oscillations of the particle (5).

Fig.1 (circle 3) gives a visual impression of the growth of the amplitude of the betatron oscillations and the displacement of the closed orbit due to multiple Coulomb scattering and ionization energy loss after interaction with a target.

The final expression for the amplitude of the particle oscillations at the first deflector asimuth after a single interaction with a target is:

$$A_{rd\Sigma} = \left[\frac{\beta_d}{\beta_t} \cdot \left(A_{0rt} + \psi_t \cdot \frac{\Delta p}{p_0}\right)^2 + \beta_d \cdot \beta_t \cdot \Phi^2\right]^{1/2}; \qquad (13)$$

The resulting growth of the amplitude after n interactions is important for the non–resonant slow extraction. The grows is expressed by the formula:

$$\Delta A_{rd\Sigma} \sim \left[n \cdot (\beta_d \cdot \beta_t) \cdot \Phi^2 + 2n \cdot \left(\frac{\beta_d}{\beta_t} \right) A_{0rt} \cdot \psi_t \cdot \frac{\Delta p}{p_0} \right]^{1/2}; \qquad (14)$$

The maximum displacement of a particle at the first deflector asimuth is:

$$\Delta r_{dmax\Sigma} = (\Delta A_{rd\Sigma} \pm n \cdot \Delta r_d) = \left[n \cdot (\beta_d \cdot \beta_t) \cdot \Phi^2 + 2n \cdot \left(\frac{\beta_d}{\beta_t} \right) A_{0rt} \cdot \psi_t \cdot \frac{\Delta p}{p_0} \right]^{1/2} \pm n \cdot \psi_d \cdot \frac{\Delta p}{p_0} \qquad (15)$$

The last term of expression (16) has a sign depending on the coordinate of the septum–magnet (inside or outside of the closed orbit). For example, for ED–106, which is placed inside, the growth is added to $\Delta A_{rd\Sigma}$ while for the SM–18 which is placed outside, oppositely, the grows is subtracted.

2 Practical realization.

2.1 Extraction in "low intensity" mode.

Extraction to the SWD setup.

This mode of extraction was not planned for this setup since the proton beam extraction to the SWD had to be done by another method – a deflection to beam line 22 by a bent crystal [7] of a part of the beam which is extracted towards beam line 8 during the resonant slow extraction [8]. Fault in realisation of this mode of extraction makes one to search the nonresonant slow extraction possibility for the SWD. After investigations the extraction of the proton beam intensity of $10^7 \div 10^9$ ppc was done [9].

Extraction to the SPHINX setup.

At the beginning (see, e.g. [10]) it was planed to obtain proton beams with the intensity of $10^6 \div 10^9$ ppc by diffraction scattering of protons of primary intensity of $5 \cdot 10^{11}$ ppc on the target installed in the beam line bending magnets with further forming of the beams obtained.

In spite of the fact that this technique solves the impose problem it still had the essential disadvantages:

- The very large planned losses of intensity inevitably result in unjustified irradiation of the magnetic optical elements;

- so high levels of induced radiation that makes the maintenance and repair of this equipment quite problematic.

The NRSE allowed one to avoid such losses. As it turned out this technique has a number of advantages in the case of the IHEP accelerator. The most important among them are the possibility to extract proton beam of intermediate intensities ($10^6 \div 10^{11}$ ppc), to provide high quality time structure and to work simultaneously with several (3–4) internal targets.

As is seen from fig.2 the target of beam line 4 (T_2) installed on the coordinates of $r \approx 35 - 45$ mm w.r.t. the central orbit may be a sorce of elastically scattered protons for beam line 21. The production angle for protons is 5–7 mrad wich causes the beam intensity of $\sim 5 \cdot 10^6$ ppc which allows to carry some experiments.

This operating mode seems rather attractive because it does not require the SM power supplies to be switched on, no need for local bumps and what is more, the consumption efficiency for the beam from the internal target becomes considerably higher. Under such operational conditions one can provide the duration of the beam extraction towards beam line 21 equal to that of the secondary beam extraction which is up to now is not possible because of SM–18.

NRSE without septum–magnets has been tested and was used successfully at the end of the last IHEP accelerator run (spring of 1999) when we had one of the septum–magnets of the extraction chain broken. Thanks to this mode of extraction the experiment of the SPHINX setup was finished successfully.

2.2 Extraction of higher intensity for the Tagged Neutrino Facility (TNF).

During the investigations it was desirable to study a principle possibility for extraction of a beam of intensity $\sim 10^{10}$ ppc for a new experiment. Earlier, when NRSE of protons was made towards channel 22 such an intensity was not obtained. The next stage of the research was to obtain the maximum intensity is possible to reach when all deflectors of the slow extraction system of A–70 were used.

Two schemes were used for beam extraction. The first one, where beam extraction is done by a series of magnetic deflectors (septum–magnets) with gradually increasing septum thicknesses which are installed in straight sections (SS) 18, 20, 22, 26 has already been used at IHEP [3,11]. The first deflector of the scheme which is responsible for the extraction efficiency is the septum–magnet SM–18 with a "mechanical" thickness of the septum of ~ 0.5== [12].

The second scheme has, in addition to the septum–magnets arrangement of the first one, a first element that is an electrostatic deflector (ED). The septum is installed at SS–106 to separate the circulating and extracted beam. It consists of 0.1mm diameter W–Re alloy wires, the dimension of the gap between the anode and cathode of the deflector is 20 mm and the length of the deflector is 3 = [13].

Beam extraction by scheme 1 (with SM–18).

NRSE was done in series with high intensity resonant slow extraction (RSE) which used the first half of the flat top of the magnetic cycle and in parallel with generation of secondaries by internal targets for experimental setups for beam lines 2, 4, 5N. Duration of the extracted beam pulse was about 0.7 s. After tuning the septum and magnets, the extracted intensity was optimized by special bumps to displace the beam towards septums SM–18, 20. Those bumps were formed with currents in additional windings on magnetic blocks 15, 21 and 16, 22 [3]. With a current of $\sim$ 95A in 15–21 bump, the extracted intensity was optimized with bump 16–22. It was programmed in such a way that the change of the gap between the circulatig beam and the septums at SM–18, 20 due to absorption of particles by the target was compensated. Simultaneously, the displacement of the extracted particles towards septum inside of the septum–magnet apertures due to energy loss in the target was decreased. So, the above gaps are kept constant during the extaction time. Extracted itensity of $\sim 10^{10}$ppc was reached. After the target of the TNF there are $\sim 10^6$ K–mesons. This flux was enough to start physics research for the new program.

Beam extraction by scheme 2 (with ED–106).

In this case proton extraction continues for the duration of the accelerated beam interaction with the internal targets ($\geq$ 1.7 s). Scattered particles are thrown into the ED–106 aperture, then further into SM–18 (with a current reduced by 20%) and extracted towards channel 23 along the well known trajectory through the septum–magnets of SS–20, 22, 26. The full scheme of extraction is shown in fig.2 where the beam orbit distortion in the region of the ED–106, septum–magnets and targets (T_1, T_2) is presented with curve $1, 1'$ while the trajectory of the extracted beam is shown with the curve $2, 2'$.

At this experiment an extracted beam intensity of $\geq 5 \cdot 10^{10}$ ppc was obtained and at the TNF $\sim 5 \cdot 10^6$ K–mesons was achieved. After obtaining practically full duration of extraction and an intensity of the proton beam high enough to start the research program, the factor which determines the statistical efficiency of new experiment is a quality of the extracted beam. Beam quality is illustrated by fig.3 where oscillogramms of the time structure of the secondary particle beams extracted towards beam lines 14, 4 and 18

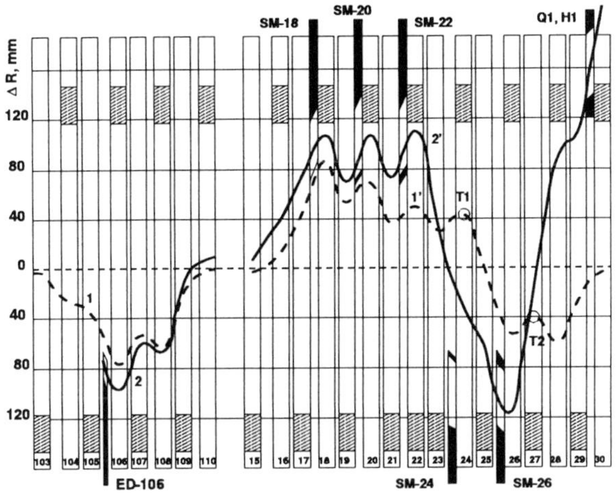

Figure 2: Full proton extraction scheme: ED–106,SM–18,20,22,26 are the electrostatic and magnet deflectors of the extraction system; Q_1, H_1 are the first elements of the beam transport system; T_1, T_2 are internal targets of beam lines 2(14) and 4, respectively.

(three upper traces) and primary protons extracted towards beam line 23 (the lower trace) are presented.

Here we stress only that due to the "thin" carbon target [14] forming such distribution of particles into the beam that allows the bump systems, having the limited band of frequencies, to facilitate the beam steering. As a result, there are no high frequency ripples in the extracted beams (for example, the frequency of the beam revolution $\sim 200KHz$). Simulteneously low frequencies (from one to hundred of Hz) to be in the working region for the beam steering systems are significantly suppressed. The level of the ripple is not higher than $7 - 12\%$ of the amplitude of the appropriate signals.

3 Evaluation of the extraction efficiency.

The extraction efficiency for NRSE was evaluated by taking account of the intensity of the primary beam that interacted inelastically with internal targets, the number of particles scattered by targets as well as the intensity measured in the beam line. All these data are summarized in the next table.

$^{*)}$Limit of measurements due to saturaton of the ionization chamber that was used.

Explanation of the table:

$I_{\Sigma A-70}$ is the accelerator intensity; ΔI_i is the intensity of the beam interact-

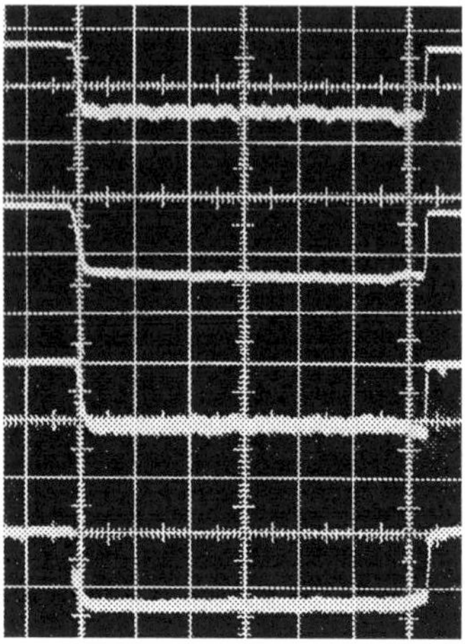

Figure 3: Oscillograms of the time structure of secondary particle and primary proton beams extracted simultaneously towards beam lines 14, 4, 18 and 23.

Table 1: Distribution of accelerated intensity($\times 10^{11}$ppc)

I_Σ	Intensity for internal targets				ΔI	ΔI
A–70	ΔI_{24}	ΔI_{27}	ΔI_{35}	ΔI_Σ	NRSE	EXTR
24.0	11.0	10.0	1.0	22.0	2.0	$\geq 0.5^{*)}$

ing with targets (i is the magnetic block number where respective target is placed); $\Delta I_{NRSE}, \Delta I_{EXTR}$ are intensities of particles scattered by targets and extracted, respectively.

The extraction efficiency was evaluated as follows:

$$\varepsilon = \frac{\Delta I_{EXTR}}{\Delta I_{NRSE}} \geq \frac{5 \cdot 10^{10}}{2 \cdot 10^{11}} \geq 25\%.$$

The error of the reported value is not more than 10%. It is expected that with nominal current for SM–18 an extraction intensity $\geq 10^{11}$ ppc can be obtained. With the high quality of the extracted beam time structure it meets the requirements of the new research program for physics facilities of beam lines 21, 22, 23.

The results of numerical estimates of the particle step size into aperture due to deflection from scattering by the Be target ($D_t = 3.0$cm) of beam line 2(14) after 1,4 and 6 interceptions respectively are given below. Analysis of the data of the estimates allows one to explain the difference of extracted intensities for the two cases mentioned above (see sections 2.2.1 and 2.2.2).

We have, at extraction through the SM–18, a particle step size after a single target pass of $\Delta r_{dmax\Sigma} = 2.36$mm. After 4 and 6 passes $\Delta r_{dmax\Sigma}$ is 3.94 and 4.40 mm. After a number of passes through the target a particle undergoes elastic scattering which results in kicking it into the the deflector aperture [3] with a high level of probability. For extraction through SM–18, taking into account the septum thickness of 0.5 mm we see that the real step size of particles in the aperture of a septum–magnet is 1.86–3.90 mm. Considering septum–magnet allignments errors, orbit instabilities, beam divergence and the decrease of intensity of the beam interacting with a target one can see that the possibility of extraction of a high intensity beam through the SM–18 is limited.

In the case of extraction a beam through the ED–106 we have $\Delta r_{dmax\Sigma} =$ 3.07, 6.90, and 8.87 mm for 1, 4 and 6 intersections of a target, respectively. This means that if ED–106 is powerful enough to deflect a beam over the SM–18 septum only, the gap between circulating and extracted beams at the SM–18 asimuth will be increased considerably. Obviously, this means there will be a decrease of particle losses on the septum of SM–18. As a result there will be an increase of the intensity of the extracted beam.

Based on the above estimates the increase of the effective gap between the circulating and extracted beams more than by 1.5–2 times due to the use of an electrostatic deflector have led to an increase of the extraction intensity by almost an order of magnitude. For scheme 1 this amounts to $\leq 10^{10}$ at extraction while for scheme 2 it results in $\geq 5 \cdot 10^{10}$ ppc extracted towards the setup TNF.

4 Beam parameters.

Results of calculations of particle distributions thrown due to the scattering in the internal target into the aperture of the septum–magnet SM–18 are shown in the fig.4. The impact parameter of the particle depends on the target parameters and the number of intersection of it by the beam. Under optimally selected parameters of the target one can form the beam phase space to be small enough that facilitates the seach of the variants of beam focusing in the head part of the beam line.

Phase ellipses of the extracted from the accelerator beam in $r-$ and $z-$ planes are shown in fig. 5. The phase acceptance of the beginning part of beam line 8 (in SS–30) and phase ellipces (on $r-$space is shown with the broken line) which serve as the optical source to calculate the beam focusing are shown in the figure.

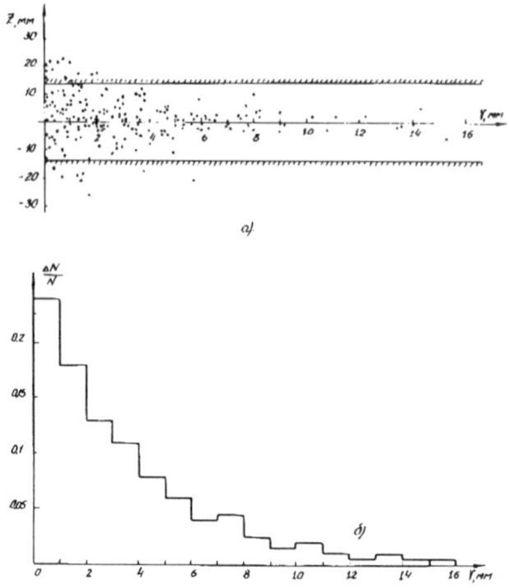

Figure 4: Distribution of particles in the SM–18 aperture: a)– on the plane $r-,z-$ across the orbit; b)– the radial distribution.

Other characteristics of the extracted from A–70 beam are illustrated by figures 6–8.

Fig.6 shows the angular distribution of particles inside the SM–18 aperture for Be and Al targets. Here the parts of the beam due to elastic scattering are also shown.

The relation of number of nuclear scattered protons to the all extracted protons in dependence of the mass number of the target material is shown in fig.7.

It follows from the last figures that the maximum of angular distribution for the nuclear scattering is around the region of angles of 0.5–0.6 mrad (with subtraction of the r.m.s. value of angular divergence of the circulating beam ~ 0.2 mrad). The share of such particles in the case of Be target is $\sim 45\%$.

This data represent the independent interest since in the regarded region of the scattering angles the appearence the partial polarization of protons is forecasted [15–17]. The polarization is due to the interferention of the hadron amplitude without the spin revolution with the electromagnetic one with the spin revolution for a magnetic momentum of the particle.

The dependence of the polarization P_{pp} from the radial projection of angles calculated by R.A.Rzaev [3] for the proton–proton scattering at the 70 GeV is shown in fig. 8.

At transition to the nucleis (A,Z) the values of the maximal polarization P_0^A and the corresponding transmitted momentum t_0^A are changed by formulas:

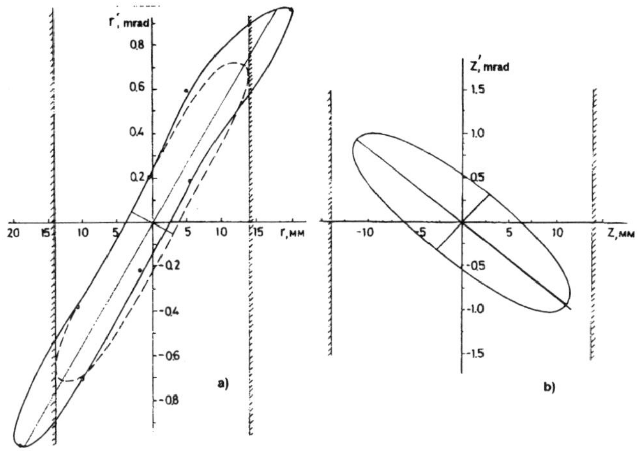

Figure 5: The phase space of the beam at the accelerator output: a)– in the horisontal plane; b)– in the vertical plane. The shadowed lines show the boundaries of the vertical aperture of the first lens.

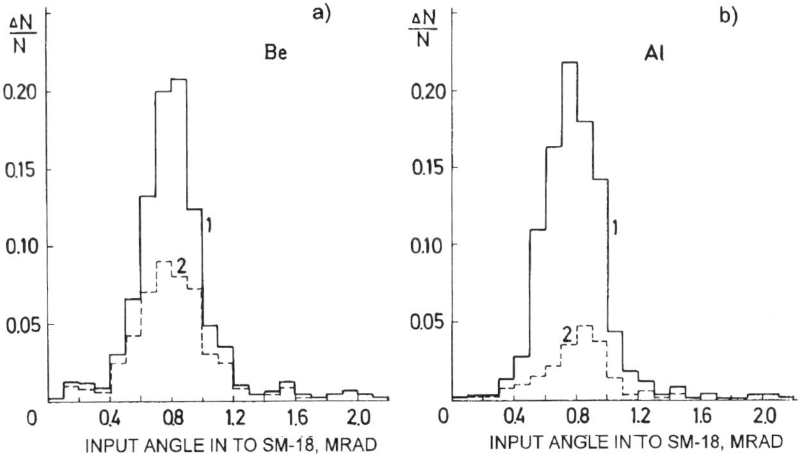

Figure 6: Angular distribution of particles inside the SM–18 aperture at the interaction of proton beam with the Be (a) and Al (b) targets: 1 is for all the particles undergone the scattering; 2 is for the particles undergone the elastic nuclear scattering.

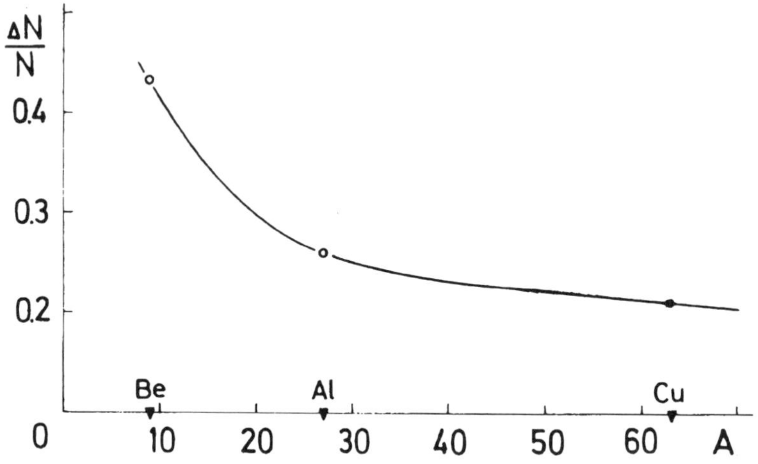

Figure 7: The share of the elastic nuclear scattering versus of the mass number of a target material.

$$P_0^A = \sqrt{\frac{Z}{A^{2/3}}} \cdot P_{pp}; t_0^A = \frac{Z}{A^{2/3}} \cdot t_0^{pp}. \tag{16}$$

For the case of the Be nuclei we have:

$$P_0^{Be} = 0.96 P_{pp}; t_0^{Be} = 0.92 t_0^{pp}. \tag{17}$$

From the fig. 8 it follows that the scattering angles of 0.5–0.6 mrad on the target correspond to the maximum of the polarization curve and the beam of protons obtained by this mode may turn to be partly polarized with the rate of the polarisation of 2–3%.

5 Conclusion

Use of an electrostatic deflector as the first element of the extraction system in the non–resonant slow extraction mode allowed one to reduce particle loss on septums. Even with the reduced current of SM–18 the extracted beam intensity $\geq 5 \cdot 10^{10}$ ppc was obtained. It is about an order of magnitude higher the intensity obtained when the SM–18 was as the first deflector of a system. Evaluations show that it is possible to extract $\geq 10^{11}$ ppc keeping the high quality time structure for the extracted beams. Beams of a high quality and intensity of $\geq 10^{10}$ ppc are needed also for the setups FODS–2, SWD of channel 22 to generate the electron beams of high energy.

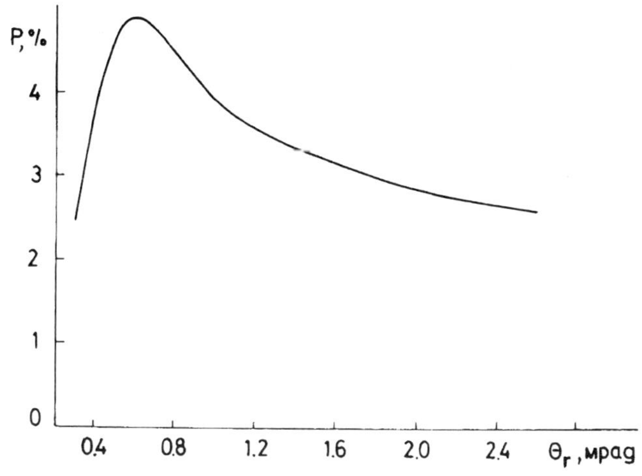

Figure 8: Dependence of the polarization rate vs the scattering angle for the Be target.

Non–resonant slow extraction is the only method at the A–70 so far to extract high quality proton beams of the intensity $\sim 10^{11}$ ppc in parallel with extraction of secondaries for other experiments. Extraction of particle beams of such intensity, long duration and high quality time structure as well as stability of other parameters had not yet succeeded with other well–established methods.

References

1. Ch.Steinbach, EJ/27, CERN/MPS–CO/70–7, Geneva, 1970.
2. CERN MC/60, Geneva, 1970. A design of the European research facilities.
3. Yu.M.Ado, A.A.Asseev et al., IHEP 85–23, Serpukhov, 1985.
4. Hereward H.G., Ranft J., Richter W., CERN 65–1, Geneva, 1965.
5. A.N.Kalinovsky, N.V.Mokhov, Yu.P.Nikitin, Propagation of high energy particles through the matter, Moscow, Energoatomizdat, 1985.
6. N.D.Malitsky, Yu.P.Severgin, I.A.Shukeylo et al., Proceedings of the XI All-Union Part. accel.conf., Dubna, 1989, v.2, p.170.
7. A.A.Arkhipenko et al., IHEP 90–91, Protvino, 1990.
8. K.P.Myznikov et al., IHEP 70–51, Serpukhov, 1970.
9. A.A.Asseev et al., Proceed. of the 3–d EPAC, Berlin, 1992, v.2, p. 1486.
10. A.A.Batalov et al., IHEP 87–116, Serpukhov, 1987.
11. A.A.Asseev et al., J. of Techn. Phys. N9, v.60(1990)70, Leningrad.
12. Yu.M.Ado, E.A.Ludmirsky, IHEP 87-30, Serpukhov, 1987.
13. A.G.Afonin et al., Proceed. of the XII All-Union Conference on charged

part. accel., Dubna, 1992, v.1, p.371.
14. Yu.M.Ado, A.A.Asseev et al., IHEP 88–9, Serpukhov, 1988.
15. B.Z.Kopeliovich, L.I.Lapidus, Nucl. Phys. 19(1974)218.
16. C.Bourelly, J.Soffer, D.Wray, Nucl.Phys. B77(1974)386.
17. S.Nurushev, A.Penzo, CERN/WAG/Int. 3/78, 1978.
18. L.I. Lapidus, Journ. of Element.Part. and Nuclei, 9(1978)84.

H. Terazawa

GLUONS, GLUEBALLS, ODDERONS, POMERONS, AND "COLOR-BALLS"

H. TERAZAWA

Institute of Particle and Nuclear Studies, HEARO (Tanashi)
Midori-cho, Tanashi, Tokyo 188, Japan

Various pictures of glueballs in a generic sense, the hadrons consisting of gluons only, are reviewed and discussed in some detail in the light of their relevance to hadron-hadron, photon (gauge-boson)-hadron, and (hadronic) photon (gauge-boson)-photon (gauge-boson) scatterings at high energies. In particular, it is proposed that the Pomeron is a "color-ball", the color-singlet complex object consisting of an arbitrary number of gluons, which can be approximately described in QCD either by the extended Nambu-Goto action or by the extended Polyakov action for the hadronic membrane or for the hadronic bundle. The contents of this talk include

1. Introduction,
2. Glueballs,
3. Odderons,
4. Pomerons,
5. "Color-Balls",

and 6. Conclusions and Future Prospects.

Preface

I would like to dedicate this talk to Professor Bjorn H. Wiik, the late Director of DESY, who together with his collaborators has discovered the existence of gluons by observing three-jet events in one-photon annihilation of electron-positron pairs into hadrons by PETRA, the e^+e^- collider at DESY.

1 Introduction

Mesons and baryons are hadrons, the particles with strong interactions. From the long and detailed theoretical and experimental studies in hadron physics for the last more than half century, the following things have all become self-evident: mesons are composite states of a quark and an antiquark while baryons are those of three quarks. A quark has not only the flavor quantum number but also the color quantum number [1]. The strong interaction of quarks can be well described by quantum chromodynamics (QCD) [2], the Yang-Mills gauge theory of color SU(3), in which a color-octet of gluons mediate the interaction. If color-confinement is exact, all hadrons are color-SU(3)-singlet states. If we take the gluon also as a constituent of hadrons, to be more precise, we must, however, define hadrons as color-singlet states consisting of quarks, antiquarks and/or gluons; mesons (and exotic mesons) as those consisting of a quark, an antiquark, and an arbitrary number of gluons (and quark-antiquark pairs); and baryons (and exotic baryons) as those consisting of three quarks and an arbitrary number of gluons (and quark-antiquark pairs). In addition, we can imagine that there may exist another type of hadrons, color-singlet states consisting of gluons only. Let us call them "glueballs" in a generic sense. The simplest one of glueballs is a color-singlet state consisting of two gluons, which we call "glueball" in a narrow sense. The second simplest one is a color-singlet state consisting of three gluons, which is so-called "odderon". In this talk, I am going to discuss "glueballs" in a generic sense as they not only appear as extremely interesting purified hadrons but also seem to play an important role in hadron-hadron, photon (gauge-boson)-hadron, and (hadronic) photon (gauge-boson)-photon (gauge-boson) scatterings at high energies[3,4]. This talk consists of the following contents:

2. Glueballs,
3. Odderons,
4. Pomerons,
5. "Color-Balls",

and 6. Conclusions and Future Prospects.

2 Glueballs

In this Section, let me concentrate on glueballs in a narrow sense, the color-singlet bound states consisting of two gluons. The ground state $I^G(J^{PC}) = 0^+(0^{++})$ glueball mass has been extensively calculated for the last more than two decades in many different ways such as the lattice QCD [5] and the $1/N_c$ expansion [6]. The best predicted values all lie in the range of 1.5-1.7 GeV. Experimental search for glueballs has also been extensively made for the last more than two decades in many different ways. The best candidate for the ground state is $f_J(1710)$ with $I^G(J^{PC}) = 0^+$ (even^{++}), $m = 1712 \pm 5$ MeV, and $\Gamma = 133 \pm 14$ MeV [7]. It seems, however, that there are still some problems in identifying which one of observed resonances as the ground state. In fact, although a few years ago the BES Collaboration [8] found $f_J(1710)$ with $m = 1781 \pm 8^{+10}_{-31}(1696 \pm 5^{+9}_{-34})$ MeV and $\Gamma = 85 \pm 24^{+22}_{-19}(103 \pm 18^{+30}_{-11})$ MeV for $J = 0(2)$ in $e^+e^- \to J/\psi \to \gamma f_J \to \gamma K^+K^-$ reactions at BEPS, very recently the Crystal Barrel Collaboration [9] has found $f_0(1770)$ with $I^G(J^{PC}) = 0^+(0^{++})$, $m = 1770 \pm 12$ MeV, and $\Gamma = 220 \pm 40$ MeV in $\bar{p}p \to f_0\pi^0 \to \eta\eta\pi^0$ reactions at LEAR and argued that the $f_J(1710)$ observed by the BES Collaboration is not seen in their experiment. For higher and excited glueball states with $J^{PC} = 2^{++}, 1^{+-}$, etc., both theoretical and experimental investigations have been made. However, what seem more relevant here are not these physical glueballs but the "Reggeized glueballs".

Already in 1975, Low and Nussinov [3] pointed out that the exchange of two gluons may dominate hadron-hadron scatterings at high energies. Then, Kuraev, Lipatov, Fadin, and Balitskii [4] formulated the exchange of two gluons interacting with each other, based on the DGLAP evolution equation [10], and calculated the intercept of the so-called "BFKL Pomeron" in perturbative QCD. In 1986, Lipatov [11] found a quasi-classical solution for the BFKL Pomeron equation and, a year later, Landshoff and Nachtmann [12] clarified that the "two-gluon Pomeron" has all the properties that the Pomeron should possess. Since then until now, many theoretical works have been made not only for further clarifying the BFKL Pomeron [13] but also for modifying or extending it to more sophisticated models such as the "two-Pomeron model" by Donnachie and Landshoff [14] and the "running BFKL Pomeron model" by Nikolaev and Zoller [15]. However, there still remain many questions including what the Pomeron really is, whether it is "soft" or "hard", etc.. I shall discuss these problems later in Section 4.

3 Odderons

In this Section, let me concentrate on "odderons", the color-singlet bound states consisting of three gluons. The ground state $I^G(J^{PC}) = 0^-(1^{--})$ odderon mass has not been precisely calculated. Suppose that the "constituent gluon mass" m_g^* is well defined and estimated as $m_g^* \sim \frac{1}{2}m_{f_0} \sim 0.8 - 0.9$ GeV. Then, the odderon mass can be simply estimated as $m_O \sim 3m_g^* \sim 2.4 - 2.7$ GeV. As experimental search for odderons has not been seriously made, there is still no serious candidate for a physical odderon. However, what seems more relevant here is not the physical odderon but the "Reggeized odderon".

Concerning hadron-hadron scatterings at high energies, the simplest model for the odderon such as a C-odd state of three non-interacting gluons [16] or a non-perturbative Regge pole at $J = 1$ [17] have been discussed in the literature. The second simplest model is a C-odd state of three gluons which interact pairwise with a potential. The equation for such an idealized odderon state was first written down by Bartels and by Kwiecinski and Praszalowicz in 1980 [18] and had been analyzed to find the odderon intercept by Lipatov [19,20] and by Fadeev and Korchemsky [20]. More recently, several groups [21,22,23,24] have tried to solve the difficult mathematical equation. Very lately, Janik and Wosiek [25] have found that the energy of the lowest state gives the intercept of the odderon, $\alpha_O(0) = 1 - 0.2472\alpha_s N_c/\pi$, which Braun, Gauron, and Nicolescu [26] have reproduced numerically. In any case, however, it seems quite certain that the odderon intercept lies slightly below unity and that the contributions of odderons to the total, elastic, and diffractive cross sections for hadron scatterings are therefore suppressed (compared to those of the Pomeron whose intercept is larger than unity) and difficult to observe at high energies. However, it seems also very natural to suppose that the real Pomeron contains not only a component of glueballs but also another one of odderons. I shall discuss such a complex picture of the Pomeron in Sections 4 and 5.

4 Pomerons

What is the Pomeron? It must be a kind of object which has the vacuum quantum numbers ($I^G = 0^+$ and $J^{PC} = J^{++}$) and whose t-channel exchange between hadrons universally dominates the cross section for hadron scatterings at high energies. If it is a Regge pole, its intercept is larger than unity, which leads to the increase of hadron cross sections at high energies, and its residue is rather universal to all hadrons. Is it the "soft Pomeron" whose intercept is about 1.1 or the "hard Pomeron" whose intercept is as large as 1.4? Jenkovszky, Paccanoni, and company [27] have been insisting on the uniqueness of their "extended hard Pomeron" while Donnachie, Landshoff, and company[14,28] have been insisting on their two-Pomeron model consisting of a conventional non-perturbative soft Pomeron with the intercept $\alpha_{\rm soft} \sim 1.08$ and a hard Pomeron with the intercept $\alpha_{\rm hard} \sim 1.40$. In addition, Nikolaev and Zollar [15] have been advocating their "running BFKL Pomeron". However, from now on I would rather stay out of this hot debate, and concentrate on the following question: What is the Pomeron made of?

The simplest model is, of course, the "two-gluon-exchange model" for the Pomeron in which the Pomeron consists of two gluons and in which the Pomeron exchange can be approximated by the exchange of two gluons[3]. The second simplest model is the "BFKL Pomeron model" in which the Pomeron consists of two gluons interacting with each other and that the Pomeron exchange can be approximately described by the BFKL Pomeron equation[4] based on the DGLAP evolution equation [10]. In other words, the Pomeron is similar to the Reggeized glueballs. The third simplest picture is that the Pomeron consists of two or three gluons and that the Pomeron exchange can be better approximated by the sum of the exchanges of two and three gluons. In other words, the Pomeron is similar to the sum or mixture of the Reggeized glueballs and Reggeized odderons. Thus, we must go on and finally find the more realistic picture of the Pomeron in which the Pomeron consists of an arbitrary number of gluons and in which the Pomeron exchange is equivalent to the sum of the exchanges of any number of gluons. This "color-ball model" for the Pomeron is similar to the "color-singlet gluon cluster model" very lately proposed by Meng, Rittel, Zhang, and company [29]. Note also that it not only extends the now classic Nambu-Susskind hadronic string model but also revives the celebrated Freund-Harari hadron duality [which asserts that resonances in a s-channel corresponds to Regge poles in a t-channel while background (or continuum) states in the s-channel to the Pomeron in the t-channel].

5 "Color-Balls"

After the long introductory reviews on glueballs, odderons, and Pomerons in the previous Sections, I have come to the main point in this talk that the Pomeron must be a "color-ball", the color-singlet complex object consisting of an arbitrary number of gluons. It seems very difficult to describe such complex objects in quantum field theory and to evaluate their effects on hadron scatterings. In fact, Meng, Rittel, Zhang, and company [29] have adopted a statistical approach based on the formulation of complex systems by Bak, Tang, and Wiesenfeld [30] in describing the "color-singlet gluon clusters" with "self-organized criticality" and used an optical-geometrical method in examining the space-time properties of such objects. Here, however, let me discuss how to describe "color-balls" in an ordinary particle-theoretical way.

In 1979, Nambu and, independently, Tetsuaki Matsumoto[31] showed that, if a meson can be described by the path-ordered phase factor sandwiched between a quark and an antiquark in QCD and if the gluon flux is bunched along the path when quarks are largely separated, the meson can be approximated by the Nambu-Goto massless string with quarks at their end points. It was the first field-theoretical demonstration in which the Nambu-Susskind hadronic string is realized in QCD. From their demonstration for the realization of hadronic strings it seems natural to imagine that, if the gluon flux is not bunched but diversed along the path between quarks, it can be approximated by a "hadronic (two-dimensional) membrane" or better by a "hadronic (three-dimensional) bundle". In either case, the color-ball can be approximately described by the extended Nambu-Goto action of

$$S_{NG} = -\frac{1}{2\pi\alpha'}\int d^n\sigma[-\det\,(g_{\mu\nu}\partial_\alpha X^\mu \partial_\beta X^\nu)]^{1/2} F^{-(n-2)/2}$$

or by the extended Polyakov action of

$$S_P = -\frac{1}{2\pi\alpha'}\int d^n\sigma\sqrt{-h}(\frac{1}{2}h^{\alpha\beta}g_{\mu\nu}\partial_\alpha X^\mu \partial_\beta X^\nu - \frac{n-2}{2}F),$$

where $\sigma_\alpha(\alpha = 0, 1, \cdots, n-1)$ and $X^\mu(\mu = 0, 1, 2, 3)$ are the world-sheet parameters and the space-time coordinates of the "hadronic membrane" for $n = 3$

or the "hadronic bundle" for $n = 4$, respectively, $g_{\mu\nu} = \text{diag}(-1,1,1,1)$, $h = \det h_{\alpha\beta}$, and α' and F are constants.

6 Conclusions and Future Prospects

In this talk, I have reviewed and discussed in some detail various pictures of glueballs in a generic sense, the hadrons consisting of gluons only, in the light of their relevance to hadron-hadron, photon (gauge-boson)-hadron, and (hadronic) photon (gauge-boson)-photon (gauge-boson) scatterings at high energies. In particular, I have proposed that the Pomeron is a color-ball, the color-singlet complex object consisting of an arbitrary number of gluons. I have also shown that it can be approximately described in QCD either by the extended Nambu-Goto action or by the extended Polyakov action for the hadronic membrane or for the hadronic bundle. In either case, to proceed further in this way for evaluating the effects of color-balls on hadron scatterings seems very difficult, if not impossible, since it would inevitably make us be involved in some mathematical complexity.

However, in this picture of the Pomeron as a color-ball we may be able to understand the following properties of the Pomeron found in hadron scatterings at high energies without any manipulations: 1) The Pomeron cannot be taken as a simple Regge pole but a hadronic membrane or bundle, which is an extension of the Nambu-Susskind hadronic string, or a linear combination of an infinite number of Regge poles and Regge cuts, which revives the Freund-Harari hadron duality. 2) The slope of the Pomeron trajectory α'_P is much smaller than that of a ordinary Regge trajectory α'_R, i.e. $\alpha'_P \sim \frac{1}{4}\alpha'_R$, since the "membrane or bundle tension" is much larger than the string tension. 3) The Pomeron couplings are universal to any flavors of quarks (and antiquarks) and, therefore, their factorizability holds since the color-balls are not only color-blind but also flavor-insensitive. The experimental fact that the Pomeron intercept is larger than unity, i.e. $\alpha_P(0) > 1$, seems to be very dynamical and cannot be explained intuitively even in this picture of the Pomeron.

In conclusion, I wish to emphasize the importance of not only further theoretical works but also future experimental investigations on the real picture of the Pomeron by checking various things including the ever increasing total, elastic, and diffractive cross sections, the universality and factorizability such as the now classic relations of $\sigma(\bar{p}p) \cong \sigma(pp)$, $\sigma(\pi p) \cong \frac{2}{3}\sigma(pp)$, and $\sigma(\gamma\gamma \to \text{hadrons}) \cong \sigma(\gamma p \to \text{hadrons})^2/\sigma(pp)$ [32], and the triple-Pomeron and Pomeron-Reggeon-Reggeon vertices based on the Brandt-Preparata and Mueller diagrams [33] in hadron-hadron, photon (gauge-boson)-hadron, and (hadronic) photon (gauge-boson)-photon (gauge-boson) scatterings at high energies [34]. After all, they will eventually clarify one of the most fundamental problems in high energy physics, the origin of finite and non-vanishing hadron sizes.

Acknowledgments

The author would like to thank Professor V.A. Petrov (Chairman), Professor A.V. Prokudin (Scientific Secretary), and all the other organizers of the International Conference on Elastic and Diffractive Scattering (VIIIth 'Blois Workshop'), Protvino, Russia, June 28-July 2, 1999 for inviting him to the Conference and for their warm hospitality extended to him during his stay at Institute for High Energy Physics for two weeks from June 20 to July 4, 1999. He also wishes to thank Professor Bjorn H. Wiik, the late Director of DESY, to which this talk has been dedicated, for the long friendship and invitation to DESY.

References

1. M.-Y. Han and Y. Nambu, *Phys. Rev.* B **139**, 1006 (1965); Y. Miyamoto, *Prog. Theor. Phys. Suppl. Extra Number* , 187 (1965).
2. Y. Nambu, in Preludés in Theoretical Physics, edited by A. de Shalit, H. Feshbach, and L. Van Hove (North-Holland, Amsterdam, 1966), p. 133; H. Fritzsch and M. Gell-Mann, in Proc. XVI Int'l Conf. on High Energy Physics, Chicago and Batavia, 1972, edited by J.D. Jackson and A. Roberts (NAL, Batavia, 1973), Vol. 2, p. 135.
3. F.E. Low, *Phys. Rev.* D **12**, 163 (1975); S. Nussinov, *Phys. Rev. Lett.* **34**, 1286 (1975).
4. E.A. Kuraev, L.N. Lipatov, and V.S. Fadin, *Phys. Lett.* B **60**, 50 (1975); *Sov. Phys. JETP* **44**, 443 (1976); **45**, 199 (1977); Ya. Ya. Balitskii and L.N. Lipatov, *Sov. J. Nucl. Phys.* **28**, 822 (1978).
5. See, for example, C. Amsler and F. Close, *Phys. Lett.* B **353**, 385 (1995); *Phys. Rev.* D **53**, 295 (1996); J. Sexton et al., *Phys. Rev. Lett.* **75**, 4563 (1995); W. Lee and D. Weingarten, IBM Report No. IBM-HET-98-1, hep-lat/9805029, and references therein.
6. See, for example, S. Dalley and B. van de Sande, *Phys. Rev. Lett.* **82**, 1088 (1999), and references therein.
7. Particle Data Group, C. Caso et al., *Eur. Phys. J.* C **3**, 1 (1998).
8. See, for example, J.Z. Bai et al. (BES Collaboration), *Phys. Rev. Lett.* **77**, 3959 (1996).

9. See, for example, A.V. Anisovich et al. (Crystal Barrel Collaboration), *Phys. Lett.* B **449**, 154 (1999).
10. V.N. Gribov and L.N. Lipatov, *Sov. J. Nucl. Phys.* **15**, 438 (1972); G. Altarelli and G. Parisi, *Nucl. Phys.* B **126**, 298 (1977); Yu. L. Dokshitzer, *Sov. Phys. JETP* **46**, 641 (1977).
11. L.N. Lipatov, *Sov. Phys. JETP* **63**, 904 (1986).
12. P.V. Landshoff and O. Nachtmann, *Z. Phys.* C **35**, 405 (1987).
13. See, for example, V.S. Fadin and L.N. Lipatov, *Phys. Lett.* B **429**, 127 (1998).
14. A. Donnachie and P.V. Landshoff, *Phys. Lett.* B **437**, 408 (1998); K. Golec-Biernat and M. Wüsthoff, Report No. DTP-98-50.
15. N.N. Nikolaev and V.R. Zoller, *JETP Lett.* **69**, 103 (1999); **69**, 187 (1999).
16. I.F. Ginzburg and D. Yu. Ivanov, *Nucl. Phys.* B **388**, 376 (1992); J. Czyzewski, J. Kwiecinski, L. Motyka, and M. Sadzikowski, *Phys. Lett.* B **398**, 400 (1997) R. Engel, D. Ivanov, R. Kirschner, and L. Szymanowski, *Eur. Phys. J.* C **4**, 93 (1998); L. Motyka and J. Kwiencinski, hep-ph/9802278.
17. W. Kilian and O. Nachtmann, hep-ph/9712371, to be published in Eur. Phys. J.C; M. Rueter, H.G. Dosch, and O. Nachtmann, hep-ph/9806342.
18. J. Bartels, *Nucl. Phys.* B **175**, 365 (1980); J. Kwiecinski and M. Praszalowicz, *Phys. Lett.* B **94**, 413 (1980).
19. L.N. Lipatov, *Sov. Phys. JETP* **63**, 904 (1986); *Phys. Lett.* B **251**, 284 (1990); B **309**, 394 (1993).
20. L.N. Lipatov, *JETP Lett.* **59**, 571 (1994); L.D. Faddeev and G.P. Korchemsky, *Phys. Lett.* B **342**, 311 (1994); G.P. Korchemsky, *Nucl. Phys.* B **443**, 255 (1995); B **462**, 333 (1996).
21. P. Gauron, L.N. Lipatov, and B. Nicolescu, *Z. Phys.* C **63**, 253 (1994).
22. N. Armesto and M.A. Braun, Report No. US-FT/9-94, hep-ph/9410411; *Z. Phys.* C **63**, 709 (1997).
23. P.A. Janik, *Acta. Phys. Pol.* B **37**, 1819 (1996); J. Wosiek and R.A. Janik, *Phys. Rev. Lett.* **79**, 2935 (1997); R.A. Janik and J. Wosiek, Report No. TPJU-2/98, hep-th/9802100.
24. M. Praszalowicz and A. Rostworowski, hep-ph/9805245.
25. R.A. Janik and J. Wosiek, *Phys. Rev. Lett.* **82**, 1092 (1999).
26. M.A. Braun, P. Gauron, and B. Nicolescu, *Nucl. Phys.* B **542**, 329 (1999).
27. See, for example, M. Bertini, M. Giffon, L.L. Jenkovszky, F. Paccanoni, and E. Predazzi, *Rivista Nuovo Cimento* **19**, 1 (1996); P. Desgrolard, L. Jenkovszky, and F. Paccanoni, in Proc. 14-th Int'l Conf. on Strong

Interaction at High Energies (HADRONS-98), Parthenit, Crimea, 1998, edited by L.L. Jenkovszky (Bogolyubov Institute for Theoretical Physics, Kiev, 1998), p. 78.

28. A. Donnachie, H.G. Dosch, and M. Rueter, *Phys. Rev.* D **59**, 074011 (1999).
29. T. Meng, R. Rittel, and Y. Zhang, *Phys. Rev. Lett.* **82**, 2044 (1999). See also C. Boros, T. Meng, R. Rittel, and Y. Zhang, hep-ph/9704285; T. Meng, R. Rittel, K. Tabelow, and Y. Zhang, to be published; and C. Boros, T. Meng, R. Rittel, K. Tabelow, and Y. Zhang, hep-ph/9905318.
30. P. Bak, C. Tang, K. Wiesenfeld, *Phys. Rev. Lett.* **59**, 381 (1987); *Phys. Rev.* A **38**, 364 (1988). For a review, see, for example, P. Bak, How Nature Works (Springer-Verlag, New York, 1996).
31. T. Matsumoto and Y. Nambu, papers contributed to the 19-th Int'l Conf. on High Energy Physics, Tokyo, 1978 (Contribution No. 369 and 956, respectively). For reviews, see K. Kikkawa and B. Sakita, in Proc. 19th Int'l Conf. on High Energy Physics, Tokyo, 1978, edited by S. Homma, M. Kawaguchi, and H. Miyazawa (Physical Society of Japan, Tokyo, 1979), p. 507 and 921, respectively. For more details, see T. Matsumoto, UT Report No. UT-Komaba 78-10 (unpublished) and Y. Nambu, *Phys. Lett.* B **80**, 372 (1979).
32. S.J. Brodsky, T. Kinoshita, and H. Terazawa, *Phys. Rev.* D **4**, 1532 (1971). For a latest work, see Donnachie et al. in Ref. [28]. For a review on the two-photon process, see H. Terazawa, *Rev. Mod. Phys.* **45**, 615 (1973).
33. H. Terazawa, a contributed paper to the 1971 Int'l Symp. on Electron and Photon Interactions at High Energies, Cornell, 1971, Report No. CLNS-160 (LNS, Cornell Univ., 1971), in Proc. 1971 Int'l Symp. on Electron and Photon Interactions at High Energies, Cornell, 1971, edited by N.B. Mistry (Lab. of Nucl. Studies, Cornell Univ., Ithaca, N.Y., 1972), p. 327. For a more recent work, see, for example, L.L. Jenkovszky, E.S. Martynov, and F. Paccanoni, in Proc. XII-th Workshop on "Soft" Physics (HADRONS-96), Novy Svet, Crimea, 1996, edited by G. Bugrij, L. Jenkovszky, and E. Martynov (Bogoliubov Institute for Theoretical Physics, Kiev, 1996), p. 159.
34. For a latest review, see S. Donnachie, CERN Courier Vol. 39, No. 3, p. 29 (1999).

E. R. Berger

SOFT HIGH ENERGY SCATTERING IN NONPERTURBATIVE QCD

E. R. Berger
LAPTH, Universite Paris-Sud, Batiment 211,
F-91405 Orsay Cedex, France
E-mail: berger@th.u-psud.fr

In this report diffractive high energy reactions are discussed in a functional integral approach where hadronic amplitudes are calculated from vacuum expectation values of lightlike Wegner-Wilson loops. In the first part we calculate elastic differential cross sections for high energy and small momentum transfer elastic proton-proton (pp) scattering which are in reasonable agreement with the experimental data. In the second part we consider exclusive π^0 production in ep-scattering. At high energies photon and odderon exchange contribute to this reaction. We show that odderon exchange leads to a much larger inelastic than elastic π^0 production cross section, dominating the γ contribution by orders of magnitude. Observing our process at HERA would establish the soft odderon.

1 Elastic pp scattering

We will consider here some results of [1] for elastic differential cross sections $d\sigma/dt$ in elastic pp(p$\bar{\text{p}}$) and of [2] for exclusive π^0 production in ep scattering, both at high center of mass (cm) energies $\sqrt{s} \gtrsim 20$ GeV and low momentum transfer squared $|t| \lesssim O(1\,\text{GeV}^2)$ calculated within a functional approach [3,4].

We start with elastic pp scattering. For large energies the amplitude $T_{\text{pp}}(T_{\text{p}\bar{\text{p}}})$ is known to be dominated by $C = P = +1$ (pomern $\mathcal{P}$) exchange, $C = P = -1$ (odderon $\mathcal{O}$) exchange[a] is small or absent. In our approach, where the p is treated as a wave packet of its constituent quarks, this is achieved by considering the proton to be a quark-diquark (q-qq) system [6], what we will do in the following. Then the result for T_{pp} is:

$$T_{\text{pp}} = (-2is) \int d^2 b_T \exp(i\vec{q}_T \vec{b}_T) \cdot$$
$$\int d^2 x_T\, d^2 y_T\, |\Psi^p(x_T)|^2 |\Psi^p(y_T)|^2 \tilde{J}(\vec{x}_T, \vec{y}_T, \vec{b}_T); \qquad (1)$$

The scattering amplitude (1) is obtained by first considering the scattering of quarks (antiquarks) on a fixed gluon potential G. Travelling through it the quarks (q) and diquarks (qq) pick up non-abelian phase factors. To ensure

[a]The soft (nonperturbative) odderon is introduced in elastic hadron-hadron scattering as the $C = P = -1$ partner of the pomeron [5].

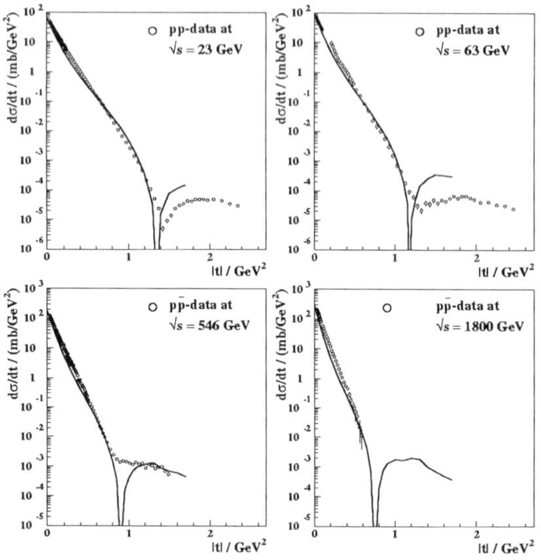

Figure 1: Differential elastic cross sections for c.m. energies $\sqrt{s} = 23, 63, 546$ and 1800 GeV.

gauge invariance the phase factors for q and qq of the same p are joined at large positive and negative times, yielding lightlike Wegner-Wilson loops $W_{\pm}$. The parton scattering amplitude $\tilde{J}$, given by the functional integral over G (indicated with barackets $\langle\,\rangle_G$),

$$\tilde{J}(\vec{x}_T, \vec{y}_T, \vec{b}_T) = \left\langle W_+(\frac{1}{2}\vec{b}_T, \vec{x}_T) W_-(-\frac{1}{2}\vec{b}_T, \vec{y}_T) - 1 \right\rangle_G, \tag{2}$$

is the central object in our approach. The transverse separation between the centres of the loops (impact parameter) is given by $\vec{b}_T$. The vectors $\vec{x}_T$ and $\vec{y}_T$ give the extensions and orientations of the loops in transverse space. The resulting parton scattering amplitude has to be integrated over all extensions and orientations of the loops in transverse space with a measure given by the q-qq densities, $\Psi^p \sim exp(-x_T^2/(4S_p^2))$ where $S_{\rm p}$ is the p extension parameter. To calculate (1) a Fourier transform over $\vec{b}_T$ has to be done finally.

The parton amplitude $\tilde{J}$ is calculated in nonperturbative QCD, using the minkowski version of the model of the stochastic vacuum (MSV). In terms of a matrix cumulant expansion the result is [1]:

$$\tilde{J} = [\frac{2}{3}\cos(\frac{1}{3}\chi) + \frac{1}{3}\cos(\frac{2}{3}\chi) - 1] + i[-\frac{2}{3}\sin(\frac{1}{3}\chi) + \frac{1}{3}\sin(\frac{2}{3}\chi)]. \tag{3}$$

The real function χ is odd under space inversion or the replacement of all partons by its antipartons, for example $\chi(-\vec{x}_T, \vec{y}_T, \vec{b}_T) = -\chi(\vec{x}_T, \vec{y}_T, \vec{b}_T)$. For

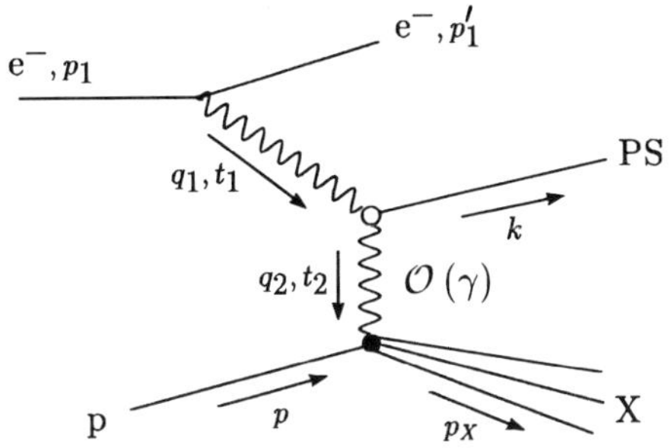

Figure 2: Feynman diagrams for pseudoscalar meson production in *ep* scattering at high energies with odderon (photon) exchange.

this reason $\tilde{J}$ contains odd and even contributions. But since the q-qq densities in (1) are invariant under space inversion the odd contributions to $T_{\rm pp}$ average out consistent with experiment[b].

χ is calculated from an integral over the non-local gluon condensate, for which the MSV makes an ansatz [4,1]. The parameters that occur are the gluon condensate G_2, the correlation lenght a and κ, a parameter measuring the non-abelian content of the condensate. In this way high energy scattering is related to the QCD vacuum charactereised by these parameters.

Leaving G_2, κ, a and $S_{\rm p}$ in (1) as free parameters we determine them in high energy scattering, using as input experimental data at $\sqrt{s} = 23$ GeV. We obtain [1]: $G_2 = (529\,\text{MeV})^4, \kappa = 0.75, a = 0.32\,\text{fm}$, consistent with existing lattice measurements [7], and a loop extension parameter $S_{\rm p}$ in the range of the electromagnetic proton radius, $S_{\rm p} = 0.87$ fm.

σ_{tot} increases with increasing extension parameter. Following [4] we introduce an energy dependence of (1) by making S_p energy dependent, fixed by requiring that our model reproduces the pomeron part of σ_{tot} in the DL parametrisation (The vacuum parameters should not depend on the energy).

In Fig. 1 we show our results for $d\sigma/dt$. A major point is, that our results for $d\sigma/dt$ depend crucially on $\kappa \neq 0$ which implies in terms of the MSV a string tension $\rho \neq 0$ and so confinement. A detailed discussion of this point can be found in [1].

2 Exclusive π^0 production in ep scattering

In this section we discuss exclusive π^0 production in high energy ep scattering. The π^0 is produced (see Fig. 2) by $\gamma\mathcal{O}$-fusion and by $\gamma\gamma$-fusion, but can not

---

[b] A consequence is, that $T_{\rm pp}$ and $T_{\rm p\bar{p}}$ are equal in the diquark limit.They contain no real part which is due to a second cumulant approximation [1]. As we will see, this results in infinitly deep dips in $d\sigma/dt$

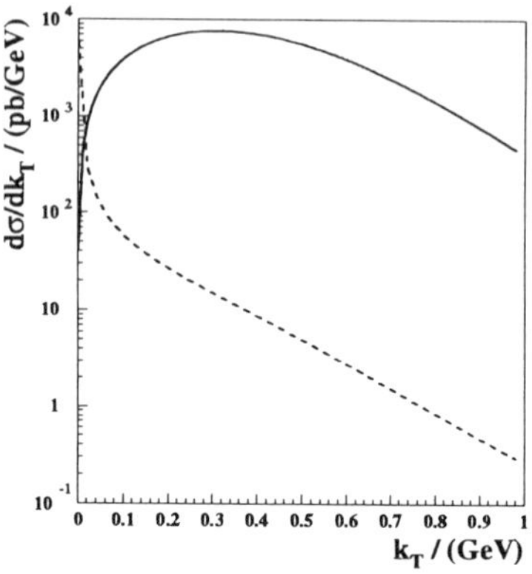

Figure 3: The k_T distribution in pion production from the 2P resonance channels for odderon exchange (solid line) compared to the complete electromagnetic result (dashed line). Interference contributions are not taken into account.

be produced by $\gamma\mathcal{P}$-fusion since the π^0 has positive C-parity.

In Fig. 2 X stands for a proton or resonances or a sum over resonances. Here we treat the very small $Q^2 := -q_1^2$ range where the ep cross section can be calculated by folding the γp cross section with the equivalent photon spectrum of the electron.

To calculate the $\mathcal{O}$ exchange we look at γp-scattering in the c.m. system. In the following we discuss the cases, (i) that the proton, stays intact or (ii) gets diffractively exited into the resonances $N(1520)$ with $J^P = \frac{3}{2}^-$ and $N(1535)$ with $J^P = \frac{1}{2}^-$, described as exited quark-diquark systems.

When considering unpolarised cross sections, summed over both resonances in case (ii), the quark spin degree of freedom becomes irrelevant and the calculation reduces to one where a spinless state stays intact or is exited to a 2P resonance. The amplitudes are again given by folding $\tilde{J}$ with approperiate wave functions. For (ii) the helicity amplitudes are [2]

$$
\begin{aligned}
T(s_2,t_2)_{\lambda,\lambda_\gamma} &= -2is_2 \int d^2b_T\, e^{i\vec{q}_{2T}\vec{b}_T}\, \hat{J}_{\lambda,\lambda_\gamma}(\vec{b}), \\
\hat{J}(\vec{b})_{\lambda,\lambda_\gamma} &= \int \frac{d^2x_T}{4\pi} \int \frac{d^2y_T}{4\pi} \\
&\times \sum_{f,h_1,h_2} \Psi^{*\,\pi^0}_{fh_1h_2}(\vec{x}_T)\Psi^{\gamma}_{\lambda_\gamma,fh_1h_2}(\vec{x}_T) \\
&\times \Psi^{*\,2\mathrm{P}}_{\lambda}(\vec{y}_T)\Psi^{p}(\vec{y}_T)\;\; \tilde{J}(\vec{x}_T,\vec{y}_T,\vec{b}_T).
\end{aligned} \tag{4}
$$

Here $\lambda_\gamma(\lambda)$ is the helicity and $\Psi^\gamma(\Psi^{2P})$ the wave function of the photon (2P state). Now, when the proton stays intact there occurs in (4) the quark-diquark density ($\Psi_\lambda^{2P} \to \Psi^{P}$). But, since this density is symmetric under a parity transformation whereas the odderon coupling changes sign, there is a cancelation when we integrate over all angles. On the other hand when the proton gets exited to a negative parity state like the resonances $N(1520)$ and $N(1535)$ there is no cancellation (the overlap is odd under parity) and the odderon couples to the nucleon without any restriction [2,8]. Thus in our model the odderon couples only if breakup of the proton occurs.

The photoproduction cross section calculated from (4) is independent of s_2 and so the EPA conversion to electroproduction can be achieved by simply multiplying it with a constant $c_{\rm EPA} = 0.0136$.

In the following we consider only ep scattering. An experimentally prefered observable is the k_T spectrum, the transverse momentum distribution of the π^0 with respect to the beam direction. This is displayed in Fig. 3. We also show there the distribution of the pion's k_T summed over the elastic and all inelastic channels (with invariant mass $M_X \leq 2$ GeV) for γ exchange. As we can see γ exchange is larger than $\mathcal{O}$ exchange only for very small k_T. For $k_T \gtrsim 0.1$ GeV the $\mathcal{O}$ exchange dominates by orders of magnitude.

3 Conclusions

The functional integral approach gives a reasonable description of soft elastic pp scattering at high energies, relating it to fundamental parameters of the QCD vacuum. The values determined from high energy scattering compare well with lattice results. In π^0 production for the p breakup into negative parity hadronic final states, a large odderon cross section arises which should be observable at HERA. This would establish the soft odderon as an exchange-object in high energy scattering on an equal footing with the soft pomeron.

References

1. E. R. Berger, O. Nachtmann, Eur. Phys. J. C7, 459 (1999)
2. E. R. Berger, A. Donnachie, H. G. Dosch, W. Kilian, O. Nachtmann, M. Rueter, Eur. Phys. J. C9, 491 (1999);
3. O. Nachtmann, Ann. Phys. 209 (1991) 436.
4. H. G. Dosch, E. Ferreira, A. Krämer, Phys. Rev. **D50** (1994) 1992.
5. L. Lukaszuk, B. Nicolescu, Nuov. Cim. Lett. **8**, 405 (1973); D. Joynson, E. Leader, C. Lopez, B. Nicolescu, Nuov. Cim. **A 30**, 345 (1975)
6. M. Rueter, H. G. Dosch, Phys. Lett. B 380, 177 (1996)
7. E. Meggiolaro, Phys. Lett. B451, (1999) 414
8. M. Rueter , H.G. Dosch , O. Nachtmann, Phys. Rev. D59, 014018, 1999

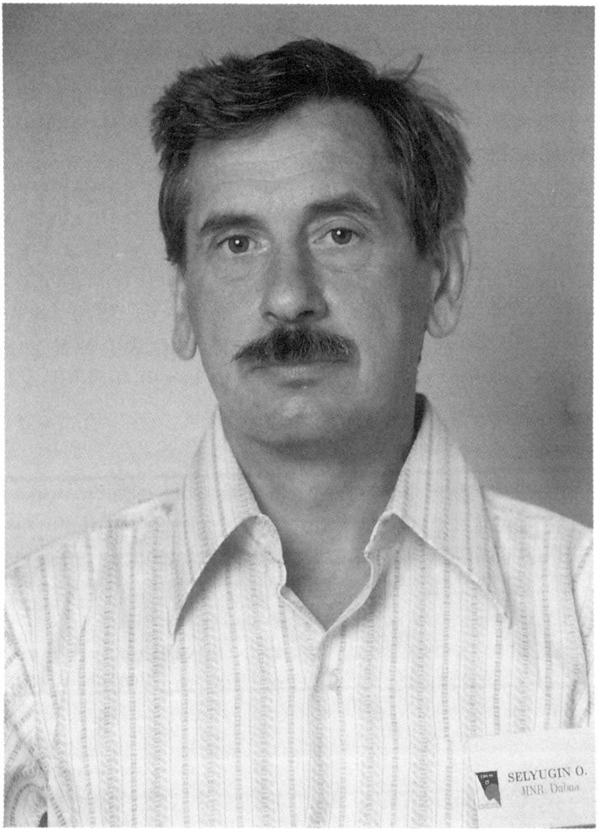

O. V. Selyugin

ODDERON CONTRIBUTION IN DIFFRACTION REACTIONS

O. V. SELYUGIN
Bogoliubov Laboratory of Theoretical Physics, JINR, Dubna, Russia
E-mail: selugin@thsun1.jinr.ru

The odderon coupling and a possible contribution of the odderon in single pseudoscalar meson production are analyzed.

Some approaches to diffractive processes are connected with the t-channel point of view. The charge conjugation partner of Pomeron with $C = -1$ - Odderon exists. For the maximal Odderon [1] the Intercept odderons equals the intercept of Pomeron. The recent calculations of the odderon intercept give it as follows [2]:

$$\alpha_{odd}(0) \;=\; 1 \;-\; (9\alpha_s/2\pi)\varepsilon, \tag{1}$$

where ε is the odderon energy. Note that in the case of Pomeron this calculation gives the pomeron intercept significant large than 1, but the calculation of the next to leading Logarithmic Approximation [3] has a different sign and reduces the pomeron intercept. May be that this calculation in the case of odderon will lead to increasing the intercept and finally we will have both the intercepts near 1.

In the early 90-ies it was noted that the odderon would play an important role in the pseudoscalar meson production [4]. Now we have some works which analyze this contribution and its interference with a competitor-process $\gamma^*\gamma^* \to PSV$-meson [5,6,7]. In work [5] the calculations were made under some assumption on the odderon intercept ($\alpha(0) = 0.8;\ 1.;\ 1.08$). The coupling constant was estimated from the supposition that $\rho_{\bar{p}p} - \rho_{pp} < 0.05$ as $\sqrt{s} > 100\ GeV$.

Let us try to improve these estimation. We have the experimental data for pp and $\bar{p}p$-elastic scattering at $\sqrt{s} = 52.8\ GeV$. For this energy we have

$$\Delta\rho = \rho_{\bar{p}p} - \rho_{pp}; \quad \Delta\sigma_{tot} = \sigma_{tot}^{\bar{p}p} - \sigma_{tot}^{pp}$$

and

$$\Delta d\sigma/dt(-t_{min}) = d\sigma/dt(-t_{min})|_{\bar{p}p} - d\sigma/dt(-t_{min})|_{pp}$$

is the difference of differential cross sections at the point of the diffraction minimum. Let us use these data to obtain the bound on the odderon coupling. Our analysis includes not only the contribution of pomeron and odderon but also the second reggeon. If an analysis is carries out without the reggeon contribution, the picture will be contradictory between these three cases. So, into our examination we include two energies $\sqrt{s} = 9.78\ GeV$ and $\sqrt{s} = 52.8\ GeV$ with specific pictures of the differential cross sections.

Let us start with the data on σ_{tot}.

$$\sigma_{tot}^{pp} = \sigma_{tot}^{pom} + \sigma_{tot}^{r+} - \sigma_{tot}^{odd} - \sigma_{tot}^{r-}; \quad \sigma_{tot}^{\bar{p}p} = \sigma_{tot}^{pom} + \sigma_{tot}^{r+} + \sigma_{tot}^{odd} + \sigma_{tot}^{r-}, \tag{2}$$

where $r+$ is denotes the cross-even contribution of the second reggeon and $r-$ is noted a cross-odd contribution. The pomeron and odderon conributions and a second reggeon $\sigma_{tot}^{r+(-)}$ (plus and mines correspond to cross-even and cross-odd parts of second reggeon)

$$\sigma_{tot}^{pom} = Ps^{\alpha+}; \quad \sigma_{tot}^{odd} = Ds^{\alpha_- - \delta}, \quad \sigma_{tot}^{r+} = as^{-0.5}, \quad \sigma_{tot}^{r-} = bs^{-0.5}, \tag{3}$$

where δ is the value on which the intercept of the real part of odderon grows more fast than its imaginary part. The difference of $\sigma_{tot}^{+(-)}$ at s_1 and s_2 is

$$\Delta\sigma_{tot}(s_1) = 2\ (bs_1^{-0.5} + Ds_1^{\alpha_- - \delta}); \quad \Delta\sigma_{tot}(s_2) = 2\ (bs_2^{-0.5} + Ds_2^{\alpha_- - \delta}); \tag{4}$$

And for b and D

$$b = \frac{\Delta\sigma_{tot}(s_1)s_2^{-0.5} - \Delta\sigma_{tot}(s_2)s_1^{-0.5}}{s_2^{\alpha_- - \delta}s_1^{-0.5} - s_1^{\alpha_- - \delta}s_2^{-0.5}}, \quad D = \frac{\Delta\sigma_{tot}(s_2)s_1^{-0.5} - \Delta\sigma_{tot}(s_1)s_2^{-0.5}}{s_2^{\alpha_- - \delta}s_1^{-0.5} - s_1^{\alpha_- - \delta}s_2^{-0.5}} \tag{5}$$

The arithmetic mean difference of $\Delta\sigma_{tot}$ is

$$\Delta\sigma_{tot}(s_1) = 5.73\ mb; \qquad \Delta\sigma_{tot}(s_1) = 1.34\ mb.$$

For the maximal odderon the real part grows with $\alpha_- = 0.08$. In this case the imaginary part either does not grows or grows very slowly. Take for the first examination $\alpha_- - \delta = 0$. As a result,

$$ImF_{odd}(s_2)ImF_{pom}(s_2) = k(\beta_{odd})^2 s_2^{\alpha_-}(\beta_{pom})^2 s_2^{\alpha+} = 0.004, \tag{6}$$

where k is obtained using local dispersion relations $k = (\pi/2)\ /\ \ln(s_2/s_0)$, with $s_0 = 1\ GeV^2$.

$$C_{0dd} = (\beta_{odd})^2(\beta_{pom})^2 \ = \ 0.02. \tag{7}$$

Defined that $C_{odd}^s(s) = C_{odd}s^{\alpha_-}$, one takes account of the energy dependence $\alpha_- = 0.08$ and denotes $\sqrt{s_3} = 250\ GeV$

$$C_{odd}^s(s_2) \ = 0.037; \qquad C_{odd}^s(s_3) \ = 0.043. \tag{8}$$

If one takes the "minimal" odderon with the energy dependence of imaginary part with $\alpha_- - \delta = -0.2$, one obtains $b = 23.8$ and $D = 1.07$; Hence

$$ImF_{odd}(s_2)/ImF_{pom}(s_2) = k\ (\beta_{odd})^2 s_2^{-0.1}/(\beta_{pom})^2 s_2^{0.08}, \tag{9}$$

and

$$C_{0dd} = (\beta_{odd})^2/(\beta_{pom})^2 = 0.106. \tag{10}$$

If one takes account of the energy dependence $\alpha_- = -0.1$ $C^s_{odd}(s_2) = 0.047$; $C^s_{odd}(s_3) = 0.035$. We obtained that the ratio of the odderon contribution to reggion at $\sqrt{s} = 52.8\ GeV$ is

$$R_r^{odd} = (0.107(s_2)^{-02})/(23.8(s_2)^{-0.5}) = 0.47 \tag{11}$$

Note, that this values is the ratio of imaginary parts of the odderon and reggion exchanges.

For the "middle" odderon with the energy dependence of imaginary part with $\alpha_- - \delta = -0.1$, we obtain $C_{0dd} = (\beta_{odd})^2/(\beta_{pom})^2 = 0.041$. In that case the real part of odderon practically does not change with energy and $C^s_{odd}(s_2) = 0.041$; $Cs_{odd}(s_3) = 0.041$. The ratio of the odderon contribution to reggion at $\sqrt{s} = 52.8\ GeV$ is

$$R_r^{odd} = (0.416(s_2)^{-0.1})/(25.44(s_2)^{-0.5}) = 0.41 \tag{12}$$

Hence, the odderon contribution is one-half or one-third of the cross-odd contribution to the total cross section in the case of maximal odderon and is practical the same in the case of "minimal" odderon:

Now let us treat the second set of values. The ratio of real to imaginary parts of the scattering amplitude at one energy in some approximation at high energies is

$$\Delta\rho \sim 2\ \frac{ReF_{cross-odd}}{ImF_{cross-even}} = 2\ \frac{ReF_{r-} + ReF_{odd}}{ImF_{pom} + ImF_{r+}} \tag{13}$$

We can considere two cases. In the first, as in work[8]

$$ImF_{r+} \sim -ReF_{r-}. \tag{14}$$

The other case (2_b) was obtained in work[9]

$$ImF_{r+} \sim -3ReF_{r-}. \tag{15}$$

The imaginary part $ImF_{pom}(s_2)$ of pomeron is connected with $\sigma^{pp}_{tot}(s_2)$

$$ImF_{pom}(s_2) \sim \sigma^{pp}_{tot}/(4\pi 0.39); \quad ImF_{pom}(s_1) = ImF_{pom}(s_2)s^{-\alpha_+(0)}. \tag{16}$$

Again our analysis for the upper examined cases gives the odderon coupling which are showed on Table.

It is clearly understood as in this case the odderon contribution is determined by the real part of the odderon which is 5 times as larger as the Imaginary part at $\sqrt{s} = 52.8 GeV$. So, it coincides with the results obtained by the first method from the $\Delta\sigma_{tot}$ if we take into account the ratio of imaginary to real parts of odderon amplitude.

Now let us examine the region of diffraction minimum. In this place the imaginary part of scattering amplitude changes his sign and the diffraction minimum is filled by the real part of amplitude. The differential cross sections for proton-proton and proton-antiproton elastic scattering at the energy s_1 and s_2 are a very different picture. The $\bar{p}p$ -scattering at $\sqrt{s_1} = 9.78\ GeV$ has a very sharp minimum and the measured $\rho_{\bar{p}p}(s_1) \simeq 0 \div -0.05$. The pp-scattering has a "shoulder" and $\rho_{pp}(s_1) \simeq -0.16$. At $\sqrt{s_2} = 52.8\ GeV$ the behavior of the differential cross section becomes opposite.

$$\begin{aligned}(ReF_+(t_{min}) + ReF_-(t_{min}))^2 &= 4.8\ 10^{-5}\ mb/GeV^2 \\ (ReF_+(t_{min}) - ReF_+(t_{min}))^2 &= 2.2\ 10^{-5}\ mb/GeV^2 \end{aligned} \qquad (17)$$

The contributions of cross-even and cross-odd parts can have the different sign. In result one can obtain the eight solutions for pomeron and odderon contributions.

We can obtain the separately contribution of pomeron, odderon and reggions. But for obtaining the ratio of odderon-pomeron coupling it is need extend our results to point $t = 0$. Note, that in point of diffraction minimum the energy dependence of different parts of scattering amplitude are depended on the one hand the whole energy dependence at $t = 0$ and the other hand, the energy dependence of the real parts in the point of diffraction minimum.

Real part of odderon amplitude one take in the same form as imaginary part of pomeron amplitude proposed by Landshoft [10].

$$Re(F_{odd}(s,t)) = \frac{h_{odd}}{\sqrt{0.39}\ \pi}\ (\alpha' s)^{(\epsilon_0^{odd} + \alpha' t)}\ G(t)^2. \qquad (18)$$

The real part of pomeron amplitude is obtained by using the local dispersion relations [11,12]

$$Re(F_{pom}(s,t)) = \frac{h_{pom}}{\sqrt{0.39}\ \pi}\ (\epsilon_0 + \alpha' t)\ G(t)^2, \qquad (19)$$

where

$$G(t) = \frac{4m^2 - 2.8t}{4m^2 - t}\left(\frac{1}{1 - t/0.71}\right)^2 \qquad (20)$$

with $\epsilon_0 = 0.08$, and $\alpha' = 0.25$

In first we choose the solutions with the negative sign of real part of pomeron in the point of the diffraction minimum. As using the local dispersion relations we obtain the first zero in real part of pomeron amplitude at small t and second after the second diffraction maximum. Some different phenomenological models give the same result, for example [9]. The four solutions satisfy this criteria.

amplitude

As other criteria for chosen one solution one take the corresponding the ratio of real-imaginary parts of pomeron amplitude at point $t = 0$. Two of the obtaining solutions lead to the small real part of pomeron amplitude and, correspondingly, to the very small magnitute of $\rho(t = 0)$ and to the large odderon amplitude. Other two solutions lead to the real part of pomeron amplitude which give the $\rho(t = 0) = 0.08$ and to the small odderon amplitude with practically the same magnitude $H_{odd} = 0.2\ mb^{1/2}/GeV$. Our criteria satisfy the solutions with plus or mines sign in the root at $\sqrt{s} = 9.78\ GeV$ and with the negative signs of the both roots at $\sqrt{s} = 52.8\ GeV$ It gives for maximal odderon

$$C_{odd} = 0.038 \quad C^{s}_{odd}(s_1) = 0.07 \quad C^{s}_{odd}(s_1) = 0.09$$

For three cases $(3_a, 3_b, 3_c)$ which are, respectively, "maximal" odderon, "middle" odderon and "minimal", odderon the result is reported in Table 1.

All our calculations are gathered in Table. Of course, the errors of the obtained values are large. But an essential result of this work consists in that they do not used any fitting procedures and involve as little theoretical assumptions as possible. The results of calculation by three methods are very close to each other.

We defined the odderon coupling in the nucleon-nucleon scattering. The case of single pseudoscalar meson production in $\gamma * N$ scattering requires a careful analysis. Now there are minimum five models of pomeron interaction. For the odderon the situation is more complicated. One can there regard the interaction of odderon with quarks constituents of γ^* or, on the contrary, the interaction of gamma with the constituents of odderon. In the case of creation of π-meson, we have a very non-perturbative object at small q $\gamma*(q) \rightarrow (q\bar{q})$ as $r_\perp \sim 1/m$ is very large. In this case it can be treated as interaction of a π-meson with a nucleon. If

$$(\beta_{odd,\pi}\ \beta_{odd,N})/(\beta_{odd,N})^2 \sim (\beta_{pom,\pi}\ \beta_{pom,N})\ (\beta_{pom,N})^2 = 2/3, \qquad (21)$$

the coefficient $C_{odd}(\pi N)$ for the creation of a single π-meson will be obtained by multiplying the calculated value of $C_{odd}(NN)$ from the table by factor 2/3.

Table 1: The $C^s_{0dd}(s)(NN)$

Var.	C_{0dd}	$C^s_{0dd}(\sqrt{s} = 52.8\ GeV)$	$C^s_{0dd}(\sqrt{s} = 250\ GeV)$
1_a	0.02	0.037	0.043
1_b	0.04	0.04	0.04
1_c	0.106	0.047	0.035
2_{1a}	0.027	0.05	0.065
2_{1b}	0.054	0.054	0.054
2_{1c}	0.133	0.06	0.044
2_{2a}	0.03	0.05	0.07
2_{2b}	0.06	0.06	0.06
2_{2c}	0.15	0.07	0.05
3_a	0.04	0.07	0.09
3_b	0.065	0.065	0.065
3_c	0.12	0.05	0.04

Besides, one should mention, the importance of t dependence of the odderon amplitude, since the relative maximum of the interference between $\gamma^*\gamma^*$ and γ^* *odderon* amplitudes will be in the region of $p^2_\perp$ where both amplitude will be equal. In the case of large effective α'_- this region will be that where both amplitudes small and the effect is very difficult to observe. So, the researches are necessary which can reveal the possible bounds on the t-dependence of odderon amplitude.

References

1. L. Lukaszuk, B. Nicolescu, *Nuovo Cim. Lett.*, **8**, 405 1973; P. Gauron, L. Lukaszuk, B. Nicolescu, *Phys. Lett.* **B 294**, 298 (1992).
2. B. Nicolescu, Talk at XXIX Int. Conf. on High Energy Physics, Vancouver, Canada, July 23-30,1998, preprint LPTPE/UP6/98-13, hep-ph 9810465.
3. V. S. Fadin, Talk at XX Int. workshop on "Hadrons-98", Krim, June 23-30,1998.
4. B.V. Struminski, A.N. Shelkovenko, Yad. Phys. bf 53, p.788 (1991).
5. W. Killian, O. Nachtmann,
6. Kwestinsky
7. Ellis
8. P. Gauron, E. Leader, B. Nicolescu, Phys.Rev.Lett. **52**, p.1952 (1984).
9. S. V. Goloskokov, S. P. Kuleshov and O. V. Selyugin, *Z. Phys.* **C 50**, 455 (1991).
10. P.V. Landshoff, preprint DAMP 96/48 (1996).
11. J.B. Bronzan and G. Kane, *Phys.Lett.*, **B 49**, 272 (1974).
12. V.P. Gerd, V.I. Inozemtzev and V.A. Mescheryakov, *Yad.Fiz.* **24** 176 (1976).

Basarab Nicolescu

RECENT ADVANCES IN ODDERON PHYSICS

BASARAB NICOLESCU
LPNHE- LPTPE, Université Pierre et Marie Curie, 4, Place Jussieu, 75252 Paris Cedex 05, France*
E-mail: nicolesc@ipno.in2p3.fr

We present in this talk the phenomenological and theoretical advances in Odderon physics which occured since the last EDS Blois Workshop (Seoul, 1997).

1 Introduction

The elastic and diffractive scattering is controlled by two singularities located near $J = 1$ in the complex J-plane : the Pomeron[1], in the even-under-crossing amplitude F_+ and the Odderon[2], in the odd-under-crossing amplitude F_-.

The Odderon idea was longtime forced to stay in the Purgatory, because it contradicted the belief, founded on the dominant approach of the 70's - the Regge-pole model, that F_- is dominated by singularities located near $J = 1/2$ (ρ and ω Regge poles and their cuts). In spite of its rediscovery in QCD in the 80's[3], and of its solid theoretical foundation in the framework of asymptotic theorems[2,4] and derivative relations[5], the Odderon continued to be considered as an heretical concept. As late as in October 1990, here, in Dubna, André Martin did not hesitate to associate the words "revolution" and "Odderon"[6]. The complete theoretical legitimacy of the Odderon came only in the last few years from the calculation of the Odderon intercept in QCD. In the first part of my talk I will discuss the last results in this field.

From the phenomenological point of view, the interest clearly shifted from the somewhat biased study of Odderon effects in $\bar{p}p$ and pp scattering towards HERA Odderon physics. The surprisingly rich activity in Odderon phenomenology in the last two years will be described, before drawing conclusions, in the second part of the talk.

2 Calculation of the Odderon intercept in QCD

The most important recent result is, beyond any doubt, the discovery by Janik and Wosiek[7] and, independently, by Lipatov[8], of an exact solution for the Odderon intercept in LLA.

*Unité de Recherche des Universités Paris 6 et Paris 7, Associée au CNRS

In QCD the Odderon is a C-odd state of 3 reggeized gluons which interact pairwise with a well-defined potential (Fig. 1). The problem is to find an operator $\hat{q}_3$,

$$\hat{q}_3^2 = -r_{12}^2 r_{23}^2 r_{31}^2 p_1^2 p_2^2 p_3^2, \tag{1}$$

which commutes with the Odderon hamiltonian H,

$$[\hat{q}_3^2, H] = 0 \tag{2}$$

and has a much simpler form than H.

The exact solution of the problem is formulated in terms of the eigenequation

$$\hat{q}_3 f = q_3 f \tag{3}$$

where

$$f = \left(\frac{\rho_{12}\rho_{13}\rho_{23}}{\rho_{10}^2\rho_{20}^2\rho_{30}^2}\right)^{\mu} \Phi(z), \tag{4}$$

with $\mu = h/3$, h being the conformal weight. The function Φ satisfies a third order linear equation

$$a(z)\frac{d^3}{dz^3}\Phi(z) + b(z)\frac{d^2}{dz^2}\Phi(z) + c(z)\frac{d}{dz}\Phi(z) \\ +d(z)\Phi(z) = 0, \tag{5}$$

where

$$a(z) = z^3(1-z)^3, \tag{6}$$

$$b(z) = 2z^2(1-z)^2(1-2z), \tag{7}$$

$$c(z) = z(z-1)\left(z(z-1)(3\mu+2)(\mu-1) + 3\mu^2 - \mu\right), \tag{8}$$

$$d(z) = \mu^2(1-\mu)(z+1)(z-2)(2z-1) - iq_3 z(1-z), \tag{9}$$

$$z = \frac{\rho_{12}\rho_{30}}{\rho_{10}\rho_{32}}, \rho_k = x_k + iy_k (k = 1,2,3), \rho_{ij} = \rho_i - \rho_j. \tag{10}$$

In Ref. 7, the numerical solution

$$q_3 = -0.20526\, i \tag{11}$$

was found, corresponding to the Odderon energy

$$\epsilon = 0.16478. \tag{12}$$

The relation between the Odderon energy ϵ and the Odderon intercept $\alpha_O(0)$ is given by the equation

$$\alpha_O(0) = 1 - (9\alpha_s/2\pi)\epsilon. \tag{13}$$

For realistic values of α_s ($\alpha_s \simeq 0.19$) one gets

$$\alpha_O(0) = 0.94. \tag{14}$$

In collaboration with M.A. Braun and P. Gauron we recently performed a direct calculation of the lower bound for the Odderon intercept[9] in the framework of the variational approach we formulated earlier in collaboration with L. Lipatov[10]. In this variational approach the Odderon energy is defined as

$$\epsilon = E/D \tag{15}$$

In Eq.(15) D is a normalization constant and the energy functional is given by

$$E = \sum_{n=-\infty}^{\infty} \int_{-\infty}^{\infty} d\nu \epsilon_n(\nu) |\alpha_n(\nu)|^2, \tag{16}$$

where

$$\epsilon_n(\nu) = 2Re[\psi(\frac{1+|n|}{2} + i\nu) - \psi(1)], \tag{17}$$

$$\alpha_n(\nu) = \int_0^{\infty} dr r^{-2-2i\nu} \int_0^{2\pi} d\phi e^{-in\phi}$$
$$\left(i\nu + \frac{n+1}{2} + re^{i\phi}(h - i\nu - \frac{n-1}{2})\right)(i\nu - \frac{n-1}{2})$$
$$(-\tilde{h} + i\nu - \frac{n-1}{2}) Z(r,\phi), \tag{18}$$

$$h = 1/2 + n/2 - i\nu, \tilde{h} = 1/2 - n/2 + i\nu,$$
$$-\infty < \nu < \infty, n = \ldots, -1, 0, 1, \ldots, \tag{19}$$

$$Z(z, z^*) = |z(1-z)|^{2h/3}\Psi(z, z^*). \tag{20}$$

The function Ψ in Eq. (20) is invariant under the transformations $z \to 1-z$ and $z \to 1/z$ (Bose symmetry in the 3 gluons).

The following trial functions Ψ were used in Ref. 9 :

$$\Psi = \sum_{k=1}^{N-N_1} c_k a^{k/2-1/6} + \sum_{k=1}^{N_1} d_k a^{k-1/6} \ln a, \tag{21}$$

where

$$a = \frac{r^2 r_1^2}{(1+r^2)(1+r_1^2)(r^2+r_1^2)}. \tag{22}$$

The result is

$$\epsilon = 0.22269, \tag{23}$$

a value which has to be compared with Eq. (12). The corresponding Odderon intercept is

$$\alpha_O(0) = 0.96. \tag{24}$$

By comparing the values (24) and (14), one sees that there is only a 2% difference between the "exact" result and the variational one.

We draw from this section the conclusion that the Odderon intercept is very close to 1, i.e. is much higher than the 1/2 (ρ,ω) intercept. We therefore expect important Odderon effects at high energy.

The LLA result shows that the gap $\alpha_O(0) - 1$ is surprisingly small and therefore very sensitive to higher order corrections. The crucial problem of knowing if $\alpha_O(0)$ is bigger than 1, smaller than 1 or just equal to one, is therefore still an open problem. The last case in the one I favor. The best way of solving this problem is, in my opinion, the study of the non-perturbative Odderon. Promising results on this line were already obtained[11].

Let me add, before closing this section, that very recently, G.P. Korchemsky and J. Wosiek obtained a new representation for the Odderon wave function[12], which is in agreement with the results of Refs. 7 and 8 and which allows identification of a new quantum number - *triality* - associated with the Odderon.

An intriguing problem is if there is or not a Pomeron-Odderon exchange-degeneracy, like in the non-leading reggeon $\rho - \omega - f - A_2$ sector. The study of the C-even state of 3 reggeized gluons is crucial in this context[13].

3 The Odderon phenomenology

3.1 *What is the problem with the Odderon ?*

A quarter of century after its birth the Odderon has still an uncertain existential status. Of course, there are some interesting experimental indications of its existence :

- the difference between pp and $\bar{p}p$ differential cross-sections in the dip-shoulder region at ISR energies[14] ;

- the unusual shape of the polarisation in $\pi^- p \to \pi^o n$ at low energies[15] (indicating a ρ-type Odderon, distinct from the ω-type Odderon as required in LLA) ;

- the extraction of the semi-theoretical ρ-parameter from the dN/dt UA4/2 $\bar{p}p$ data in the presence of oscillations at very small t and high energies or of a more complicated phase of the forward scattering amplitude[16].

However, these experimental indications are either isolated or controversial.

The real problem with the Odderon is the paradoxical scarcity of the high-energy data in hadron-hadron scattering, leading to an excessive focus on the pp and $\bar{p}p$ scattering. In other words we fit high-energy parameters by using mainly low-energy data and, even worse, we draw conclusions about the Odderon based only on pp and $\bar{p}p$ scattering. The folklore about the "suppression" of the Odderon has its source in these facts. The recent shift in attention from pp and $\bar{p}p$ scattering towards ep scattering is, in this context, very positive.

3.2 *HERA-Odderon phenomenology*

The most active team in the Odderon phenomenology in the last two years is, of course, the Heidelberg group (H.G. Dosch, O. Nachtmann, E.R. Berger) and its associates (A. Donnachie, P.V. Landshoff, W. Kilian, M. Rueter). In a series of papers[17],[18], they produced impressive results with a solid theoretical ground.

For example, the pseudoscalar meson production

$$e^{\pm}p \to e^{\pm}pPS, \text{ where } PS = \pi^o,\ \eta,\ \eta',\ \eta_c \quad (25)$$

at HERA ($\sqrt{s}$ = 300.6 GeV) constitutes a direct probe for the Odderon. The Odderon is here in competition with the photon only (see Fig. 2) : the Odderon contribution is not obscured by the huge Pomeron contribution as in the hadron-hadron reactions.

By taking, as a toy model, the Odderon as a Regge pole located near $J = 1$,

$$\eta_O \beta(t) \left(\frac{s}{s_0}\right)^{\alpha_O(t)-1} \xi(t) \quad (26)$$

where $\alpha_O(0) \simeq 1$, $\xi(t)$ is the signature fractor, $\beta|t|$-the Odderon residue and $\eta_O = \pm 1$ (due to the absence of a positivity property for the Odderon contribution), one gets important Odderon effects in a variety of observables. I give just one example in Fig. 3 : the $p_\perp$ distribution for pion production in the photoproduction region. One can see from Fig. 3 (where c_O is proportional to η_O) that dramatic effects are induced by the presence of the Odderon. Moreover, one can see that the effects for the case $\eta_O = +1$ are drastically different from the ones for $\eta_O = -1$: the HERA data can give important indications on the sign of $\Delta\sigma$ - the difference between hadron-hadron and hadron-antihadron total cross-sections.

An important theoretical ingredient in several works of the Heidelberg group[18] is the Stochastic Vacuum Model (SVM)[19] which established a very interesting connection between the topological Y-shape of the baryons and the coupling of the Odderon. If the angle between two sheets of the baryon is very small (i.e. the baryon has a diquark-quark structure) the Regge-pole Odderon is suppressed. An interesting process is the photoproduction of pions with single dissociation (breakup of the target proton), because this process is independent of the particular structure of the baryon. Its cross-section is $\simeq$ 300 nb, i.e. 50 times larger for the corresponding one in the elastic photoproduction.

Of course, taking the Odderon as a Regge-pole could be too simplistic : for example, perturbative QCD indicates a more complicate singularity structure in the complex J-plane. Moreover, we showed longtime ago[20] that the Regge-pole Odderon induces an overall shift of the low-energy data for the real parts which are already very well described by the Pomeron and the secondary Regge poles. Therefore, the Regge-pole Odderon is the worst case to be considered as a possible Odderon singularity. However, as a toy model, it can be still used as an illustration of typical Odderon effects, present even if

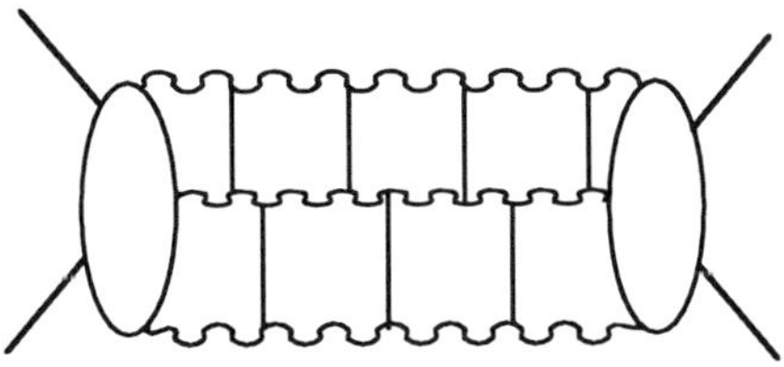

Fig. 1. The Odderon in QCD as a C-odd state of 3 reggeized gluons.

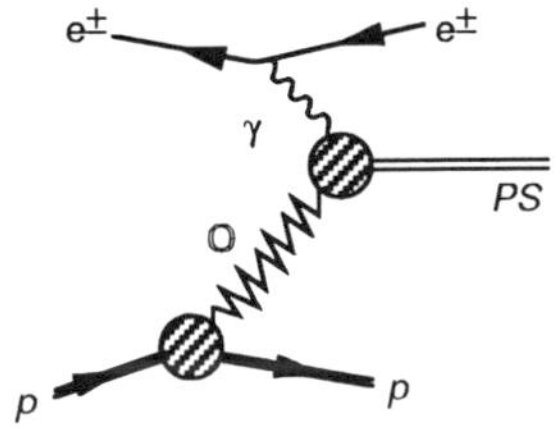

Fig. 2. The Odderon exchange in $e^{\pm}p \to e^{\pm}pPS$.

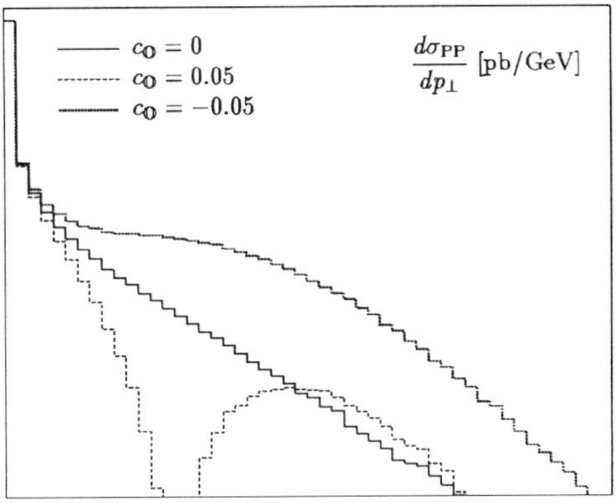

Fig. 3. The $p_\perp$ distribution for pion production in the photoproduction region (Ref. 17).

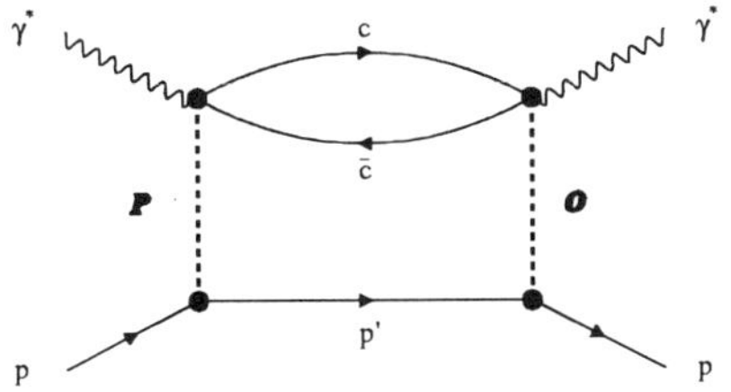

Fig. 4. The Pomeron-Odderon interference in $\gamma p \to c\bar{c}p'$ (Ref. 21).

the Odderon coupling is very much suppressed as compared with the Pomeron one.

A very interesting Odderon effect was recently studied by S.J. Brodsky et al.[9] : the assymetry in the fractional energy of charm versus anticharm jets in the diffractive photoproduction $\gamma p \to c\bar{c}Y$ at HERA. This assymmetry is very sensitive to the Pomeron-Odderon interference (Fig. 4) : it measures the Odderon amplitude linearly. Namely

$$A(t, M_x^2, z_c) \simeq \epsilon \frac{\sin[(\pi/2)(\alpha_O - \alpha_P)]}{\cos((\pi/2)\alpha_O)} \left(\frac{s_{\gamma p}}{M_x^2}\right)^{\alpha_O - \alpha_P} \tag{27}$$

where $z_c = E_{c(\bar{c})}/E_\gamma$ and $\epsilon = \pm 1$. It can be seen from eq.(27) that the sign of the asymmetry is controlled by the gap $(\alpha_O - 1)$ and the energy dependence by the gap $(\alpha_O - \alpha_P)$. By taking, as a numerical illustration, $\alpha_P = 1.13$ and $\alpha_O = 0.95$, Brodsky et al. get a lower bound for the asymmetry equal to 15%.

Before closing this section, let me mention other interesting phenomenological studies of the Odderon : diffractive $C = +$ neutral meson production from virtual photons[22], exclusive η_c photo- and electroproduction[23], exclusive f_2 leptoproduction[24] and single-spin asymmetries for small-angle pion production in hadron colisions[25].

4 Conclusions : the Odderon in the future

As we understood from the talk of S. Weisz at this Conference[26], the TOTEM experiment at LHC will not be very helpful for the Odderon physics : at least in the initial stage of the experiment, priority will be given to just mesuring the pp total cross-section.

Great hopes arise from the talk of S.B. Nurushev[27] concerning the R7 experiment at RHIC[28] with a higher luminosity that the UA4/2 experiment at CERN. Several measurements concern directly the Odderon physics : extraction of the ρ parameter with a precision $\delta\rho = 0.01$; research for oscilation at very small t ($|t| \simeq (1-4) \cdot 10^{-4}$ GeV^2 ; the evolution of the dip in $d\sigma/dt$; the polarisation parameters. In connection with this last measurement, E. Leader and T.L. Trueman showed in a very recent paper[29] the high sensitivity of the spin dependence of pp scattering, particularly the parameter A_{NN}, to the Odderon.

We had a nice surprise learning that the $\bar{p}p$ option at RHIC is realistic. That would allow the detection of the Odderon through the measurement of the difference $\Delta\sigma$ of the $\bar{p}p$ and pp total cross-sections : the Regge-pole model predicts $\Delta\sigma = 40\ \mu b$ at RHIC while the maximal-Odderon approach[2] predicts[14] $\Delta\sigma = -2.4$ mb. The expected precision at RHIC being $\Delta\sigma = 0.5$ mb, one can clearly detect the presence of the Odderon through $\Delta\sigma$.

Let me also stress the importance of the systematic study of the energy-dependence of $\sigma_T(s)$ in the RHIC range 50 GeV $\leq \sqrt{s} \leq$ 500 GeV. A huge gap in the high-energy hadron data will be filled.

However, we have not to wait till the next millenium in order to get a long awaited evidence of the Odderon. The results of the H_1 experiment on the pseudoscalar meson production at HERA[30] will be soon available.

Acknowledgments

It is a great pleasure to thank Prof. Vladimir Petrov and his young and dynamic staff for the remarkable organization of this Conference.

References

1. I.Ya. Pomeranchuk, Zh. Eksp. i Teor. Fiz. **34**, 725 (1958) ; G.F. Chew and S.C. Frautschi, *Phys. Rev. Lett.* **7**, 394 (1961).
2. L. Lukaszuk and B. Nicolescu, *Nuovo Cimento Lett.* **8**, 405 (1973).
3. J. Bartels, *Nucl. Phys.* B **175**, 365 (1980) ; J. Kwiecinski and M. Praszalowicz, *Phys. Lett.* B **94**, 413 (1980).
4. D. Joynson, E. Leader, C. Lopez and B. Nicolescu, *Nuovo Cimento* **30A**, 345 (1975).
5. K. Kang and B. Nicolescu, *Phys. Rev.* D **11**, 2461 (1975).
6. A. Martin, talk given at CERN-Dubna-Serpukhov Triangle Seminar, Dubna, 1990, CERN-TH. 5630/90 preprint.
7. R.A Janik and J. Wosiek, *Phys. Rev. Lett.* **82**, 1092 (1999).
8. L.N. Lipatov, talk given at IPN Orsay, France, March 1997 (unpublished).
9. M.A. Braun, P. Gauron and B. Nicolescu, *Nucl. Phys.* B **542**, 329 (1999).
10. P. Gauron, L.N. Lipatov and B. Nicolescu, *Phys. Lett.* B **260**, 407 (1991).
11. P. Desgrolard, M. Giffon, E. Martynov and E. Predazzi, hep-ph/9907451 ; E. Predazzi, talk at this Conference.
12. G.P. Korchemsky and J. Wosiek, hep-ph/9908304.
13. L.N. Lipatov, private communication.
14. A. Breakstone *et al.*, *Phys. Rev. Lett.* **54**, 2180 (1985) ; S. Erhan *et al.*, *Phys. Lett.* B **152**, 131 (1985) ; A. Donnachie and P.V. Landshoff, *Nucl. Phys.* B **244**, 322 (1984) ; P. Gauron, E. Leader and B. Nicolescu, *Phys. Rev. Lett.* **54**, 2656 (1985).
15. V.P. Apokin *et al.*, *Z. Phys.* C **15**, 293 (1982) (Prof. S.B. Nurushev was part of this experiment) ; P. Gauron, E. Leader and B. Nicolescu, *Phys.*

Rev. Lett. **52**, 1952 (1984).

16. UA4/2 Coll., C. Augier *et al.*, *Phys. Lett.* B **316**, 448 (1993) ; P. Gauron, B. Nicolescu and O.V. Selyugin, *Phys. Lett.* B **397**, 305 (1997) ; O.V. Selyugin, talk at this Conference ; V. Kundrat, talk at this Conference.
17. W. Kilian and O. Nachtmann, *Eur. Phys. J.* C **5**, 317 (1998) ; P.V. Landshoff and O. Nachtmann, hep-ph/9808233 ; E.R. Berger, A. Donnachie, H.G. Dosch, W. Kilian, O. Nachtmann and M. Rueter, *Eur. Phys. J.* C **9**, 491 (1999) ; E.R. Berger, hep-ph/9905519.
18. M. Rueter and H.G. Dosch, *Phys. Lett.* B **380**, 177 (1996) ; *Phys. Rev.* D **57**, 4097 (1998) ; E.R. Berger and O. Nachtmann, *Eur. Phys. J.* C **7**, 459 (1999) ; E.R. Berger, *Nucl. Phys. Proc. Suppl.* **74**, 96 (1999) ; M. Rueter, H.G. Dosch and O. Nachtmann, *Phys. Rev.* D **59**, 014018 (1999) ; E.R. Berger, talk at this Conference.
19. H.G. Dosch and Y.A. Simonov, *Phys. Lett.* B **205**, 339 (1988).
20. P. Gauron, E. Leader and B. Nicolescu , *Phys. Lett.* B **238**, 406 (1990) ; P. Gauron and B. Nicolescu, *Phys. Lett.* B **292**, 448 (1992).
21. S.J. Brodsky, C. Merino and J. Rathsman, hep-ph/99044280 ; SLAC-PUB-8272 and SLAC-PUB-8273, October 1999 ; S.J. Brodsky, hep-ph/9907346.
22. R. Engel, D. Yu. Ivanov, R. Kirschner and L. Szymanowski, *Eur. Phys. J.* C **4**, 93 (1998).
23. J. Czyzewski, J. Kwiecinski, L. Motyka and M. Sadzikowski, *Phys. Lett.* B **398**, 400 (1997) ; B **411**, 402 (1997).
24. M.G. Ryskin, *Eur. Phys. J.* C **2**, 339 (1998).
25. A. Ahmedov, I.V. Akushevich, E.A. Kuraev and P.G. Ratcliffe, hep-ph/9902418.
26. TOTEM Coll., W. Kienzle *et al.*, CERN/LHCC 99-7 ; S. Weisz, talk at this Conference.
27. S.B. Nurushev, talk at this Conference.
28. R7 Coll., N. Akchurin *et al.*, Proposal, September 1995.
29. E. Leader and T.L. Trueman, hep-ph/9908221.
30. H1 Coll. S. Tapprogge *et al.*, Proc. of the Int. Conf. on the Structure and the Interactions of the Photon (Photon 97) including the 11th Int. Workshop on Photon-Photon Collisions, ed. F.C. Erne and A. Buijs (World Scientific, Singapore, 1997) ; S. Tapprogge, talk at the Pomeron-Odderon Workshop, Heidelberg, March 1998 ; S. Tapprogge, PhD Thesis, Heidelberg University, HD-IHEP-96-19, December 1996.

M. N. Mnatsakanova

ABSOLUTE UPPER BOUNDS ON $\pi^0\pi^0$ TOTAL CROSS SECTIONS AT ARBITRARY ENERGIES

M. N. MNATSAKANOVA
Institute for Nuclear Physics of Moscow State University, Russia

YU.S. VERNOV
Institute for Nuclear Research, Russian Academy of Sciences, Russia

Absolute upper bounds on the imaginary part and on the modulus of $\pi^0\pi^0$ elastic scattering amplitude are obtained for arbitrary physical energies as the consequence of analyticity, unitarity and crossing symmetry.

One of the remarkable achievements of axiomatic quantum field theory is the derivation of absolute (numerical) bounds on $\pi^0\pi^0$ elastic scattering amplitude. These bounds are rigorous consequences of general principles of quantum field theory only. Precisely they are consequences of analyticity in s and t planes, crossing symmetry and unitarity. The distinguished role of $\pi^0\pi^0$ scattering follows from the full symmetry of its elastic scattering amplitude $F(s,t,u)$ in s,t and u-variables. According to the quark structure of hadrons $\pi^0\pi^0$ total cross section $\sigma_{tot}^{\pi^0\pi^0}$ has the same order as $\sigma_{tot}^{\pi N}$ and σ_{tot}^{NN}.

The first bound of this kind has been derived by Martin [1]. The improved bounds were obtained in paper of Lukashúk and Martin [2] and then in papers [3-8]. Above mentioned bounds are bounds in nonphysical region, where $|s| < 4m_\pi^2$, $|t| < 4m_\pi^2$ and $|u| < 4m_\pi^2$ or bounds on some integrals over low energies intervals. As it has been shown by A.Martin [1], such bounds, in fact, imply the restrictions on the physical amplitudes, precisely, on the integrals of the scattering amplitude over all physical region.

In our paper [9] we obtained that absolute bounds lead to Froissart-Martin bound [2,10] without unknown constant and so to the bound which is valid at arbitrary physical energies. We also derived an absolute upper bound on modulus of $F(s,t)$ at physical s and $t > 0$ inside Martin's ellipse [9, 11]. In this report we generalize our previous results.

Let's recall that Froissart-Martin bound is the following inequality:

$$\sigma_{tot}(s) < \frac{\pi}{m_\pi^2} ln^2 \frac{s}{s_0} \qquad (1),$$

where s_0 is an unknown constant and strictly speaking this inequality is valid at $s \to \infty$ only.

We obtain our bounds in two steps:
1. Using dispersion relations (DR) and absolute bounds we obtain upper bounds on amplitude at physical s and nonphysical t_0, belonging to Martin's ellipse, precisely, for $0 < t_0 < 4m_\pi^2$.
2. Then we use these bounds to derive upper bounds on $A(s,t)$ at physical s and $t < t_0$.

In accordance with the result of Jin and Martin [12], if $t > 0$ and belongs to Martin ellipse, DR require only two subtractions. It is convenient to use the variable $\omega \equiv s - 2m_\pi^2 + t/2$:

$$F(\omega,t) = F(0,t) + \frac{2\omega^2}{\pi}\int_{\omega_0}^{\infty}\frac{A(\omega',t)d\omega'}{\omega'(\omega'^2-\omega^2)}, \quad A(\omega,t) \equiv ImF(\omega,t) \qquad (2),$$

Thus

$$F(\omega_2,t_0) - F(\omega_1,t_0) = \frac{2(\omega_2^2-\omega_1^2)}{\pi}\int_{\omega_0}^{\infty}\frac{A(\omega',t_0)\omega'\,d\omega'}{(\omega'^2-\omega_1^2)(\omega'^2-\omega_2^2)} \qquad (3)$$

Let $0 < \omega_{1,2} < \omega_0$ and $\omega_2 > \omega_1$:

$$\int_{\omega}^{\nu\omega}\frac{A(\omega',t_0)\omega'\,d\omega'}{(\omega'^2-\omega_1^2)(\omega'^2-\omega_2^2)} < \frac{\pi}{2}\,\frac{F(\omega_2,t_0)-F(\omega_1,t_0)}{\omega_2^2-\omega_1^2} \qquad (4),$$

where ν is an arbitrary number. Using the low of the mean, we have:

$$\int_{\omega}^{\nu\omega}\frac{A(\omega',t_0)d\omega'}{\omega'^3} = \frac{A(\bar\omega,t_0)}{\bar\omega}\,\frac{1}{\omega}\,\frac{\nu-1}{\nu} \qquad (5)$$

$$\frac{A(\bar\omega,t_0)}{\bar\omega} < \frac{\pi}{2}\,\frac{\nu}{\nu-1}\,\omega\,\frac{F(\omega_2,t_0)-F(\omega_1,t_0)}{\omega_2^2-\omega_1^2} \qquad (6)$$

To simplify our formulae let's consider the case $\omega \gg \omega_0$. (In fact it means $s > 1\,Gev^2$). In our case $A(\omega,t_0)$ can be presented as the following expansion:

$$A(\omega,t_0) = 2\sum_{\substack{l=0\\ even}}^{\infty}(2l+1)a_l(\omega)P_l\left(1+\frac{2t_0}{\omega}\right) \qquad (7),$$

where due to unitary condition

$$0 \le a_l(\omega) \le 1 \qquad (8)$$

Let's recall that for $\pi^0\pi^0$-scattering $a_{2l+1}(\omega) = 0$.

Now we have to find $max A(\bar{\omega}, t)$ with the fixed value of $A(\bar{\omega}, t_0)$. Let's consider $t \geq 0$. It is easy to prove (see, e.g. [13]), that since $P_{l_2}(x)/P_{l_1}(x)$, $l_2 > l_1$ increases with the growth of x if $x > 1$, this maximum is realized by the following well-known set of partial amplitudes:

$$a_l(\bar{\omega}) = \begin{cases} 1 & l \leq L \\ \eta \leq 1 & l = L+2 \\ 0 & l \geq L+4 \end{cases} \tag{9}$$

At high energies we can neglect the contribution of one partial wave with $l = L + 2$. So,

$$A(\bar{\omega}, t) = 2 \sum_{\substack{l=0 \\ even}}^{L} (2l+1) P_l(x) \qquad x = \cos\theta = 1 + \frac{2t}{\omega} \tag{10}$$

This set can be easily calculated as

$$(2l+1)P_l(x) = P'_{l+1}(x) - P'_{l-1}(x)$$

Thus,

$$A(\bar{\omega}, t) = 2P'_{L+1}(x) \cong 2P'_L(x), \quad x = 1 + \frac{2t}{\bar{\omega}} \tag{11}$$

It has been obtained by Vernov and Mnatsakanova [13], that

$$P'_l(1 + \frac{2t}{\omega}) \cong \frac{e^{\gamma}\sqrt{\gamma}}{4\sqrt{2\pi}} \frac{\omega}{t} \text{if} \qquad \gamma > 1, \qquad \gamma = 2L\sqrt{\frac{t}{\omega}} \tag{12}$$

At $\gamma \to \infty$ eq. (12) is the known asymptotic expression for $P'_l(x)$. So

$$A(\bar{\omega}, t) = 2P'_L(1 + \frac{2t}{\bar{\omega}}) = \frac{e^{\gamma}\sqrt{\gamma}}{2t\sqrt{2\pi}}\bar{\omega} \tag{13}$$

If we recall formula (6) for $\frac{A(\bar{\omega}, t_0)}{\bar{\omega}}$ and substitute the last equation in it, we obtain:

$$\frac{e^{\gamma_0}\sqrt{\gamma_0}}{t_0\sqrt{2\pi}} < \pi \frac{\nu}{\nu - 1} \omega \frac{F(\omega_2, t_0) - F(\omega_1, t_0)}{\omega_2^2 - \omega_1^2}, \qquad \gamma_0 = 2L\sqrt{\frac{t_0}{\bar{\omega}}} \tag{14}$$

We derive the upper bound, changing sign $<$ on $=$ in the last equation.

Let's find the decision of the equation:

$$e^{\gamma_0}\sqrt{\gamma_0} = B\bar{\omega} \qquad \text{at} \quad \gamma_0 > 1 \tag{15}$$

As $\gamma_0 > 1$ we can easily solve it by the method of successive approximations.

$$\gamma_0^{(1)} = ln(B\omega), \qquad \gamma_0^{(n+1)} = \ln\left[(B\,\omega)/\sqrt{\gamma_0^{(n)}}\right]$$

It is easy to see that $\gamma_0^{(2n+1)} > \gamma_0$, $\gamma_0^{(2n)} < \gamma_0$. Thus we have to use $\gamma_0^{(2n+1)}$, but yet $\gamma_0^{(2)}$ practically coinsides with $\gamma_0^{(3)}$ (see Table below).

For $t = 0$ in accordance with (9) we obtain the following formula for σ_{tot}^{max}:

$$\sigma_{tot}^{max} = \frac{16\pi}{\bar{\omega}} \sum_{\substack{l=0 \\ even}}^{L} 2l \cong \frac{8\pi}{\bar{\omega}} L^2 = \frac{2\pi}{t_0}\gamma^2, \qquad L^2 = \frac{\gamma^2}{4}\frac{\bar{\omega}}{t_0} \tag{16}$$

Owing to the calculations made above

$$\sigma_{tot}^{max} < \frac{2\pi}{t_0}\left(\gamma_0^{(2n+1)}\right)^2 < \frac{2\pi}{t_0}\, ln^2(B\omega) \tag{17}$$

Similarly we can obtain the absolute bound on $A(\bar{\omega}, t)$, $t > 0$. We consider only the case $\gamma \gg 1$, in which eq.(12) is valid and after simple calculations we find:

$$\frac{A(\bar{\omega}, t)}{\bar{\omega}} < \frac{\sqrt{\gamma_0}\, e^{\gamma_0\sqrt{\frac{t}{t_0}}}}{2\sqrt{2\pi}\, t^{\frac{3}{4}}\,(t_0)^{\frac{1}{4}}} \tag{18}$$

To obtain the absolute bound on $A(\bar{\omega}, t)$, $t < 0$ we use well-known inequality: $|P_l(\cos\theta)| < [2/(\pi\, l\, \sin\theta)]^{\frac{1}{2}}$ if $|\cos\theta| < 1$. Then we can use our reasons given above to obtain the inequality:

$$A(\bar{\omega}, t) < \frac{4\sqrt{2}}{3\sqrt{\pi}\,\sin\theta} L^{\frac{3}{2}} = \frac{2}{3\sqrt{\pi}\,\sin\theta}\gamma_0^{\frac{3}{2}}\left(\frac{\bar{\omega}}{t_0}\right)^{\frac{3}{4}} \tag{19}$$

To obtain the numerical bounds let's use the results of Lopez and Mennessier [7] ($m_\pi^2 = 1$): $|F(1,3,0)| < 3.3$, $|F(2,3,0)$. After trivial calculations we have:

$$e^{\gamma}\sqrt{\gamma} = 210\frac{\nu}{\nu - 1}\omega \tag{15$'$}$$

Here and below we omit the subscript 0. We obtain the decision of eq.(15′) at various ω and ν to the third approximation.

The results are presented in the following table:

	$\nu = 1.1$			$\nu = 2$			$\nu = 10$		
$\sqrt{s}$(Gev)	γ_1^2	γ_2^2	γ_3^2	γ_1^2	γ_2^2	γ_3^2	γ_1^2	γ_2^2	γ_3^2
1	134	107	108	100	78	79	88	68	69
10	261	218	219	220	182	183	194	159	160
100	431	370	372	367	312	314	345	293	294
10^4	899	800	802	805	713	715	772	682	684
10^6	1536	1396	1397	1410	1277	1279	1398	1266	1268
10^8	2343	2159	2160	2190	2014	2016	2135	1962	1964
10^{12}	4465	4190	4191	4250	3982	3984	4176	3911	3913
10^{16}	7267	6893	6895	7000	6635	6637	6534	6184	6186
10^{20}	10750	10274	10276	10400	9934	9936	10300	9836	9839

So we come to the following conclusions: as the dependence of ν is weak, especially, at high energies, we have obtained "almost" local bounds; upper bounds slowly increase with ω and γ depends very weakly on $B\omega$ if $\gamma \gg 1$.

References

1. A. Martin, High-energy physics and elementary particles (IAEA, Vienna, 1965), p. 165.
2. L.Łukashúk and A. Martin, Nuovo Cim. **52A** (1967) 122.
3. B. Bonnier and R. Vinh-Mau, Phys. Rev. **165** (1968) 1923.
4. J.B. Healy, Phys. Rev. **D8** (1973) 1904.
5. G. Auberson, L. Epele, G. Mahoux and F.R.A. Simaõ, Nucl. Phys. **B73** (1974) 314; Nucl. Phys. **B94** (1975) 311.
6. C. Lopez, Nucl. Phys. **B88** (1975) 358; Nuovo Cim. Lett. **13** (1975) 69; **B96** (1975) 515; C. Lopez and G. Mennessier, Nucl. Phys. **B96** (1975) 515.
7. C. Lopez and G. Mennessier, Nucl. Phys. **B118** (1977).
8. A.D. Gangal and S.M. Roy, Nuovo Cim. **76A** (1983) 52.
9. M.N. Mnatsakanova and Yu.S. Vernov, "Absolute upper bound on $\pi^0\pi^0$ total cross sections at arbitrary energies ". In: Proc. of the XIth Workshop on High Energy Physics and Quantum Field Theory, Sankt-Petersburg, 1996, P. 385.
10. M. Froissart, Phys. Rev. **123** (1961) 1053;
    A. Martin, Phys. Rev. **129** (1963) 1432.
11. M.N. Mnatsakanova and Yu.S. Vernov, "Absolute upper bounds ..." In: Proc. of the Xth Intern. Seminar "Quarks'98", Suzdal, in print.
12. Y.S. Jin and A. Martin, Phys. Rev. **B135** (1964) 1375.
13. Yu.S. Vernov and M.N. Mnatsakanova, Teor. Mat. Fiz. **52** (1982) 199 (in Russian).

Yu. S. Vernov

WHEN DISPERSION RELATIONS CAN BE REPLACED BY THEIR DERIVATIVE ANALOGS?

M. N. MNATSAKANOVA
Institute for Nuclear Physics of Moscow State University, Russia
E-mail: mnatsak@theory.npi.msu.su

YU.S. VERNOV
Institute for Nuclear Research, Russian Academy of Sciences, Russia
E-mail: vernov@theory.npi.msu.su

Local relations between real and imaginary part of elastic scattering amplitude have been obtained. They are valid with the good accuracy at high, but finite energies, if scattering amplitude has not strong oscillations. These relations coincide with the first term of differential dispersion relations.

Differential dispersion relations (DDR) are point-to-point correlations between the real and the imaginary part of the scattering amplitudee. They were obtained in paper [1]. These relations are in a good agreement with experiment, but, in general, they are not the rigorous consequences of dispersion relations (DR) [2,3]. Detailed investigation of DDR was done in papers of P. Kolář and J. Fischer (see [4] and references herein). Recently interesting results in this direction were obtained in papers [5,6].In our work [7] it was proved that DDR (precisely speaking the first term of them) are rigorous consequences of DR at asymptotic energies for some class of amplitudes. The amplitudes which admit such substitution (DR – DDR) have to be weakly oscillating ones (precise definition see below).

In this report we extend our results to finite, but high energies and also obtain new rigorous equations, which are the basis for derivation of local relations between imaginary and real part of the scattering amplitude.

We restrict ourself by the most important case of forward scattering amplitude.

Let's represent elastic forward amplitude $F(E)$ on the interval $(\alpha E, \alpha^{-1})$ in the form:

$$F(E) = F_\beta(E)\,\varphi(E) \qquad (1),$$

where $F_\beta(E)$ and $\varphi(E)$ have the same analytical properties as $F(E)$, $\varphi(E)$ changes slowly on the interval under consideration and $F_\beta(E) = C_\beta\, E^\beta$. $C_\beta = C\,exp(-i\pi\beta/2)$ in symmetric case, $C_\beta = i\,C\,exp(-i\pi\beta/2)$ in antisymmetric; $C \in \mathbb{R}$.Thus in both cases $\varphi(E)$ has the same crossing properties with symmetric amplitude. Precisely we suppose that

$$\left|\frac{Im\,\varphi\,(E') - Im\,\varphi\,(E)}{Im\,\varphi\,(E)}\right| \ll 1, \qquad \left|\frac{\psi\,(E') - \psi\,(E)}{\psi\,(E)}\right| \ll 1,$$
$$\psi\,(E) = E\,\frac{d\,Re\,\varphi\,(E)}{dE}, \qquad \alpha\,E < E' < \alpha^{-1}\,E \qquad (2),$$

where α is some constant, which have to be sufficiently small, in fact if $\alpha \sim 1/3$, the corrections terms in derived equations are small (see below eqs.(8), (9)). If $E \to \infty$ the similar conditions can be obtained by the substitution: $\ll 1$ by $\to 0, \alpha$ is any fixed number. In this case the corrections in derived equations are negligible. At finite energies these equations are only approximate ones.

Let's use dispersion relation and inverse dispersion relation:

$$\varphi\,(E) = \frac{2E^2}{\pi}\int\limits_1^\infty \frac{Im\,\varphi\,(E')\,dE'}{E'(E'^2 - E^2)} \qquad (3)$$

$$\varphi\,(E) = \frac{2}{i\pi}\sqrt{E^2-1}\int\limits_1^\infty \frac{E'\,Re\,\varphi\,(E')\,dE'}{\sqrt{E'^2-1}(E'^2 - E^2)} \qquad (4)$$

(In fact we passed to the function $\varphi\,(E) - \varphi\,(0)$, but additional term plays no role, thus we don't change our notations, we also denote the beginning of the cut by one.) Equate the right parts of eqs.(3) and (4) in a pure imaginary point $i\,E$ we come to the relation:

$$-E^2\int\limits_1^\infty \frac{Im\,\varphi\,(E')\,dE'}{E'(E'^2 + E^2)} = \sqrt{E^2+1}\int\limits_1^\infty \frac{E'\,Re\,\varphi\,(E')\,dE'}{\sqrt{E'^2-1}(E'^2 + E^2)} \qquad (5)$$

Let's differentiate this relation. Below we consider interval $(\alpha\,E,\ \alpha^{-1}\,E)$, satisfying the condition $\alpha\,E \gg 1$. So we can present the right part of eq.(5) in the form:

$$\sqrt{E^2+1}\int\limits_1^\infty \frac{E'\,Re\,\varphi\,(E')\,dE'}{\sqrt{E'^2-1}(E'^2 + E^2)} \cong C\,(E_0, E) + E\int\limits_{E_0}^\infty \frac{Re\,\varphi\,(E')\,dE'}{E'^2 + E^2} \qquad (6),$$

where $1 \ll E_0 \ll \alpha\,E$. Thus we can neglect the term $C\,(E_0, E)$. After differentiation we use integration by parts and come to the equation:

$$-2\,E\int\limits_1^\infty \frac{E'\,Im\,\varphi\,(E')\,dE'}{(E'^2 + E^2)^2} = \int\limits_{E_0}^\infty \frac{\psi\,(E')\,dE'}{E'^2 + E^2} \qquad (7)$$

Here and below we omit the terms $\sim E_0/E$. To estimate the left part of eq.(7) let's divide interval $(1,\infty)$ on three parts: $(1,\alpha E)$, $(\alpha E),\alpha^{-1} E)$, $(\alpha^{-1} E,\infty)$. Using the low of the mean and introducing the function $\mu(E') = Im\,\varphi(E') - Im\,\varphi(E)$ we obtain after simple calculations that:

$$-2E\int\limits_1^\infty \frac{E'\,Im\,\varphi(E')\,dE'}{(E'^2+E^2)^2} = -\frac{1}{E}\left[Im\,\varphi(E)\left(1+\frac{\mu(\bar{E}_2)+\mu(\bar{E}_3)}{2\,Im\,\varphi(E)}\right)+ \right. \qquad (8),$$
$$\left. +\alpha^2\left(Im\,\varphi(\bar{E}_1)+Im\,\varphi(\bar{E}_4)-2\,Im\,\varphi(E)\right)\right]$$

where $1\le \bar{E}_1\le \alpha E$, $\alpha E\le \bar{E}_2\le E$, $E\le \bar{E}_3\le \alpha^{-1}E$, $\alpha^{-1}E\le \bar{E}_4<\infty$. We omit the terms with higher powers of α. From the condition (2) it follows that:

$$\left|\frac{\mu(\bar{E}_2)+\mu(\bar{E}_3)}{2\,Im\,\varphi(E)}\right| \ll 1.$$

Let's point out that if $Im\,\varphi(E')$ is a monotonic function in the interval $(\alpha E,\alpha^{-1}E)$, then $\mu(\bar{E}_2)$ and $\mu(\bar{E}_3)$ have different signs. This leads to additional smallness of the main correction term. The same is also true for the term proportional to α^2 if $Im\,\varphi(E)$ is a monotonic function in the interval wider then $\alpha E,\alpha^{-1}E)$. Estimating the right part of eq. (8) by the similar calculations we obtain:

$$\int\limits_{E_0}^\infty \frac{\psi(E')\,dE'}{E'^2+E^2} = \frac{1}{E}\left[\frac{\pi}{2}\psi(E)+(\frac{\pi}{4}-\alpha)\left(\nu(\bar{E}_2)+\nu(\bar{E}_3)\right)+ \right. \qquad (9),$$
$$\left. +\alpha\,\psi(\bar{E}_1)+\psi(\bar{E}_4)-2\,\psi(E)\right]$$

$\nu(E') = \psi(E')-\psi(E)$, $E_0\le \bar{E}_1\le \alpha E$, $\alpha E\le \bar{E}_2\le E$, $E\le \bar{E}_3\le \alpha^{-1}E$, $\alpha^{-1}E\le \bar{E}_4<\infty$. From the condition (2) it follows that only the first term in the right part of (9) is essential.

Finally we obtain the relation:

$$Im\,\varphi(E) \cong -\frac{\pi}{2}E\frac{d\,Re\,\varphi(E)}{d\,E} \qquad (10)$$

In conclusion let's stress that in the most likely case of monotonic behavior of $\varphi(E)$ at high energies eq.(10) has to be fulfilled in a good accuracy.

References

1. J. B. Bronzan, *Argonne Nat. Lab. Report, N ANL/HEP 7327* (1973); J. B. Bronzan, G. L. Kane and H. P. Sukhatme, *Phys. Lett.* B **249**, 272 (1974).
2. G. K. Eichman, Y. Dronhers, *Phys. Lett.* B **52**, 428 (1974).

3. J. Heidrich, E. Kares, *Lett. al. N. Cimento* **12**, 365 (1975).
4. P. Kolář, J. Fischer, *J. Math. Phys.* **25**, 2538 (1984);
J. Fischer, P. Kolář, *Czech. J. Phys.* B **37**, 297 (1987).
5. M. J. Menon, A. E. Motter, B. M. Pimentel, *Phys. Lett.* B **451**, 207 (1999).
6. A. F. Martini and M. J. Menon, J. T. S. Paes and M. J. Silva Neto, Preprint *Unicamp, IFGW ABSTRACTA A 008-99* (1999).
7. Yu. S. Vernov, M. N. Mnatsakanova, "Consequences of analyticity for weakly oscillating amplitudes", in *Proc. of the XII Intern. Seminar on high energy physics and quantum field theory, Protvino*, (Nauka, 1990).

S. M. Troshin

UNITARITY BOUND FOR THE SINGLE-HELICITY-FLIP AMPLITUDE IN ELASTIC pp-SCATTERING

S. M. TROSHIN, N. E. TYURIN
*Institute for High Energy Physics, Protvino, Moscow Region,
142284 RUSSIA*

On the basis of the U–matrix method of the s–channel unitarization, we obtain an unitarity bound for the single-helicity-flip amplitude F_5 of elastic pp-scattering at small values of t.

The size and energy dependence of the hadron helicity-flip amplitude is an interesting problem in the study of the asymptotic properties and the role of spin-dependent interaction and it is also an important issue in the polarimetry studies based on the Coulomb-Nuclear Interference (CNI) [1]. If the hadronic part of the proton-proton interaction is helicity conserving the CNI analyzing power would be then due to the interference between a real electromagnetic helicity-flip amplitude and an imaginary hadronic helicity-nonflip amplitude.

However, the hadronic interaction may not conserve helicity in small angle scattering. Helicity conservation does not follow from QCD in a region where chiral symmetry is spontaneously broken. In Regge theory the Pomeron is usually assumed to be helicity conserving. But this is merely an assumption which seems to be wrong; recent experimental results on the central production of mesons and observation of nontrivial azimuthal dependence suggest a nonzero helicity transfer by the effective Pomeron [2]. Earlier it was shown that the unitarity generates different phases in the helicity-flip and nonflip Pomeron contributions [3] and leads to a nonzero analyzing power.

The CNI analyzing power would also certainly change if there were any nonzero hadronic single-helicity-flip amplitude.

The above issues are considered in the recent survey [1]. Among the new results, the most interesting one is the bound for the single-helicity-flip amplitude F_5 of elastic pp-scattering, i.e. the function

$$\hat{F}_5(s,0) \equiv \frac{mF_5(s,t)}{\sqrt{-t}}|_{t=0} \qquad (1)$$

cannot increase at $s \to \infty$ faster than

$$cs\ln^3 s, \qquad (2)$$

while for the helicity nonflip amplitudes there is the Froissart-Martin bound $cs\ln^2 s$.

We show that, in fact, a stronger bound for the function $\hat{F}_5(s,0)$ can be obtained if one apply an unitarity in the specific way, i.e. uses an unitarization method based on the U-matrix approach[4]. This method is based on the unitary representation for helicity amplitudes of elastic pp-scattering:

$$F_{\lambda_3,\lambda_4,\lambda_1,\lambda_2}(s,b) = U_{\lambda_3,\lambda_4,\lambda_1,\lambda_2}(s,b)+i\rho(s)\sum_{\lambda',\lambda''} U_{\lambda_3,\lambda_4,\lambda',\lambda''}(s,b)F_{\lambda',\lambda'',\lambda_1,\lambda_2}(s,b), \tag{3}$$

where $\lambda's$ are the initial and final proton's helicities. F_i are the helicity amplitudes in the standard notations. The kinematical function $\rho(s) \simeq 1$ at $s \gg 4m^2$ and will be neglected in the following. Explicit solution of Eq. (3) has the following form:

$$F_1(s,b) = \frac{\tilde{U}_1(s,b)[1-iU_1(s,b)] - i\tilde{U}_2(s,b)U_2(s,b)}{[1-iU_1(s,b)]^2 - [U_2(s,b)]^2}, \tag{4}$$

$$F_3(s,b) = \frac{\tilde{U}_3(s,b)[1-iU_3(s,b)] - i\tilde{U}_4(s,b)U_3(s,b)}{[1-iU_3(s,b)]^2 - [U_4(s,b)]^2}, \tag{5}$$

where

$$\tilde{U}_i(s,b) = U_i(s,b) + 2U_5(s,b)F_5(s,b) \tag{6}$$

and

$$F_5(s,b) = \frac{U_5(s,b)}{[1-iU_1(s,b)-iU_2(s,b)][1-iU_3(s,b)-iU_4(s,b)] - 4U_5^2(s,b)}. \tag{7}$$

For simplicity we consider the two extreme cases. First, we suppose that the helicity nonflip functions $U_1(s,b)$ and $U_3(s,b)$ are the dominant ones, i.e.

$$|U_i(s,b)| \ll |U_1(s,b)|,\ |U_3(s,b)| \tag{8}$$

for $i = 2,4,5$. In this case one can get [3]

$$F_5(s,t) = \frac{s}{\pi^2}\int_0^\infty bdb \frac{U_5(s,b)}{[1-iU_1(s,b)][1-iU_3(s,b)]} J_1(b\sqrt{-t}). \tag{9}$$

Unitarity requires that $\mathrm{Im}U_{1,3}(s,b) \geq 0$. The functions $U_{1,3}(s,b)$ could be different. For our purposes, however, it is safe to assume that they are the same $U_1(s,b) = U_3(s,b) = U(s,b)$. For the function $U(s,b)$ we use a simple form

$$U(s,b) = gs^{\Delta}e^{-\mu b}, \tag{10}$$

where the parameter $\Delta > 0$ guarantees the rise of the total cross-section. This is a rather general parameterization for $U(s,b)$ which provides correct analytical properties in the complex t–plane, i.e. it is consistent with the spectral representation for the function $U(s,b)$ [5]:

$$U(s,b) = \frac{\pi^2}{s}\int_{t_0}^{\infty}\omega(s,t)K_0(b\sqrt{t})dt. \tag{11}$$

The results do not depend on the particular functional form of the function $U(s,b)$, e.g. they are valid for the function $U(s,b)$ in the form

$$U(s,b) = gs^{\Delta}e^{-b^2/a(s)}, \tag{12}$$

where $a(s) \sim \ln s$. We do not consider this function since it does not provide correct analytical properties in the complex t-plane. Thus we do not use here model features and detailed structure of the helicity functions U_i, but consider the reasonable arguments of a general nature. To get an upper bound we should maximize the function $U_5(s,b)$. According to the Eq. (8) it can be done taking it in the form $U_5(s,b) = aU(s,b)$ where $|a| < 1$. Then from the Eq. (9) it follows that at $s \to \infty$:

$$|\hat{F}_5(s,0)| \leq cs\ln^2 s. \tag{13}$$

This means that the magnitude of the ratio

$$r_5(s,0) \equiv 2\hat{F}_5(s,0)/\mathrm{Im}[F_1(s,0) + F_3(s,0)] \tag{14}$$

cannot increase with energy and will not exceed a constant as $s \to \infty$.

This result is of a general nature and is retained when the function $U_5(s,b)$ is a dominant one. We consider now this case in some detail. We have for the amplitude $F_5(s,t)$ the following representation

$$F_5(s,t) = \frac{s}{\pi^2}\int_0^{\infty} bdb\frac{U_5(s,b)}{1-4U_5^2(s,b)}J_1(b\sqrt{-t}). \tag{15}$$

Using for $U_5(s,b)$ the functional dependence in the form of Eq. (10) it can be easily shown that the same bound Eq. (13) does take place for the single-helicity-flip amplitude $\hat{F}_5$.

Thus, we can state that the explicit account of unitarity in the form of U - matrix does not allow rising behavior of $|r_5(s,0)|$. The experimental data as well as most of the model predictions are consistent with this bound [6,7]. This result allow us to hope that the contribution of the single-helicity-flip amplitude could be controllable and the effective use of the CNI polarimeter would be possible.

Above results were obtained in impact parameter representation for simplicity. They can easily be reproduced using the partial wave expansion.

It is an account of the unitarity through the U-matrix representation for the helicity amplitudes leads to Eq. (13), i.e. the amplitude $F_5(s,b)$ has a peripheral dependence on the variable b at high energy and

$$|F_5(s, b=0)| \to 0 \tag{16}$$

at $s \to \infty$. This is a consequence of the explicit unitarity representation for the helicity amplitudes and it means that the assumption on $F_5(s,b) = constant$ at $b < R(s)$ [1] is a problematic one. However, it remains to be good for the helicity-nonflip amplitudes. Thus, as it was shown above, we have an asymptotic bound

$$|r_5(s,0)| \leq constant \tag{17}$$

at $s \to \infty$.

To conclude, it is worth noting that only the model-dependent estimations exist for the magnitude of the ratio $|r_5(s,0)|$. However, we believe that the rise of the function $|r_5(s,0)|$ at $s \to \infty$ may be excluded. At least, this is true for the used approach to the unitarization.

References

1. N.H. Buttimore et al., Phys. Rev. D 59 (1999) 114010-1 and references therein.
2. F. E. Close and G. A. Schuler, CERN-TH/99-28; hep-ph/9902243 and references therein.
3. S. M. Troshin and N. E. Tyurin, *Spin Phenomena in Particle Interactions*, World Scientific, Singapore, 1994.
4. A. A. Logunov, V. I. Savrin, N. E. Tyurin and O. A. Khrustalev, Teor. Math. Phys. 6 (1971) 157.
5. S. M. Troshin and N. E. Tyurin, Theor. Math. Phys. 50 (1982) 150.
6. T. L. Trueman, *CNI Polarimetry and the hadronic spin dependence of p p scattering,* Proceedings of the 12th International Symposium on High Energy Spin Physics, Amsterdam, Sept. 1996, SPIN96, Eds. C.W. de Jager, et al., World Scientific, Singapore, 1997, p. 833.
7. A. D. Krisch and S. M. Troshin, *Estimate of elastic proton-proton polarization at small P_T^2 near 1 TeV,* Proceedings of SPIN96, p.830.

EIKONAL CORRECTIONS FOR SCATTERING WITH SPIN-ORBIT INTERACTION

N. V. BONDARENKO, N. F. SHUL'GA

1 Introduction

Impact parameter representation that originally appeared in eikonal approximation of scattering amplitude [1] has proved to be so good tool for description of high-energy processes, that it caused a number of attempts to build on its base also calculations exceeding the eikonal approximation framework [2] . However, out of the eikonal framework a connection between the scattering particle wave function and transversal Fourier transformation of the amplitude

$$a(\mathbf{b}) = \int d^2 q_\perp e^{-iq_\perp b} a(\mathbf{q}_\perp) \tag{1}$$

is not transparent.

In the present paper we propose a way of high-energy scattering process consideration, in which instead of the wave function satisfying ordinary wave equation with a given scattering field from the outset another function is used satisfying a special differential equation and at large z turning just into $a(\mathbf{b})$, without any z dependence.

2 Eikonal expansion in the impact parameter representation of the scattering amplitude

Consider in general the process of one particle scattering which may be treated in teerms of its wave function $\psi(\mathbf{r})$.

The process of high-energy diffractive scattering physically consists of two successive subprocesses. First, during the incident particle fast passing through the scatterer a strong distortion of its wave function appears, and second, in the following free motion the distorted wave function splits into states with differtent momenta. At these conditions for the scattering amplitude calculation it is convenient to use the impact parameter representation, proposed in [3].

As it was shown in [3], scattering amplitude of a particle described by a wave function $\psi(\mathbf{r})$ may be expressed as

$$a(\mathbf{q}) = -\frac{ip'_z}{2\pi} \int d^2 b e^{-ip'r} (\psi(\mathbf{r}) - e^{ipr})|_{z=L \leq R}. \tag{2}$$

Formula (2) has a feature that should be minded. The z-dependence of the integrand must cancel, and it really does if behind the scatterer ψ is an exact solution of the free wave equation. However,at high-energy approximate calculations this is fulfilled only approximately. The given inconvenience can be removed as follows. Due to the expression

$$p'_z = \sqrt{p^2 - p'^2_\perp} \tag{3}$$

the exponent $e^{-ip'_z z}$ in (2) may be replaced by the operator form

$$e^{-iz\sqrt{p^2+\Delta_\perp}} \tag{4}$$

acting to the left. And by integratinion by parts its actrion can be transmitted to the right without changing the operator form

$$a = -\frac{ip'_z}{2\pi}\int d^2 b e^{\ ip'_\perp \rho} e^{\ iz\sqrt{p^2+\Delta_\perp}} (\psi(\mathbf{r}) - e^{ipr})|_{z=\infty} \tag{5}$$

or else

$$a = -\frac{ip'_z}{2\pi}\int d^2 b e^{-ip'_\perp \rho} (e^{-iz\sqrt{p^2+\Delta_\perp}}\psi(\mathbf{r}) - 1)|_{z=\infty} \tag{6}$$

So, $a(\mathbf{b})$ of (1) is proportional to the function

$$\varphi(\mathbf{r})\ |_{z=\infty} = e^{-iz\sqrt{p^2+\Delta_\perp}}\psi(\mathbf{r})\ |_{z=\infty} \tag{7}$$

In this regard it seems natural to look for the eikonal expansion not for function ψ, but directly for $\varphi(\mathbf{r})$.

If one takes the wave equation for ψ in the form

$$\left\{\Delta + p^2 - 2m\widehat{V}\right\}\psi = 0, \tag{8}$$

then with (7) the corresponding equation for φ reads

$$2i\sqrt{p^2+\Delta_\perp}\frac{\partial\varphi}{\partial z} - e^{-iz\sqrt{p^2+\Delta_\perp}} 2m\widehat{V} e^{iz\sqrt{p^2+\Delta_\perp}}\varphi + \frac{\partial^2\varphi}{\partial z^2} = 0. \tag{9}$$

This pattern with operator $\sqrt{p^2+\Delta_\perp}$ is intended especially for high energy consideration, so that two-valuedness of the square root does not play a role. The root further is supposed to be expanded over the small $\frac{1}{p^2}\Delta_\perp$ as

$$\sqrt{p^2+\Delta_\perp} = p - \frac{1}{2p}\Delta_\perp + ... \tag{10}$$

and from the point of view of expansion it does not matter much how complex the expanded function is. The main thing is that then result appears at once.

Expansion (10) goes in (9) as

$$\Big[2i\left(p - \tfrac{1}{2p}\Delta_\perp + ...\right)\tfrac{\partial}{\partial z}$$

$$-\left(1 + \tfrac{iz}{2p}\Delta_\perp + ...\right) e^{-ipz} 2m\widehat{V} e^{ipz}\left(1 - \tfrac{iz}{2p}\Delta_\perp + ...\right) + \tfrac{\partial^2}{\partial z^2}\Big]\varphi = 0. \tag{11}$$

Further neglections and transition to approximations depend on the operator $\widehat{V}$ magnitude and structure. In the case when $\widehat{V}$ is just a function of $\mathbf{r}$ we have

$$e^{-ipz}\widehat{V}e^{ipz} = \widehat{V}. \tag{12}$$

This case is well investigated [1,2,3]. But strictly $\widehat{V} = \widehat{V}(\mathbf{r})$ is only for electromagnetic scattering on a structureless centre. Various scattering problems involving nuclear forces, many-particle complex scatterers or just relativistic effects may be considered by introducing certain effective potential $\widehat{V}$, too. But in this case $\widehat{V}$ may contain important parts depending on the scattering particle momentum.

The simplest case of such momentum dependence is potential with spin-orbit interaction. It is worth to consider this case on the example of nucleon-nucleus scattering, at which the spin-orbit component has the simplest form and produces sizeable effects.

3 Spin–orbit potential and the eikonal wave function

In the case of scattering on a nuclear potential with a spin-orbit component the operator $\widehat{V}$ to be included into (8) reads

$$\hat{V} = V_c(r) + V_s(r)\sigma\cdot[\mathbf{r}\times(-i\nabla)]. \tag{13}$$

Then instead of (12) we have

$$e^{-ipz}\widehat{V}e^{ipz} = \hat{U}(\mathbf{r}) + V_s(r)\sigma\cdot[\mathbf{r}\times(-i\nabla)] \tag{14}$$

with

$$\hat{U}(\mathbf{r}) = V_c(r) + V_s(r)\sigma\cdot[\mathbf{b}\times\mathbf{p}]. \tag{15}$$

Though spin-orbit component V_s is usually much smaller than central one V_c, at large $\mathbf{p}$ terms in (15) can be both large to demand an eikonal consideration. And if the eikonal phase is large, corrections may be considerable.

Within the next-to-leading order accuracy equation (11) reads

$$\left(i\frac{\partial}{\partial z} - \frac{1}{v}\hat{U}\right)\varphi - \frac{1}{p}\left(\frac{iz}{v}\nabla_\perp \hat{U}\nabla_\perp + mV_s\sigma\cdot[\mathbf{b}\times\mathbf{p}]\right)\varphi \\ -\frac{1}{2p}\left(\frac{iz}{v}\Delta_\perp \hat{U} - \frac{\partial^2}{\partial z^2}\right)\varphi \cong 0. \tag{16}$$

Terms in the first bracket are largest ones and terms of second bracket can generate corrections of order unity, so that they should be exponentiated. The approximate expression for function φ reads

$$\varphi \approx \hat{F}e^{i\chi_0 + i\chi_1}, \tag{17}$$

with

$$\chi_0(\mathbf{r}) = \chi_c(\mathbf{r}) + \chi_s(\mathbf{r})\sigma\cdot[\mathbf{b}\times\mathbf{p}], \tag{18}$$

$$\chi_c(\mathbf{r}) = -\frac{1}{v}\int_{-\infty}^{z} dz' V_c, \qquad \chi_s(\mathbf{r}) = -\frac{1}{v}\int_{-\infty}^{z} dz' V_s, \tag{19}$$

$$\chi_1(\mathbf{r}) = \frac{1}{2p}\int_{-\infty}^{z} dz'\left(z'\frac{\partial}{\partial z'}(\nabla\chi_0)^2 + 2mV_s\mathbf{r}\cdot[\sigma\times\nabla\chi_0]\right), \tag{20}$$

$$\hat{F} = 1 + \frac{1}{2pv}\int_{-\infty}^{z} dz' z' \Delta\hat{U}. \tag{21}$$

4 The phase correction calculation

Appeal to the central symmetry of functions V_c, V_s allows to simplify considerably the expression for the phase correction:

$$\chi_1(b) = \frac{i}{2}\sigma[\mathbf{b}\times\mathbf{p}](\chi_s)^2 + \frac{1}{2pv^2}\left(1 + b\frac{\partial}{\partial b}\right)\int_{-\infty}^{\infty} dz\widehat{U}^2 \tag{22}$$

$$\hat{F} = 1 \tag{23}$$

To come to the common amplitude form

$$a = f + g(\sigma\cdot\mathbf{n}) \tag{24}$$

with

$$\mathbf{n} = \frac{[\mathbf{p}\times\mathbf{p}']}{|[\mathbf{p}\times\mathbf{p}']|} = \frac{[\mathbf{p}\times\mathbf{p}']}{q^2(1 - q^2/4p^2)} \tag{25}$$

we perform in (6) with χ_1 given by (22) integration over azimuthal angle of vector b. As a result we get

$$f = -ip \int_0^\infty dbb \left(e^{iC(b)} \cos pbS(b) - 1\right) J_0(qb), \tag{26}$$

$$g = iq \int_0^\infty dbbe^{iC(b)} \sin pbS(b) J_1(qb) \tag{27}$$

with

$$C(b) = \chi_c - \frac{1}{2pv^2}\left(1 + b\frac{\partial}{\partial b}\right) \int_{-\infty}^{\infty} dz(V_c^2 + p^2b^2V_s^2), \tag{28}$$

$$S(b) = \chi_s - \frac{i}{2}\chi_s^2 \frac{1}{pv^2}\left(1 + b\frac{\partial}{\partial b}\right) \int_{-\infty}^{\infty} dzV_cV_s. \tag{29}$$

5 Conclusions

As compared with known techniques developed for consideration of out-of eikonal high energy processes, our one may be recognized as most transparent and straigtforward. It also may give an idea for calculation of high energy scattering processes with comrlex interaction.

In the concrete case of scattering on nuclear potential with spin-orbit interaction it turns out to be easy to calculate eikonal corrections for nuclear potential scattering with spin-orbit interaction and to exponentiate larger ones. It makes a progress as compared with calculations made in [5] at suggestion that eikonal corrections are small.

References

1. R. Glauber In: Lectures in Theor. Phys. (W. E. Brittin and L. G. Dunham eds.), vol. 1, p. 135, NY, Interscience Pub. (1959)
2. S. J. Wallace, Ann. Phys. **78** (1973) 190
3. N. Bondarenko and N. Shul'ga, Theor. Math. Phys. **115** (1998) 588
4. N. Bondarenko and N. Shul'ga, Phys. Lett. B**427** (1998) 114
5. D. Waxman *et al.*, Phys. Rev. C**24** (1981) 578

Andrei Solodsky

HARD DIFFRACTION AT CDF

ANDREI SOLODSKY

The Rockefeller University, 1230 York Avenue, New York, NY 10021, USA

E-mail: solodsky@physics.rockefeller.edu

We summarize published results on diffractive dijet and W production and review new results on diffractive beauty and J/ψ production using the rapidity gap method to tag diffraction. The diffractive structure function of the proton obtained from the analysis of dijet events with a leading antiproton is discussed and compared with expectations based on results obtained in deep inelastic scattering experiments at HERA. Also we report on recent results of dijet production in double pomeron exchange.

1 Introduction

To study diffractive production at CDF we used two different methods. One is by searching for ***rapidity gaps***, which are pseudorapidity [1] regions devoid of particles, in the event topology. We applied this method to diffractive dijet [2], W [3] and heavy flavor [4] production measurements, described in sections 2 and 3. Another technique is the direct measurement of the diffractively scattered antiproton with a forward Roman pot spectrometer.

The CDF detector is described elsewhere [5]. Charged tracks in the region $|\eta| < 1.1$ are detected in the central tracking chamber (CTC) surrounded by the central and end-plug calorimeters. The microstrip silicon vertex detector (SVX) provides spatial measurement in the $r - \phi$ plane with a track impact parameter resolution of $\approx [13 + 40(\mathrm{GeV/c})/p_T]$ μm. The parts of the detector most relevant for the rapidity gap analyses are the forward calorimeters and the beam-beam counters (BBC). The fiducial region of the forward calorimeters covers the range $2.4 < |\eta| < 4.2$. The BBC consist of two arrays of eight vertical and eight horizontal scintillation counters which cover approximately the region $3.2 < |\eta| < 5.9$.

2 Diffractive Dijet and W Production

CDF searched for diffractive dijet production in a sample of dijet events with a single-vertex in which the two leading jets with transverse energy [1] $E_T >$ 20 GeV are both at $\eta < -1.8$ or $\eta > 1.8$. The single diffractive production was observed in these events by the requirement of a rapidity gap in both the forward calorimeter and the beam-beam counters (BBC) on one side of the detector.

Figure 1 shows the correlation of the BBC and forward calorimeter tower multiplicities in the η-region opposite the dijet system. The distinct excess in the (0,0) bin is associated with diffractive production. After subtracting the non-diffractive background and correcting for the losses caused by extra $\bar{p}p$ interactions in the same bunch crossing as a diffractive event, as well as for BBC and forward calorimeter live time acceptance and for the rapidity gap acceptance (0.70 ± 0.03), calculated using the POMPYT[6] Monte Carlo program with pomeron $\xi < 0.1$, the fraction of diffractively produced dijet events was found to be

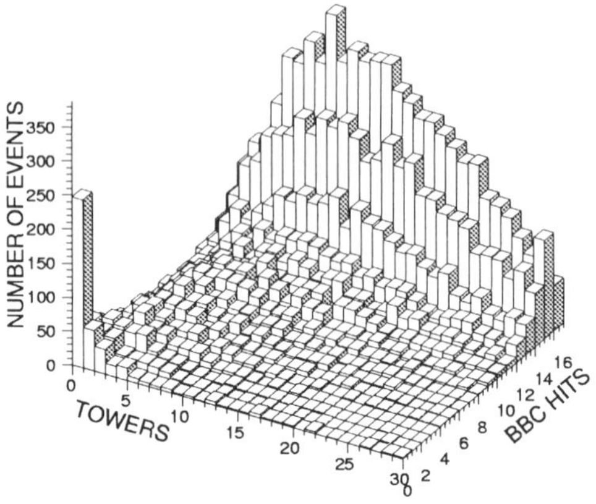

Figure 1. BBC hits versus forward calorimeter towers in region opposite a forward dijet.

$$R_{JJ} = [0.75 \pm 0.05(stat) \pm 0.09(syst)]\%.$$

In addition, CDF studied diffractive W production [3], which results predominantly from $q\bar{q}$ annihilation and probes the quark/antiquark content of the pomeron, using a sample of events with an isolated central e^+ or e^- ($|\eta| < 1.1$) of $E_T > 20$ GeV and missing transverse energy $\not{E}_T > 20$ GeV. The ratio of single diffractive to non-diffractive W production in this analysis was determined to be

$$R_W = [1.15 \pm 0.51(stat) \pm 0.20(syst)]\% \quad (\xi < 0.1).$$

Since diffractive dijet production is mostly sensitive to gluons in the pomeron, we can combine the dijet and W measurements and determine the gluon fraction in the pomeron[2], $f_g = 0.7 \pm 0.2$, which is consistent with the gluon fraction found by the ZEUS experiment [7].

3 Diffractive Heavy Flavor Production

We extended our studies of diffractive processes to diffractive heavy flavor production, charm and beauty, to probe directly the gluon content of the pomeron. Our diffractive beauty [4] production measurement is based on identifying a high transverse momentum electron with $E_T > 9.5$ GeV and $|\eta| < 1.1$, from the semi-leptonic b-quark decay, produced in single diffraction dissociation, $p + \bar{p} \to p/\bar{p} + b(\to e + X') + X$. Each event is required to have a jet consisting of at least two tracks in addition to the electron candidate.

First, we extract a diffractive signal from the obtained event sample and then estimate the b-quark fraction separately in the diffractive and total event samples.

As in our previous analyses [2,3], the diffractive signal is extracted by counting BBC hits, N_{BBC}, and adjacent forward calorimeter towers, N_{CAL}, with $E > 1.5$ GeV. Figure 2(a) shows the correlation between N_{BBC} and N_{CAL} for both the positive and negative η sides of the detector, i.e. two entries per event. The (0,0) bin contains 100 events. We evaluate the non-diffractive content of the (0,0) bin from the distribution of events along the diagonal of Fig. 2(a), with $N_{BBC} = N_{CAL}$, shown in Fig. 2(b) by extrapolating a fit to the data of bins (2,2) to (9,9) to bin (0,0). This yields 24.4±5.5 non-diffractive background events in the (0,0) bin.

Figures 2(c) and 2(d) show the electron E_T and η distribution, respectively, for the diffractive (points) and total (histogram) event samples. In Fig. 2(d), the sign of η of diffractive events with a gap at positive η was changed, so that the gap always appears at negative η. While the E_T spectra show no significant difference, the diffractive η distribution is shifted away from the gap relative to the symmetric distribution of the total event sample, in agreement with the single diffraction event topology.

In addition to events from b-quark decays, the data contain events from charm decays and background. The background is mainly due to electrons from residual photon conversions and to hadrons faking electrons.

We use two methods to extract the fraction of beauty events in the data. In the first method, we fit the electron momentum component perpendicular to the jet axis, $p_T^{e/jet}$, which depends on the mass of the parent quark, with

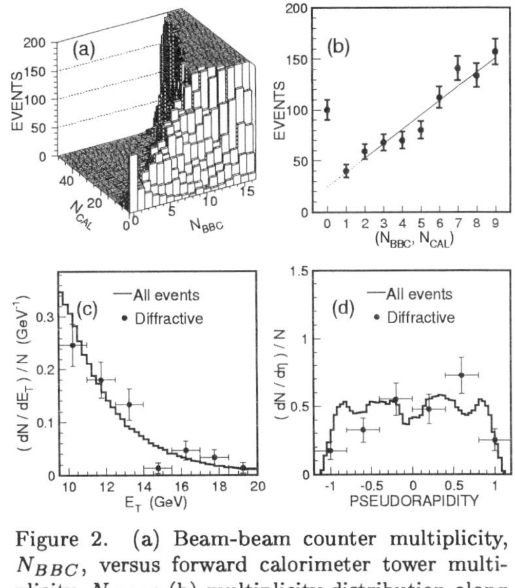

Figure 2. (a) Beam-beam counter multiplicity, N_{BBC}, versus forward calorimeter tower multiplicity, N_{CAL}; (b) multiplicity distribution along the diagonal with $N_{BBC} = N_{CAL}$ in the plot in (a); (c) electron p_T and (d) pseudorapidity for the diffractive (points) and total (histogram) event samples.

the sum of four templates: photon conversions, fake electrons from hadrons, charm and beauty. This fit yields a beauty fraction of $(42.9 \pm 0.4)\%$ $[(38 \pm 14)\%]$ for the total [diffractive] event sample. The second method uses the impact parameter of the electron track, which is defined as the minimum distance between the primary vertex and the electron track in the $r - \phi$ plane and depends on both the mass and the lifetime of the parent quark. A fit to the impact parameter distribution using four templates, as above, yields $(47.7 \pm 0.4)\%$ $[(38 \pm 14)\%]$ for our two data samples.

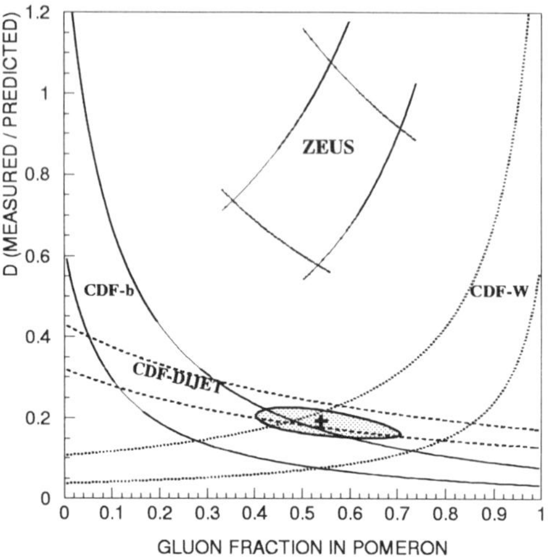

Figure 3. The ratio D of measured to predicted diffractive rates as a function of the gluon content of the pomeron.

The average of the results of both methods yields $73371 \pm 485(\boldsymbol{stat}) \pm 7774(\boldsymbol{syst})$ $[44.4 \pm 10.2(\boldsymbol{stat}) \pm 4.7(\boldsymbol{syst})]$ beauty events for the total [diffractive] event sample. The difference between the results of the two methods is assigned as systematic uncertainty. After subtracting the 24% non-diffractive background estimated from the fit in Fig. 2(b), there remain $33 \pm 10(\boldsymbol{stat}) \pm 5(\boldsymbol{syst})$ diffractive beauty events. Correcting the diffractive event yield for single-vertex selection cut efficiency (0.26 ± 0.01), and for the detector live-time acceptance (0.77 ± 0.07) due to noise or beam associated background, we obtain $165 \pm 50(\boldsymbol{stat}) \pm 29(\boldsymbol{syst})$ diffractive beauty events.

The diffractive to total b-quark production ratio obtained from the above numbers is $R_{\bar{b}b}^{gap} = [0.23 \pm 0.07(\boldsymbol{stat}) \pm 0.05(\boldsymbol{syst})]\%$. The rapidity gap acceptance for events generated using POMPYT Monte Carlo with a flat pomeron structure, which is favored by HERA measurements [7,8], and a gluon to quark ratio of 0.7 ± 0.2, as reported in ref. 2, is found to be 0.37 ± 0.02. Dividing $R_{\bar{b}b}^{gap}$ by this value yields a diffractive to total production ratio of

$$R_{\bar{b}b} = [0.62 \pm 0.19(\boldsymbol{stat}) \pm 0.14(\boldsymbol{syst})]\% \quad (\xi < 0.1).$$

POMPYT with the standard pomeron flux and a flat (hard) pomeron structure consisting of purely gluons or quarks yields $R_{\bar{b}b}$ of 10.4%(11.6%) and 0.92%(1.02%), respectively. The ratio D of the measured $R_{\bar{b}b}$ fraction to that predicted by POMPYT depends on the gluon fraction f_g of the pomeron. This dependence is shown in Fig. 3, where D is plotted as a function of f_g

along with published results from ZEUS and CDF measurements [2,3]. For each measurement the two curves show the 1σ bounds. The black cross and shaded ellipse represent the best fit and 1σ contour of a least square two-parameter fit to the three CDF results. The fit yielded $D_{CDF} = 0.19 \pm 0.04$ and $f_g^{CDF} = 0.54^{+0.16}_{-0.14}$, in agreement with the results we obtained from the W and dijet rates, namely $D = 0.18 \pm 0.04$ and $f_g = 0.7 \pm 0.2$ [2]. The value of D_{CDF} is significantly smaller than the ZEUS result. The discrepancy between the HERA and Tevatron D-values represents a breakdown of factorization. The observed discrepancy is in general agreement with predictions based on the renormalized pomeron flux model [9].

We also searched for diffractive J/ψ production in a sample of central ($|\eta| < 1.1$) dimuons. For J/ψ reconstruction we required a pair of opposite charge muons with $p_T > 2$ GeV/c and invariant mass close to the J/ψ mass. The technique we used to extract the diffractive signal is identical to that used in our previous studies. Preliminary results of this analysis, before correcting for the gap acceptance A, give a ratio of diffractive to non-diffractive J/ψ production of

$$R_{J/\psi} \times A = [0.35 \pm 0.07]\%.$$

In spite of the fact that all diffractive processes studied at CDF are differently sensitive to the quark and gluon content of the pomeron, the obtained ratios of diffractive to non-diffractive production are all of the same order of magnitude, $\sim 1\%$. This indicates that the structure of the pomeron probed in single diffraction events is not very different from the structure of the proton.

4 Study of Diffractive Structure Function

In the beginning of 1996 a forward Roman pot spectrometer (RPS) was added to CDF to detect diffractive antiprotons, which carry away a fraction x_F, measured by RPS, of their original momentum. The fraction of the $\bar{p}$ momentum taken by the pomeron is $\xi = 1 - x_F$; the momentum fraction of the parton in the pomeron, β, is given by $\beta = x/\xi$, where x is the fraction of the momentum of the $\bar{p}$ carried by the struck parton.

From an inclusive sample of single diffraction (SD) events, collected by triggering on a recoil $\bar{p}$ detected in the RPS, we selected events containing at least two jets with transverse energy $E_T^{jet} > 7$ GeV. For the non-diffractive (ND) dijet sample we used data collected with a minimum bias (MB) trigger requiring a coincidence of the BBCs in the p and $\bar{p}$ directions. From the E_T and η of the jets (including a third jet if $E_T^{jet3} > 5$ GeV) we determine the fraction x of the momentum of the antiproton carried by the struck parton,

$$x = \frac{1}{\sqrt{s}} \sum_{i=1}^{n} E_T^i e^{-\eta^i}.$$

Since the ratio $R(x)$ of the SD to ND dijet cross sections is equal to the ratio of the antiproton SD to ND structure functions, the diffractive structure function may be obtained by multiplying the known ND structure by $R(x)$. The SD dijet sample is normalized by scaling its event rate to that of the inclusive diffractive sample and using for the latter the measured [10] *soft* SD cross section. The absolute normalization of the ND dijet sample is obtained from the measured 51.2 ± 1.7 mb cross section of the BBC trigger.

The data were collected during runs of typical luminosities $\sim 3 \times 10^{29}\ cm^{-2}sec^{-1}$. We selected events with only one reconstructed vertex within $|z_{vtx}| < 60$ cm, and applied quality cuts to ensure that a reconstructed Roman pot track, assumed to originate at the primary vertex, is a diffracted $\bar{p}$. In the resulting *inclusive* sample of 1.6 million SD and 300K MB events, we observed 30,410 SD and 32,629 ND events containing at least two jets of corrected $E_T^{jet} > 7$ GeV. The E_T of a jet was defined as the sum of the calorimeter E_T within an $\eta - \phi$ cone [11] of radius 0.7.

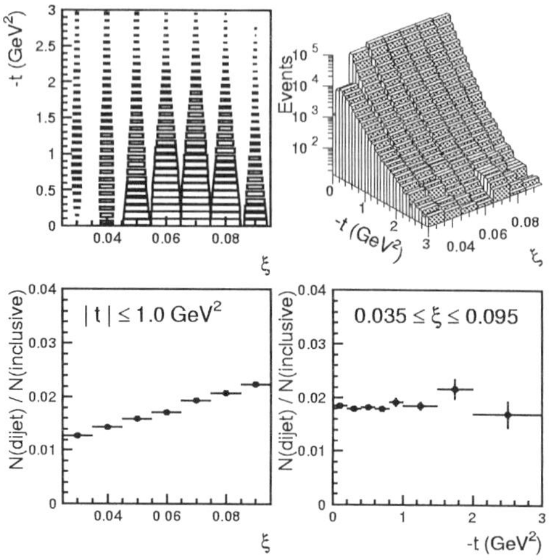

Figure 4. Distributions versus ξ and t: (a) Roman pot acceptance; (b) inclusive diffractive event sample; (c) rato of dijet to inclusive diffractive events versus ξ and (d) versus t.

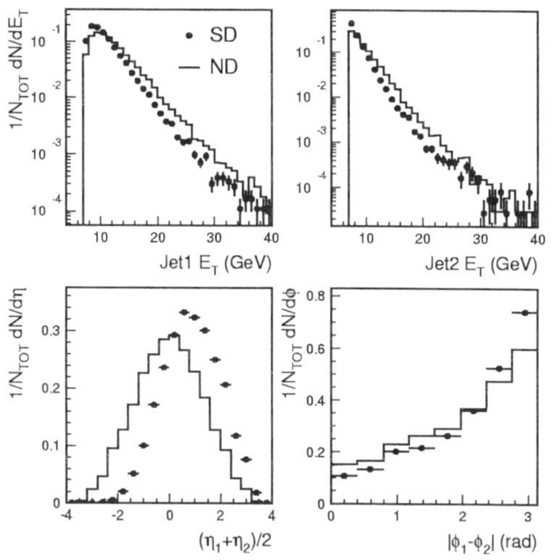

Figure 5. Comparison of diffractive to non-diffractive dijet events: (a) Leading jet E_T; (b) next to leading jet E_T; (c) mean η of dijet system; (d) azimuthal angle difference between the two leading jets.

In the jet energy correction the underlying event energy with an average

E_T of 0.54 (1.16) GeV was subtracted from each jet in SD (ND) events. These values were determined experimentally from the $\sum E_T$ of calorimeter tower energy measured within a randomly chosen η-ϕ cone of radius 0.7 in events of the inclusive data samples.

Figures 4(a) and 4(b) show, respectively, the RPS acceptance and a lego plot of the inclusive diffractive event sample as a function of ξ and t. The acceptance is large for ξ in the range of 0.04 to 0.1 and for $|t| < 1$ GeV^2. The fraction of dijet events in the inclusive sample is shown as a function of ξ in Fig. 4(c) and versus t in Fig. 4(d). We observe that while the fraction of dijet events increases linearly as a function of ξ, there is no significant t dependence in the region $0 < |t| < 3$ GeV^2, in agreement with the UA8 result [12].

Figure 5 shows distributions of E_T^{jet1}, E_T^{jet2}, $\bar{\eta} = (\eta^{jet1} + \eta^{jet2})/2$ and $\Delta\phi = \phi^{jet1} - \phi^{jet2}$ for the diffractive (points) and ND (histograms) event samples. The E_T^{jet1} and E_T^{jet2} distributions fall faster in the SD than in the ND events. This is due to the difference in SD and ND structure functions (see below). The diffractive dijets are boosted away from the leading antiproton in η (the $\bar{p}$ is at negative η). The diffractive $\Delta\phi$ distribution is narrower, jets being more back-to-back, than the ND.

The ratio $\tilde{R}(x)$ of the number of SD dijet events, corrected for Roman pot acceptance, to the number of ND dijets is shown in Fig. 6 for six bins in ξ of width 0.01 ranging from 0.035 to 0.095. The two data samples are normalized to correspond to the same luminosity. The tilde over the R indicates integration over variables (t, ξ, E_T^{jet}) for diffractive and E_T^{jet} for ND events. The results are shown in Fig. 6 for $|t| < 1$ GeV^2 and $E_T(jet1, jet2) > 7$ GeV. The data points are well fit by lines of the form $\tilde{R}(x) = R_0(x/0.0065)^{-r}$ in the region $10^{-3} < x < 0.5\xi_{min}$ for each ξ-bin. The lower x limit is imposed to minimize detector end-effects. The shape of the $\tilde{R}(x)$ distribution shows no significant ξ dependence. A fit to all the data in the region $0.035 < \xi < 0.095$ yields

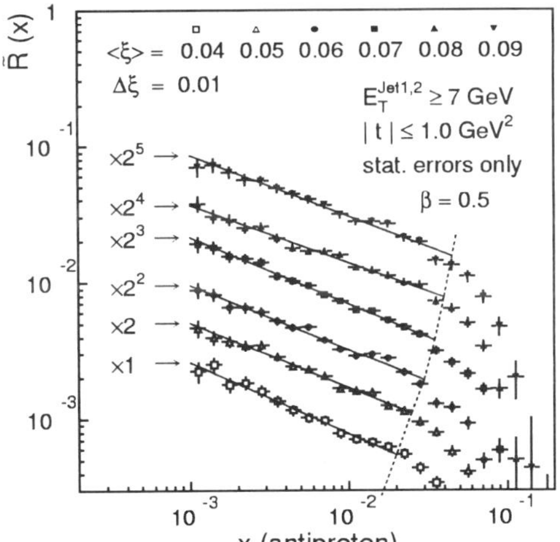

Figure 6. Ratio of diffractive to non-diffractive dijet event rates as a function of x (momentum fraction of parton in $\bar{p}$). The solid lines are fits to the form $\tilde{R}(x) = R_0(x/0.0065)^{-r}$ for $\beta < 0.5$, where $\beta = x/\langle\xi\rangle$.

$R_0 = (6.2 \pm 0.1) \times 10^{-3}$ and $r = 0.45 \pm 0.02$ with $\chi^2/\text{d.o.f.} = 0.72$.

The diffractive and ND structure functions can be written in terms of gluon and quark densities as $F_{jj}(x) = x[g(x) + \frac{4}{9}q(x)]$, where $\frac{4}{9}$ is a color factor. As mentioned above, $R(x)$ represents the ratio of the diffractive to ND structure functions of the antiproton. Thus, we can determine the diffractive structure function of the antiproton by the formula

$$\tilde{F}^D_{jj}(\beta) = \tilde{R}(x = \beta\xi) \times \tilde{F}^{ND}_{jj}(\beta),$$

where we change variables to $\beta = x/\xi$. We have evaluated $\tilde{F}^D_{jj}(\beta)$ for $|t| < 1$ GeV^2, $0.035 < \xi < 0.095$ and $E_T(jet1, jet2) > 7$ GeV using the GRV92LO parton densities [13] in $\tilde{F}^{ND}_{jj}(\beta)$. The obtained $\tilde{F}^D_{jj}(\beta)$ is shown in Fig. 7, where the solid line is a fit to the data of the form $\tilde{F}^D_{jj}(\beta) = B(\beta/0.1)^{-n}$ in the range $(10^{-3}/\xi) < \beta < 0.5$. The fit yields $B = 1.15 \pm 0.01$ and $n = 1.09 \pm 0.01$ with $\chi^2/\text{d.o.f.} = 2.2$. For our average ξ of 0.065 the value of $\beta = 0.1$, for which $\tilde{F}^D_{jj} = B$, corresponds to $x = 0.0065$, for which $\tilde{R} = R_0$. The lower and upper boundaries of the filled band surrounding the data points represent the β-distributions obtained by using only the two leading jets or up to four jets of $E_T > 5$ GeV, respectively, in the evaluation of x. The dashed (dotted) curve is obtained from fit 2 (fit 3) of the H1 diffractive structure function [8] evaluated at $Q^2 = 75$ GeV^2, which approximately corresponds to the average value of $(E_T^{jet})^2$ of our data. The measured and expected structure functions disagree both in normalization and shape. The discrepancy in normalization, defined as the integral over β of data over expectation, is $D = 0.06$ (0.05) for fit 2 (fit 3). The disagreement between our measured diffractive structure function and the expectation from diffractive DIS represents a breakdown of factorization similar to that observed [2] in comparing diffractive W-boson and dijet production rates with expectations based on ZEUS results [7] obtained from diffractive DIS and dijet photoproduction at HERA.

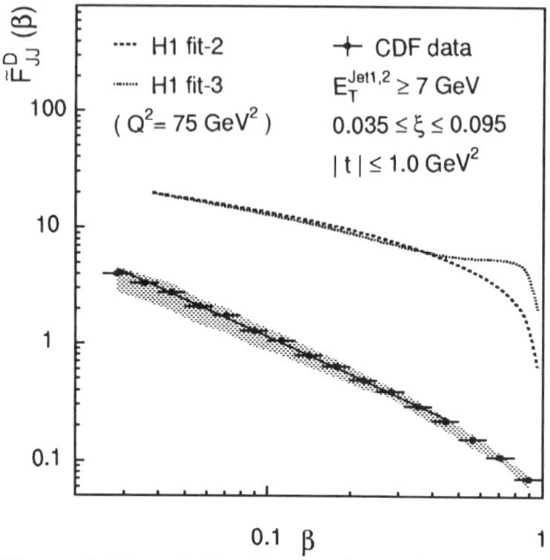

Figure 7. Data β distribution (points) compared with expectations from the parton densities of the proton extracted from diffractive deep inelastic scattering by the H1 Collaboration. The solid line is a fit to the data of the form β^{-n} in the indicated region. The dashed (dotted) lines are expectations from the H1 fit 2 (fit 3).

5 Dijets in Double Pomeron Exchange

Using the data sample obtained by triggering on an outgoing $\bar{p}$ with the RPS, we searched for dijet production by Double Pomeron Exchange (DPE). In these events the jets are produced in the central region, separated from the diffractive p and $\bar{p}$ by large rapidity gaps. From an inclusive sample of single diffractive events with detected diffractive $\bar{p}$, we selected events containing at least two jets with $E_T^{jet} > 7$ GeV and than searched for a DPE signal by looking for a rapidity gap on the p side, using the BBC array and forward calorimeters. Figure 8(a) shows a 2-dimensional plot of the BBC multiplicity, N_{BBC}, and forward calorimeter tower multiplicity, N_{CAL}, on the proton side. The excess in the (0,0) bin is attributed to DPE production of jets. We used the diagonal bins ($N_{BBC} = N_{CAL}$) of Fig. 8(a) to evaluate the contribution from low multiplicity SD events in the (0,0) bin. The diagonal multiplicity distribution is shown in Fig. 8(b).

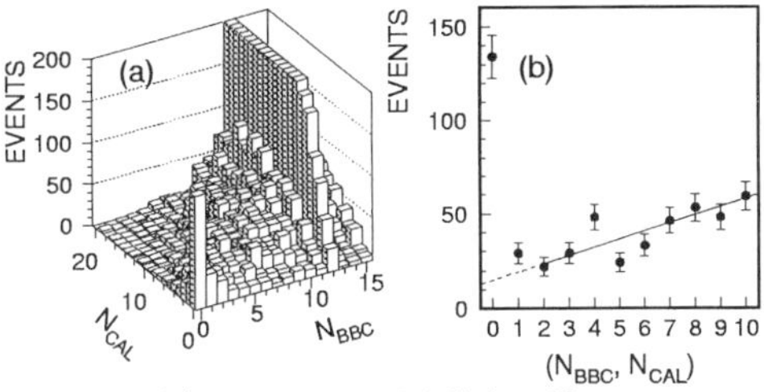

Figure 8. (a) BBC hit multiplicity, N_{BBC}, versus forward calorimeter tower multiplicity, N_{CAL}; (b) multiplicity distribution along the diagonal with $N_{BBC} = N_{CAL}$ in (a).

The measured production ratios of DPE to SD, SD to ND and DPE to ND for dijet events in the kinematic range $0.04 < \xi_{\bar{p}} < 0.095$, $|t| < 1$ GeV^2 for SD and DPE, and estimated $0.05 < \xi_p < 0.035$ for DPE, are presented in the table below:

R_{SD}^{DPE}	$0.40 \pm 0.04 \pm 0.08\,\%$
R_{ND}^{SD}	$0.31 \pm 0.004 \pm 0.03\,\%$
R_{ND}^{DPE}	$[1.3 \pm 0.1 \pm 0.4] \times 10^{-5}$

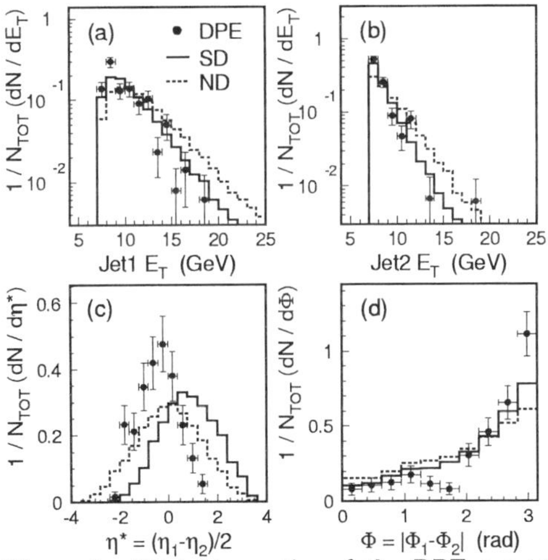

Figure 9. Dijet kinematics of the DPE events (points) compared with single diffractive (non-diffractive) dijet events shown as the solid (dashed) line: (a) Leading jet E_T; (b) next to leading jet E_T; (c) mean η of dijet system; (d) azimuthal angle difference between the two leading jets.

Figure 9 shows the distributions of E_T^{jet1}, E_T^{jet2}, $\eta^* \equiv (\eta^{jet1} + \eta^{jet2})/2$ and $\Phi \equiv \Phi^{jet1} - \Phi^{jet2}$ for the DPE (points), SD (solid histograms) and ND (dashed histograms) dijet events. The E_T^{jet} spectrum of DPE dijets is similar to that of SD jets, and slightly steeper than that of ND dijets. The η^* distribution shows that DPE dijets are boosted toward the negative η ($\bar{p}$ direction), in contrast to the SD dijets, which are boosted toward the p side, and the ND dijets, which are produced symmetrically. This behavior can be explained by the difference between ξ_p and $\xi_{\bar{p}}$ ($\xi_p < \xi_{\bar{p}}$). The DPE dijets are produced more back-to-back than the SD and ND dijets.

Acknowledgments

We thank the Fermilab staff and the technical staffs of the participating institutions for their vital contributions. This work was supported by the U.S. Department of Energy and National Science Foundation; the Italian Istituto Nazionale di Fisica Nucleare; the Ministry of Education,Science and Culture of Japan; the Natural Sciences and Engineering Research Council of Canada; the National Science Council of the Republic of China; and the A. P. Sloan Foundation; the Max Kade Foundation; and the German Landes-Ministry.

References

1. We use rapidity and pseudorapidity, η, interchangeably; $\eta \equiv -\ln(\tan\frac{\theta}{2})$, where θ is the polar angle of a particle with respect to the proton beam direction. The azimuthal angle is denoted by ϕ, and the transverse energy of a jet, E_T^{jet}, is defined as $E^{jet}\sin\theta$.
2. F. Abe *et al.*, *Phys. Rev. Lett.* **79**, 2636 (1997).
3. F. Abe *et al.*, *Phys. Rev. Lett.* **78**, 2698 (1997).
4. T. Affolder *et al.*, submitted to Phys. Rev. Lett.
5. F. Abe *et al.*, *Nucl. Instrum. Methods* A **271**, 387 (1988).
6. P. Bruni, A. Edin and G. Ingelman, Report No. DESY-95, DRAFT, ISSN 0418-9833; http://www3.tsl.uu.se/thep/pompyt/
7. M. Derrick *et al.*, *Z. Phys.* C **68**, 569 (1995); *Phys. Lett.* B **356**, 129 (1995); *Eur. Phys. J.* A **6**, 43 (1999).
8. T. Ahmed *et al.*, *Phys. Lett.* B **348**, 681 (1995); C. Adloff *et al.*, *Z. Phys.* C **76**, 613 (1997).
9. K. Goulianos, *Phys. Lett.* B **358**, 379 (1995); B **363**, 268 (1995).
10. F. Abe *et al.*, *Phys. Rev.* D **50**, 5518 (1994); D **50**, 5535 (1994); D **50**, 5550 (1994).
11. F. Abe *et al.*, *Phys. Lett.* B **45**, 1448 (1992).
12. A. Brandt *et al.*, *Phys. Lett.* B **297**, 417 (1992); B **421**, 395 (1998).
13. M. Glück, E. Reya and A. Vogt, *Z. Phys.* C **53**, 127 (1992).

PROTON INDUCED DRELL-YAN DIMUON PRODUCTION AT 800 GeV

M.A. VASILIEV[j†], T.C. AWES[i], M.E. BEDDO[h], C.N. BROWN[c], J.D. BUSH[a], T.A. CAREY[f], T.H. CHANG[h‡], W.E. COOPER[c], C.A. GAGLIARDI[j], G.T. GARVEY[f], D.F. GEESAMAN[b], E.A. HAWKER[j,f], X.C. HE[d], L.D. ISENHOWER[a], D.M. KAPLAN[e], S.B. KAUFMAN[b], P.N. KIRK[g], D.D. KOETKE[k], G. KYLE[h], D.M. LEE[f], W.M. LEE[d], M.J. LEITCH[f], N. MAKINS[b‡], P.L. McGAUGHEY[f], J.M. MOSS[f], B.A. MUELLER[b], P.M. NORD[k], V. PAPAVASSILIOU[h], B.K. PARK[f], J.C. PENG[f], G. PETITT[d], P.E. REIMER[f,b], M.E. SADLER[a], J. SELDEN[h], W.E. SONDHEIM[f], P.W. STANKUS[i], T.N. THOMPSON[f], R.S. TOWELL[a,f], R.E. TRIBBLE[j], Y.C. WANG[g], Z.F. WANG[g], J.C. WEBB[h], J.L. WILLIS[a], D.K. WISE[a], G.R. YOUNG[i]

(FNAL E866/NuSea Collaboration)

[a] *Abilene Christian University, Abilene, TX 79699*
[b] *Argonne National Laboratory, Argonne, IL 60439*
[c] *Fermi National Accelerator Laboratory, Batavia, IL 60510*
[d] *Georgia State University, Atlanta, GA 30303*
[e] *Illinois Institute of Technology, Chicago, IL 60616*
[f] *Los Alamos National Laboratory, Los Alamos, NM 87545*
[g] *Louisiana State University, Baton Rouge, LA 70803*
[h] *New Mexico State University, Las Cruces, NM, 88003*
[i] *Oak Ridge National Laboratory, Oak Ridge, TN 37831*
[j] *Texas A & M University, College Station, TX 77843*
[k] *Valparaiso University, Valparaiso, IN 46383*

Fermilab E866 has performed a precise measurement of the ratio of Drell-Yan yields in 800 GeV/c pp and pd collisions, leading to determinations of $\bar{d}/\bar{u}$ and $\bar{d}-\bar{u}$ in the proton as functions of x. The results provide valuable information regarding the origins of the $\bar{d}/\bar{u}$ asymmetry and the antiquark sea in the nucleon. The target-mass dependence of Drell-Yan dimuon production induced by an 800 GeV proton beam on targets of Be, Fe and W was also measured. Clear evidence of nuclear shadowing is observed in the Drell-Yan cross section ratios per nucleon at small x_2. The x_1 dependence of the cross section ratios provides a determination of the energy loss of ultra-relativistic quarks as they pass through cold nuclei.

1 Introduction

Fermilab E866 performed a high-precision measurement of the ratio of Drell-Yan yields per nucleon in 800 GeV/c proton-induced reactions on hydrogen and deuterium targets in order to study the light antiquark flavor asymmetry in the nucleon sea. E866 also measured the nuclear dependence of Drell-Yan dimuon production by 800 GeV/c protons on Be, Fe and W targets in order to study fast parton energy loss in nuclear matter and the effect of nuclear shadowing in Drell-Yan scattering.

1.1 Experimental Setup

E866 used a 3-dipole magnet spectrometer [1,2] employed previously in experiments E605, E772 and E789. The spectrometer was modified by the addition of new detectors with larger acceptance at the first tracking station and a new trigger system [3]. When studying the flavor asymmetry in the nucleon sea, an 800 GeV/c proton beam bombarded identical 50.8-cm long target flasks containing liquid hydrogen, liquid deuterium and vacuum that were alternated every few minutes. The cryogenic targets were replaced by solid targets of Be, Fe and W mounted in a rotating wheel for the Drell-Yan nuclear dependence studies. After passing through the target, the remaining beam was intercepted by a copper beam dump. The beam dump was followed by a thick absorber which removed hadrons produced in the target and the dump, ensuring that only muons traversed the spectrometer's detectors, consisting of four tracking stations and a momentum analyzing magnet. The trigger required two triple hodoscope coincidences having the topology of a muon pair from the target. Random muon pairs, the dominant background source, were determined from a detailed construction of random pairs using single-muon events.

2 Flavor Asymmetry in the Nucleon Sea

No known symmetry requires equality of the $\bar{d}$ and $\bar{u}$ distributions in the proton. However, until recently it had been assumed that $\bar{d}(x) \approx \bar{u}(x)$ (where x is the fraction of the proton's momentum carried by the antiquark) based on the assumption that most of the antiquark sea in the nucleon originates from gluon splitting into $q\bar{q}$ pairs. But several experiments have now given clear evidence that $\bar{d} \neq \bar{u}$ [4,5,1,6]. To date, Fermilab E866 has provided the most precise and detailed information regarding the $\bar{d}/\bar{u}$ asymmetry in the nucleon sea [1], based on measurements of the relative Drell-Yan cross sections in pp and pd collisions. Over 370,000 Drell-Yan events were recorded, using three different spectrometer settings which focused low, intermediate and high mass muon pairs and provided good acceptance from below the J/ψ to above 15 GeV. Previously results have been published [1] were based on the high-mass setting alone, $\approx$40% of the data. Here we present the results from the full data set.

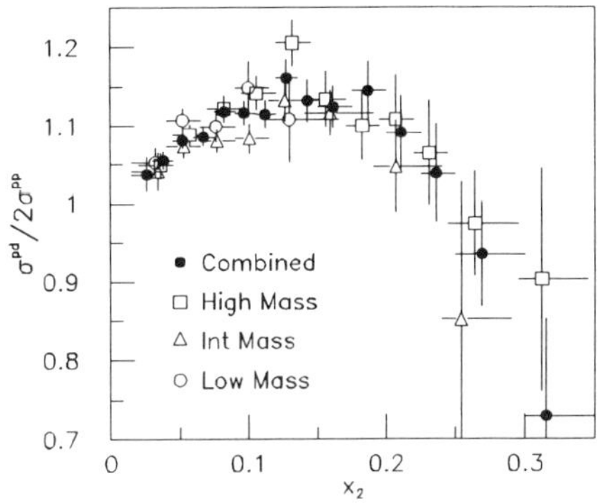

Figure 1: Preliminary ratios $\sigma^{pd}/2\sigma^{pp}$ of Drell-Yan cross sections *vs.* x_2 for the various data sets, as well as the combined result.

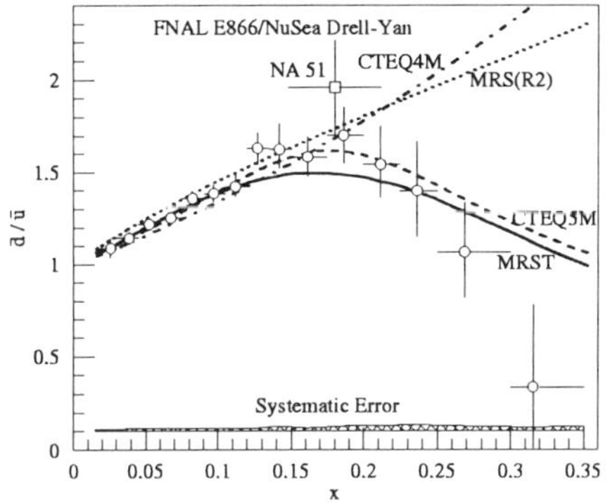

Figure 2: The ratio $\bar{d}/\bar{u}$ in the proton as a function of x, together with the predictions of several recent parton distributions.

2.1 Data Analysis

Figure 1 shows the measured ratios of the Drell-Yan cross section per nucleon as a function of x_2, the Bjorken-x of the target parton. (The Bjorken-x of the beam parton is x_1.) These ratios may be used to determine $\bar{d}/\bar{u}$ as a function of x. The acceptance of the spectrometer was largest for $x_1 > x_2$, where the Drell-Yan cross section is dominated by the annihilation of a beam quark with a target antiquark. This fact, coupled with the assumptions of charge symmetry and negligible shadowing in the deuteron, yields a simple approximation for the Drell-Yan cross section ratio:

$$\left.\frac{\sigma^{pd}}{2\sigma^{pp}}\right|_{x_1 \gg x_2} \approx \frac{1}{2} \frac{\left(1+\frac{d_1}{4u_1}\right)}{\left(1+\frac{d_1}{4u_1}\frac{\bar{d}_2}{\bar{u}_2}\right)} \left(1+\frac{\bar{d}_2}{\bar{u}_2}\right). \tag{1}$$

The subscripts 1 and 2 denote parton distributions as functions of x_1 and x_2, respectively.

Some of the data, especially at higher x_2, do not satisfy the $x_1 \gg x_2$ criterion of Eq. 1. Consequently, $\bar{d}/\bar{u}$ was extracted iteratively by calculating the leading order Drell-Yan cross section ratio using the valence, heavy quark and $\bar{d}+\bar{u}$ values from the MRST [7] parton distribution function (PDF) as input, and adjusting $\bar{d}/\bar{u}$ until the calculated cross section ratio agreed with the measured value. Leading-order and next-to-leading-order calculations of the cross section ratio give similar results, and negligible differences are found when using other recent PDF fits in place of MRST. The extracted $\bar{d}/\bar{u}$ ratios at $Q = 6.4$ GeV/c are shown in Fig. 2 along with the predictions of various

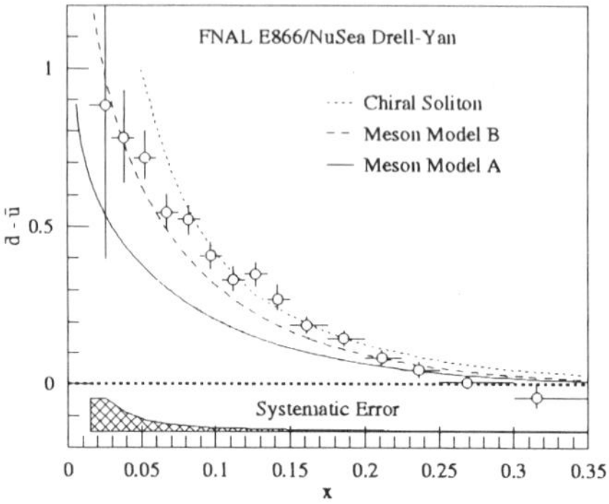

Figure 3: Comparison of the E866 $\bar{d}-\bar{u}$ results with the predictions of various models.

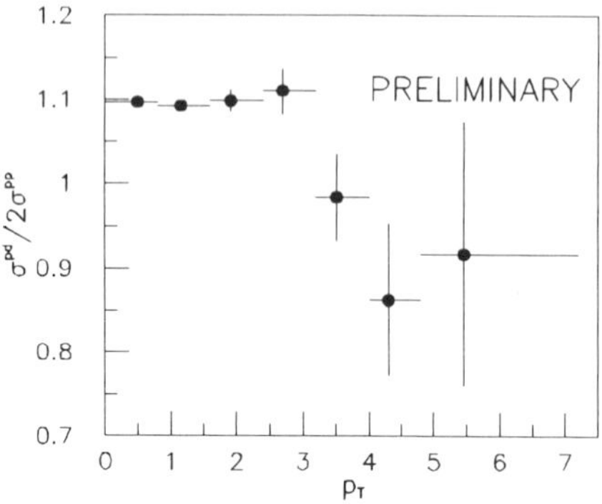

Figure 4: Preliminary ratios $\sigma^{pd}/2\sigma^{pp}$ of Drell-Yan cross sections *vs.* p_T.

PDFs. The MRST and CTEQ5M[8] PDF fits, both of which included the initial E866 results as important new inputs, provide a good description of the results, whereas previous state-of-the-art PDF fits fail to describe $\bar{d}/\bar{u}$ above $x \approx 0.18$. This demonstrates the significant impact that the E866 results have already had on global PDF fits.

The $\bar{d}/\bar{u}$ ratios measured in E866, together with the MRST values for $\bar{d}+\bar{u}$, have been used to obtain $\bar{d}-\bar{u}$ as shown in Fig. 3. As a flavor non-singlet quantity, $\bar{d}(x)-\bar{u}(x)$ has the property that its integral is Q^2-independent [9]. Furthermore, it is a direct measure of the contribution to the nucleon sea from non-perturbative processes, since perturbative QCD alone cannot generate a significant $\bar{d}, \bar{u}$ difference [10]. Integrating $\bar{d}-\bar{u}$ from E866, one finds $\int_{0.015}^{0.35}(\bar{d}-\bar{u})dx = 0.0818 \pm 0.0082 \pm 0.0049$ at $Q = 6.4$ GeV. If we use the average of MRST and CTEQ5M to estimate the contributions to the integral from the unmeasured x regions, we find $\int_0^1(\bar{d}-\bar{u})dx = 0.118 \pm 0.011$, in agreement with the NMC[4] and HERMES[6] results, but considerably more precise.

2.2 Comparison with Theoretical Models

The x dependence of $\bar{d}-\bar{u}$ provides important constraints for theoretical models [11]. At the beginning of the eighties the significant role played by virtual pions in the antiquark distributions of the nucleon sea was discussed [12,13] and it was pointed out [14] that such pions will lead to an enhancement of $\bar{d}$ relative to $\bar{u}$ via the (non-perturbative) "Sullivan process." Following the publication of the NMC result, many papers have treated virtual mesons as the origin of the asymmetry in the up, down sea of the nucleon [15]. Figure 3 compares $\bar{d}(x)-\bar{u}(x)$ with a virtual-pion model calculation [11] following the procedure described in Ref. 16. The curve labeled "Meson Model A" in Fig. 3 uses dipole form factors with $\Lambda = 1.0$ GeV for the πNN and $\pi N\Delta$ vertices, and underpredicts $\bar{d} - \bar{u}$. In contrast, much better agreement with the E866 results is obtained by reducing Λ for the $\pi N\Delta$ form factor to 0.8 GeV, as shown by the curve labeled "Meson Model B" in Fig. 3.

Several other models have also been proposed to explain the sizable $\bar{d}/\bar{u}$ asymmetry in the nucleon. In chiral field theory, the relevant degrees of freedom are constituent quarks, gluons, and Goldstone bosons. In this model, a portion of the sea comes from the couplings of Goldstone bosons to the constituent quarks, such as $u \to d\pi^+$ and $d \to u\pi^-$. The excess of $\bar{d}$ over $\bar{u}$ is then due to the additional up valence quark in the proton. However, the predicted $\bar{d}(x) - \bar{u}(x)$ from recent chiral field calculations [17] is even softer than that for "Meson Model A" in Fig. 3, suggesting that correlations between the chiral constituents must be taken into account. The chiral quark-soliton model has been used to predict $\bar{d}(x) - \bar{u}(x)$ in the large-N_c limit [18]. Figure 3 shows that this model reproduces the observed $\bar{d} - \bar{u}$ values well for $x > 0.08$, but it overestimates the asymmetry at small x. Finally, instanton models have also been proposed [19]. Their basic prediction is that the $\bar{d}/\bar{u}$ asymmetry should be primarily a large-p_T effect, but Fig. 4 shows that this is not the case. The change in cross section ratios at high p_T seen in Fig. 4 may indicate a shift in the reaction mechanism from $q\bar{q}$ annihilation to qg scattering as the major contributor to the Drell-Yan dimuon production at high p_T.

It will also be instructive to compare model predictions of $\bar{d}(x)/\bar{u}(x)$ with the E866 results. However, unlike $\bar{d} - \bar{u}$, calculations for $\bar{d}/\bar{u}$ must include the contribution to the nucleon sea of the perturbative processes $g \to u\bar{u}, d\bar{d}$, which generate a symmetric sea, in addition to the non-perturbative processes which lead to the $\bar{d}/\bar{u}$ asymmetry. Hence, the E866 results will constrain future attempts to combine the effects of perturbative and non-perturbative QCD in the structure of the nucleon.

3 Nuclear dependence of Drell-Yan production

The Drell-Yan process can be used to study the interactions of fast partons penetrating through cold nuclei. Only initial state interactions are important in Drell-Yan production since the dimuon in the final state does not interact strongly with the partons in the medium. This makes Drell-Yan scattering an ideal tool to study energy loss of fast quarks in nuclear matter, a subject of considerable theoretical interest[20,21,22,23], by comparing the observed yields from a range of nuclear targets. Drell-Yan scattering is closely related to deep-inelastic scattering (DIS) of leptons, but unlike DIS it can be used specifically

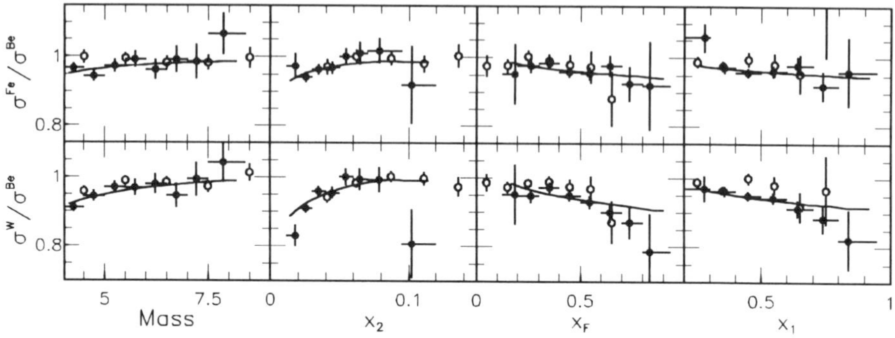

Figure 5: Ratios of the measured cross section per nucleon for Drell-Yan events from E866 are shown as solid circles and data from E772 as open circles. The solid curves are shadowing predictions as discussed in the text.

to probe antiquark contributions in target parton distributions. When DIS on nuclei occurs at $x < 0.08$ the cross section per nucleon decreases with increasing nuclear number A due to shadowing [24,25]. Shadowing should also occur in Drell-Yan dimuon production at small x_2 and should be particularly apparent at $x < 0.06$ where DIS on nuclei, like Drell-Yan, is dominated by scattering off sea quarks.

Over 130,000 Drell-Yan muon pairs, produced in 800 GeV proton-nucleus collisions on Be, Fe, and W targets, were observed with dimuon mass in the range 4.0 GeV $< M <$ 8.4 GeV, target-parton momentum fraction $0.01 < x_2 < 0.125$, x_F in the range $0.2 < x_F < 0.94$ and p_T values up to 4 GeV/c. The extended kinematic coverage of E866, reaching larger values of x_F ($\approx x_1 - x_2$), and smaller values of x_2 than the previous experiment, Fermilab E772 [26], significantly increases its sensitivity to parton energy loss and shadowing.

3.1 Shadowing

Ratios of the cross section per nucleon for Fe to Be and W to Be versus dimuon mass, x_2, x_F and x_1 are shown in Fig. 5, along with similar results from E772 for Fe to C and W to C. The figure shows very good agreement between the experiments for the cross section ratios versus x_2. The agreement versus other variables is also reasonable, given the much smaller $\langle x_2 \rangle$ and, hence, increased shadowing of the present data. (The A dependence observed here and in E772 implies the change in the cross-section ratios due to the choice of Be vs. C is small compared to the effect of the difference in $\langle x_2 \rangle$.) The reduction in the cross section per nucleon on the heavy targets, characteristic of shadowing,

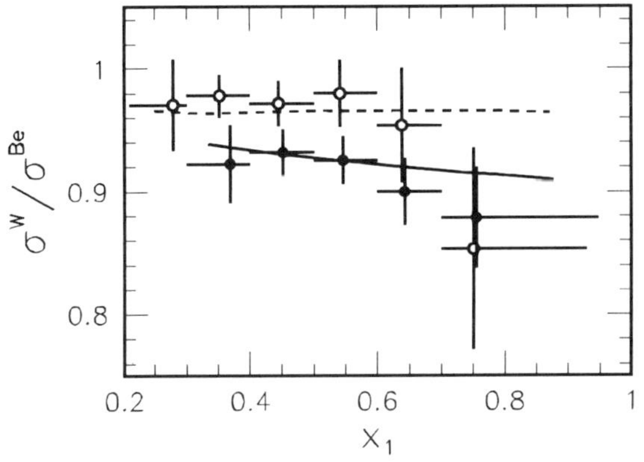

Figure 6: Ratios of the measured Drell-Yan cross section per nucleon for W/Be versus x_1. The solid (open) circles are those events with x_2 less than (greater than) 0.036, and the solid (dashed) curve is the corresponding shadowing prediction.

is apparent at small x_2. A similar reduction, part of which could be related to incident-parton energy loss, is apparent at large x_F and x_1. However, it is important to recognize that the spectrometer acceptance coupled to the intrinsic Drell-Yan cross section leads to a strong anti-correlation between x_2 and x_F for the observed events. Therefore, the events that show the cross-section reduction at large x_F and x_1 are in general the same events that appear in the shadowing region.

To identify the contributions from shadowing, Fig. 5 also shows the predicted cross-section ratios, integrated over the hidden variables, from leading-order Drell-Yan calculations using the code EKS98 [27] together with the MRST parton distribution functions [7]. Similar results are obtained using CTEQ5M or CTEQ5L parton distributions [8]. EKS98 describes the ratios $f_A(x, Q^2)/f_p(x, Q^2)$ of the various quark flavors in nucleus A, compared to those in the proton. It has been tuned to fit the shadowing observed in DIS [28] and E772 while conserving baryon number and momentum. EKS98 provides an unbiased way to separate the effects of shadowing and energy loss in the present data because it uses a single shadowing function to describe the nuclear dependence of all sea quark distributions at its initial scale ${Q_0}^2 = 2.25$ GeV^2 and only DIS results are used to constrain that function for $x \leq 0.1$.

The shape and magnitude of the ratios versus x_2 are well reproduced by the shadowing predictions. Most of the A dependence observed in the ratios versus mass, x_F and x_1 is also explained by shadowing at small x_2. As a further test of the shadowing parameterization, Fig. 6 shows the W/Be ratios vs. x_1, with the events separated into two subsets at the median x_2 value. This eliminates the correlation in acceptance between x_1 and x_2 for the large x_2 subset of the data, indicated by the flat shadowing prediction, and substantially reduces the correlation for the small x_2 subset, indicated by the reduced slope of the shadowing prediction compared to Fig. 5. The cross-section ratios differ by several percent when events at the same x_1 are observed in the two subsets of the data. EKS98 describes these differences well with no free parameters. Similar results are obtained when examining the mass or x_F dependence. This is the first experimental demonstration that the shadowing observed in Drell-Yan and DIS is quantitatively similar.

3.2 Parton Energy Loss Limits

Several groups have studied energy loss of partons in nuclei. Their results can be expressed in terms of the average change in the incident parton momentum fraction, Δx_1, as a function of target nucleus. Gavin and Milana[20] analyzed the E772 Drell-Yan data for energy loss based on the parameterization $\Delta x_1 = -\kappa_1\, x_1\, A^{1/3}$ (GM). From a comparison to the x_F dependence seen by E772 and neglecting shadowing, they concluded that the fractional energy loss of quarks passing through nuclei is $\approx 0.4\%/\text{fm}$. Brodsky and Hoyer[21] used an analogy to the photon bremsstrahlung process to obtain a form for gluon radiation leading to an initial parton energy loss $\Delta x_1 \approx -\frac{\kappa_2}{s} A^{1/3}$ (BH). They also noted that elastic scattering should make a similar contribution to the energy loss and concluded that energetic partons should lose < 0.5 GeV/fm in nuclei. The formulation developed by Brodsky and Hoyer was extended by Baier *et al.*[22,23]. They found that the energy loss of sufficiently energetic partons depends on a characteristic length and the broadening of the squared transverse momentum of the parton. For finite nuclei, both factors vary as $A^{1/3}$, so Baier *et al.* predict $\Delta x_1 \approx -\frac{\kappa_3}{s} A^{2/3}$ (B).

We have obtained empirical values for the κ's by performing simultaneous fits to the Fe/Be and W/Be Drell-Yan cross section ratios versus x_1 in Fig. 7. The solid curves are the best fit using the energy loss form (GM), and the dashed curves show the 1σ upper limits. The dotted curves show the 1σ upper limits using the energy loss forms (BH) and (B), which produce essentially identical results. When assuming the form (GM), we find that the observed fractional energy loss of the incident quarks is $< 0.14\%/\text{fm}$. For the energy loss forms (BH) and (B), the best fits imply essentially zero energy loss. We find for 1σ upper limits that the incident quarks lose energy at a constant rate of < 0.44 GeV/fm (BH) or within the model of Baier *et al* $\Delta E < 0.046\ \text{GeV/fm}^2 \times L^2$, where L is the quark propagation length through the nucleus. This is

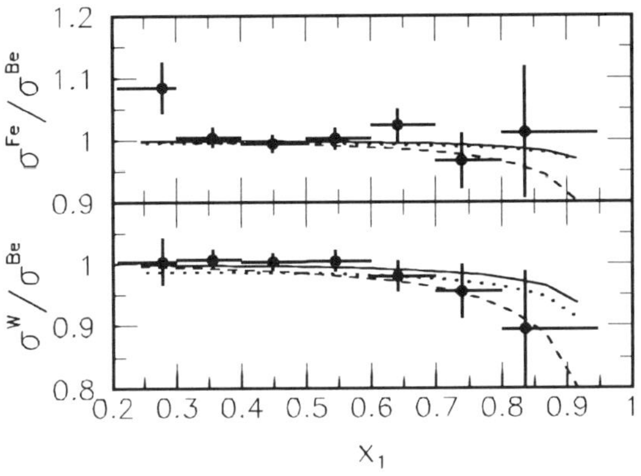

Figure 7: Ratios of the cross section per nucleon versus x_1 for Fe/Be (upper panel) and W/Be (lower panel), corrected for shadowing. The solid curves are the best fit using the energy loss form (GM), and the dashed curves show the 1σ upper limits. The dotted curves show the 1σ upper limits using the energy loss form (BH). The 1σ upper limit curves using the energy loss form (BH) are essentially identical to those using form (B).

very close to the lower value given by Baier *et al.* for cold nuclear matter.

4 Summary

E866 has provided high-precision determinations of $\bar{d}/\bar{u}$ and $\bar{d} - \bar{u}$ over the range $0.015 \leq x \leq 0.35$. The values of $\bar{d}/\bar{u}$ and $\bar{d} - \bar{u}$ for $x > 0.18$ are much smaller than predicted by pre-existing PDF parameterizations, although new global fits that include the E866 results as important additional inputs are able to reproduce them quite well. These results provide strong constraints for models that attempt to describe the antiquark sea of the nucleon. The nuclear dependence of Drell-Yan production has also been investigated. Nuclear shadowing is found to be important in the small x_2 domain, where the shadowing observed in Drell-Yan has been demonstrated to be quantitatively similar to that in DIS. The cross-section ratios versus x_1 provide direct upper limits on the energy loss of the incoming parton as it traverses a cold nucleus that are tighter than previous constraints. Shadowing and initial-state energy loss are processes that occur in both Drell-Yan production and J/ψ formation. Hence, these results should also further the understanding of J/ψ production, which is required if it is to be used as a signal for the quark-gluon plasma in relativistic heavy-ion collisions.

References

[†] On Leave from Kurchatov Institute, Moscow 123182, Russia.
[‡] Present address: University of Illinois, Urbana, IL 61801.

1. E.A. Hawker *et al.*, Phys. Rev. Lett. 80 3715 (1998).
2. G. Moreno *et al.*, Phys. Rev. D 43 2815 (1991).
3. C.A. Gagliardi *et al.*, Nucl. Instrum. Methods **A418**, 322 (1998).
4. M. Arneodo *et al.*, Phys. Rev. D 50 R1 (1994).
5. A. Baldit *et al.*, Phys. Lett. B 332 244 (1994).
6. K. Ackerstaff *et al.*, Phys. Rev. Lett. 81 5519 (1998).
7. A.D. Martin *et al.*, Eur. Phys. J. C 4 463 (1998).
8. H.L. Lai *et al.*, hep-ph/9903282.
9. A.D. Martin, W.J. Stirling, R.G. Roberts, Phys. Lett. B 252 653 (1990).
10. D.A. Ross and C.T. Sachrajda, Nucl. Phys. B 149 497 (1979).
11. J.C. Peng *et al.*, Phys. Rev. D 58 092004 (1998).
12. G.G. Arakelyan, K.G. Boreskov, A.B. Kaidalov, Sov. J. Nucl. Phys. 33 247 (1981)
13. G.G. Arakelyan and K.G. Boreskov, Sov.J. Nucl. Phys. 41 267 (1985)
14. A.W. Thomas, Phys. Lett. B 126 97 (1983).
15. S. Kumano, Phys. Rep. 303 183 (1998), and references therein.
16. S. Kumano, Phys. Rev. D 43 3067 (1991); 43 59 (1991); S. Kumano and J.T. Londergan, *ibid.* 44 717 (1991).
17. A. Szczurek, A. Buchmans and A. Faessler, Jour. Phys. G: Nucl. Part. Phys. 22 1741 (1996).
18. P.V. Pobylitsa *et al.*, Phys. Rev. D 59 034024 (1999).
19. A.E. Dorokhov and N.I. Kochelev, Phys. Lett. B **259** , 335 (1991); **304** , 167 (1993).
20. S. Gavin and J. Milana, Phys. Rev. Lett. **68** 1834 (1992).
21. S.J. Brodsky and P. Hoyer, Phys. Lett. B **298** 165 (1993).
22. R. Baier *et al.*, Nucl. Phys. B **484**, 265 (1997).
23. R. Baier *et al.*, hep-ph/9804212.
24. D.F. Geesaman, K. Saito and A.W. Thomas, Ann. Rev. Nucl. Part. Sci. **45** 337 (1995).
25. M. Arneodo *et al.*, Nucl. Phys. B **441** 3 (1995).
26. D.M. Alde *et al.*, Phys. Rev. Lett. **64**, 2479 (1990).
27. K.J. Eskola, V.J. Kolhinen and P.V. Ruuskanen, Nucl. Phys. B **535**, 351 (1998); K.J. Eskola, V.J. Kolhinen and C.A. Salgado, hep-ph/9807297.
28. M. Arneodo *et al.*, Nucl. Phys. B **481** 3 (1996); Nucl. Phys. B **481** 23 (1996).

Interaction of Two Reggeized Gluons: 2nd Order Corrections and the Bootstrap

G.P. Vacca
II Institut für Theoretische Physik, Universität Hamburg
E-mail: vacca@bo.infn.it

Our aim is to show how the reggeization of the gluon, encoded in the bootstrap property of the BFKL kernel, permits to calculate the interaction kernel in the octet colour channel in the forward and non forward direction, at least for the quark contribution, starting from the known gluon trajectory up to 2nd order. To this end an ansatz to solve the bootstrap equation is used. The obtained result, for what regard the quark contribution, has been verified by direct Feynman diagram calculations by Fadin et al..

1 Introduction

The BFKL equation and its impact in small x physics have been widely discussed subjects in the last years and recently the 2nd order corrections were calculated for the forward pomeron equation [1,2]. Generalization of this result in the non forward direction would, of course, be really interesting, not only in the singlet colour channel of the pomeron. For the other physically important case, that of the odderon, the 2nd order corrections to the non-forward interaction of two reggeized gluons in the octet colour state is needed (as well as also the form of the gluon triple interaction term).

This interaction is severely restricted by the so-called bootstrap relation which guarantees that production amplitudes with the gluon quantum number in their t channels used for the construction of the absorptive part are indeed given only by a single reggeized gluon exchange and do not contain admixture from two or more reggeized gluon exchanges [3,4]. The bootstrap relation is known to be satisfied in the lowest order in the coupling constant. The form this relation takes in the 2nd order was recently discussed by Fadin et al. [5]

The bootstrap can be expressed as a relation which involve the reggeized gluon trajectory, the interaction between two reggeized gluons and their wave function. We show here [6] how, using the bootstrap in a form stronger than in [5] and an ansatz [7] to solve it, it is possible to obtain the full second order corrections, at least due to the quark contribution, for the non-forward gluon interaction in the octet channel necessary for the odderon equation. It can also be shown that we also automatically find the corrections for the running of the coupling for the non-forward pomeron equation [6].

2 The Bootstrap relation

We find convenient to work with an operator formalism to describe the two gluon equation and the bootstrap relation. Let the two gluons with momenta $q_{1,2}$ be described by a wave function $\Psi(q_1, q_2)$. The total momentum $q = q_1 + q_2$ will always be conserved, so that in future $q_2 = q - q_1$ and the dependence on q and thus q_2 will be suppressed. The scalar product of two wave functions will be taken as $\langle\Psi_1|\Psi_2\rangle \equiv \int d^{D-2}q_1\, \Psi_1^*(q_1)\Psi_2(q_1)$, where $D = 4 + 2\epsilon$, $\epsilon \to 0$, is the dimension used for the infrared regularization.

In this metric, as a function of the angular momentum j and up to order α_s^3, the absorptive part of the scattering amplitude coming from the two-reggeized-gluon exchange is given by $\mathcal{A}_j = \langle\Phi_p|G(E)|\Phi_t\rangle$.

Here the functions $\Phi_{p,t}$ are the so called impact factors which represent the coupling of the external particles (t from the target and p from the projectile) to the two exchanged gluons and are of order $\alpha_s = g^2/4\pi$.

The Green function $G(E)$ with $E = 1 - j$ is defined as $G(E) = (H - E)^{-1}$ where H is the two gluon (Hermithean) Hamiltonian. Its explicit expression in the momentum representation is

$$\langle q_1|H|q_1'\rangle = -\delta^{D-2}(q_1 - q_1')(\omega(q_1) + \omega(q_2)) - V^{(R)}(q_1, q_1') \tag{1}$$

where $1 + \omega(q_{1(2)})$ is the Regge trajectory of the first (second) gluon and $V^{(R)}$ represents the gluonic interaction for the t-channel with the colour quantum number R.

In any colour channel the Green function $G(E)$ can be represented via the orthonormalized solutions of the homogeneous Schrödinger equation $H\Psi_n = E_n\Psi_n$. Therefore, in terms of the eigenfunctions Ψ_n the absorptive part $\mathcal{A}_j$ is

$$\mathcal{A}_j = \sum_n \frac{\langle\Phi_p|\Psi_n\rangle\langle\Psi_n|\Phi_t\rangle}{E_n - E}. \tag{2}$$

Let us consider the case with a colour quantum number of the gluon ($R = g$) in the t channel. The BFKL equation is based on the idea that physical amplitudes in this channel have the asymptotical behaviour corresponding to an exchange of a single reggeized gluon so that no contribution should come from the exchange of two or more reggeized gluons (in the leading (LO) and next-to-leading (NLO) orders). This means that the absorptive part, as a function j, should have a simple pole at $j = 1 + \omega(q)$ and no other singularities. We can therefore state the following properties valid in the **LO and NLO**:

1. the spectrum of H in the considered t channel should contain an eigenvalue $E_0 = -\omega(q)$ with the corresponding eigenfunction Ψ_0

$$(H + \omega(q))\Psi_0 = 0 \tag{3}$$

2. the expression in (2) take contributions only from this eigenfunctions. This means that both impact factors $\Phi_{p,t}$ should be orthogonal to all other eigenfunctions Ψ_n, $n > 0$. Due to the completeness of the whole set it means that both Φ_p and Φ_t should coincide with Ψ_0, up to a normalization factor, which will contain all the dependencies in the other variables (factorization).

3. there is a relation for the impact factor. In fact one considers the Mellin transform for the amplitude in (2) where only the contribution for $n = 0$ is present in the sum. Taking into account the proper signature factor and comparing to the standard form, where the coupling Γ's (particle-particle-reggeon vertices) are factorized from the contribution of the reggeized gluon, one obtains the "third bootstrap condition"

$$\Gamma(q) = q \left(\frac{q^2}{s_0}\right)^{\omega(q)/2} \frac{\langle \Phi | \Psi_0 \rangle}{\sqrt{|\sin \pi\omega(q)|}}. \tag{4}$$

Considering a perturbative approach one may write $\Psi_0 = \chi + \Psi_0^{(1)}$, $\omega(q) = \omega^{(1)}(q) + \omega^{(2)}(q)$ and $H = H^{(1)} + H^{(2)}$ and obtain at the second order for the bootstrap eq. (3) the form

$$(H^{(1)} + \omega^{(1)}(q))\Psi_0^{(1)} = -(H^{(2)} + \omega^{(2)}(q))\chi \tag{5}$$

This general relation may be projected onto the state χ (the first order wave function) to obtain the relation

$$\langle \chi | H^{(2)} + \omega^{(2)}(q) | \chi \rangle = 0 \tag{6}$$

which has been previously obtained [5] from different arguments together with the second order form of (4). We shall denote the relation (6) as a weak bootstrap condition.

3 Bootstrap at work

The main idea consists in taking one particular solution of the bootstrap relation, which describe all the inter-dependent quantities in terms of a single function. Explicitely in the LO BFKL equation every transverse momentum squares are substituted by a function of them which we call η. Explicitely taking:

$$\omega(q) = -\int d^2q_1 \frac{\eta(q)}{\eta(q_1)\eta(q_2)} \tag{7}$$

and

$$V^{(g)}(q_1, q_1') = \frac{1}{\sqrt{\eta(q_1)\eta(q_2)\eta(q_1')\eta(q_2')}} \left(\frac{\eta(q_1)\eta(q_2') + \eta(q_2)\eta(q_1')}{\eta(k)} - \eta(q) \right) \tag{8}$$

then Eq.(3) remains valid with a solution

$$\Psi_0 = 1/\sqrt{\eta(q_1)\eta(q_2)} \tag{9}$$

Therefore the strategy to determine the interaction potential for the two gluons in the octet state will be the following: *known gluon trajectory at NLO* $\rightarrow$ *determination of* η $\rightarrow$ *construction of the interaction*. In the LO case one has $\eta^{(0)}(q) = q^2/a$, where $a = \frac{g^2 N}{2(2\pi)^{D-1}}$, and in general we shall assume at the NLO the form $\eta(q) = \eta^{(0)}(q)(1+\xi(q))$.

We briefly describe the calculations leading to the full interaction at NLO for the octet state [6]. We first consider the expressions for the trajectory:

$$\omega^{(1)}(q) = -\bar{g}^2 q^{2\epsilon}\frac{\Gamma^2(\epsilon)}{\Gamma(2\epsilon)} \quad , \quad \omega^{(2)}(q) = -\bar{g}^4 q^{4\epsilon} F(\epsilon) \tag{10}$$

where

$$\bar{g}^2 = \frac{Ng^2\Gamma(1-\epsilon)}{(4\pi)^{2+\epsilon}} \quad , \quad F(\epsilon) = \frac{A}{\epsilon^2} - \frac{B}{\epsilon} - C - \frac{\pi^2}{3}A \tag{11}$$

The coefficients A, B, C contains a part proportional to N_f (the number of quark flavours) which therefore isolate the quark contribution to the reggeized gluon trajectory $\omega_Q^{(2)}(q)$, proportional on $F_Q(\epsilon)$. Thus the quark contribution to η or to the ξ functions is fixed looking only at $\omega_Q^{(2)}(q)$. Let us write this ξ_Q. Using an integral representation for $\omega_Q^{(2)}(q)$ one can write

$$2bq^2\int\frac{d^{D-2}q_1}{q_1^2q_2^2}(q^{2\epsilon}-q_1^{2\epsilon}-q_2^{2\epsilon}) = -aq^2\int\frac{d^{D-2}q_1}{q_1^2q_2^2}[\xi(q)-\xi(q_1)-\xi(q_2)] \tag{12}$$

where the $\omega_Q^{(2)}(q)$ on the LHS is matched to the form (7) which solves the bootstrap equation and expanded to the second order; b is a function of the parameter ϵ with a pole in zero.

The last relation is solved for $\xi_Q(q) = -2bq^{2\epsilon}/a$ and substituting this in the second order part of the potential (8) one easily obtains, for the quark contribution,

$$V_Q^{(g,2)}(q_1,q_1') = -b\frac{1}{\sqrt{q_1^2q_2^2{q_1'}^2{q_2'}^2}}\Big[\frac{q_1^2{q_2'}^2}{k^2}(q_1^{2\epsilon}+{q_2'}^{2\epsilon}-q_2^{2\epsilon}-{q_1'}^{2\epsilon}-2k^{2\epsilon})+$$

$$\frac{q_2^2{q_1'}^2}{k^2}(q_2^{2\epsilon}+{q_1'}^{2\epsilon}-q_1^{2\epsilon}-{q_2'}^{2\epsilon}-2k^{2\epsilon})-q^2(2q^{2\epsilon}-q_1^{2\epsilon}-q_2^{2\epsilon}-{q_1'}^{2\epsilon}-{q_2'}^{2\epsilon})\Big] \tag{13}$$

This expression (multiplied by $\sqrt{q_1^2 q_2^2 {q'_1}^2 {q'_2}^2}$ to match the integration measure) coincides with the one found later in[8] by direct methods (see Eq. (47) there).

We try to use this procedure to calculate the total gluon interaction, with also the gluonic contribution. From the relations $\xi(q)/\xi_Q(q) = \omega^{(2)}(q)/\omega_Q^{(2)}(q) = F(\epsilon)/F_Q(\epsilon)$ the answer is

$$V^{(g,2)}(q_1, q'_1) = V_Q^{(g,2)}(q_1, q'_1)F(\epsilon)/F_Q(\epsilon) \tag{14}$$

Since the full kernel K, which solves the bootstrap relation, can be written in an explicitely infrared finite way it is convenient to write the expressions using the renormalized coupling constant g_μ. In the dimensional regularization in the $\overline{\mathrm{MS}}$ scheme one has $g^2 = g_\mu^2 \mu^{-2\epsilon}\left(1 + A\bar{g}_\mu^2/\epsilon\right)$ and $\bar{g}_\mu^2 = g_\mu^2 N\Gamma(1-\epsilon)/(4\pi)^{2+\epsilon}$. This means that we can also renormalize the η or the ξ functions. One has a finite $\xi_\mu(q) = \xi(q) - A\bar{g}_\mu^2/\epsilon$ because the poles in $1/\epsilon$ cancel. Therefore we are allowed to put $\epsilon = 0$ in the infrared finite expression for the non forward kernel in the octet channel. The simple form we present here refers to a metric modified in accord to $\Psi(q_1) \to \Psi(q_1)/\sqrt{\eta(q_1)\eta(q_2)}$. For the second order contribution (as usual a superscript indicates the order) we obtain

$$K^{(2)}\Psi = \frac{3\alpha_s}{4\pi}\Big[BK^{(1)}\Psi + 6A\alpha_s \int \frac{d^2q'_1 q_1^2}{(2\pi)^2 k^2}\left(\frac{\Psi(q'_1)}{{q'_1}^2} - \frac{\Psi(q_1)}{{q'_1}^2 + k^2}\right)\ln\frac{q_1^2\mu^2}{{q'_1}^2 k^2}$$
$$+(1 \leftrightarrow 2) - 6A\alpha_s \int \frac{d^2q'_1 q^2 \Psi(q'_1)}{(2\pi)^2 {q'_1}^2 {q'_2}^2}\ln\frac{q^2\mu^2}{{q'_1}^2 {q'_2}^2}\Big]. \tag{15}$$

Acknowledgments

The results shown have been obtained in collaboration with M. Braun[6]. The author is grateful for the support from the Alexander von Humboldt Stiftung.

References

1. V.S.Fadin, L.N.Lipatov, *Phys. Lett.* B **429**, 127 (1998).
2. G.Camici, M.Ciafaloni, *Phys. Lett.* B **395**, 118 (1997).
3. L.N.Lipatov, *Yad. Fiz.* **23**, 642 (1976).
4. J. Bartels, *Nucl. Phys.* B **151**, 293 (1979).
5. V.S.Fadin, R.Fiore, *Phys. Lett.* B **440**, 359 (1998).
6. M.A.Braun and G.P.Vacca, *Phys. Lett.* B **454**, 319 (1999).
7. M.A.Braun, *Phys. Lett.* B **345**, 155 (1995); *Phys. Lett.* B **348**, 190 (1995).
8. V.S.Fadin, R.Fiore, A.Papa *Phys. Rev.* D **60**, 074025 (1999).

Victor T. Kim

The Next-to-Leading BFKL Pomeron with Optimal Renormalization [a]

Victor T. Kim[‡&], Lev N. Lipatov[‡]
and
Grigorii B. Pivovarov[§&]

[‡] *: St.Petersburg Nuclear Physics Institute, 188350 Gatchina, Russia*
[§] *: Institute for Nuclear Research, 117312 Moscow, Russia*
[&] *: International Institute of Theoretical and Applied Physics, Iowa State University, Ames, IA 50011, USA*

The next-to-leading order (NLO) corrections to the BFKL equation in the BLM optimal scale setting are briefly discussed. A striking feature of the BLM approach is rather weak Q^2-dependence of the Pomeron intercept, which might indicate an approximate conformal symmetry of the equation. An application of the NLO BFKL resummation for the virtual gamma-gamma total cross section shows a good agreement with recent L3 data at CERN LEP2 energies.

The Balitsky-Fadin-Kuraev-Lipatov (BFKL) [1,2] resummation of energy logarithms is anticipated to be an important tool for exploring the high-energy limit of QCD. Namely, the highest eigenvalue, ω^{max}, of the BFKL equation [1] is related to the intercept of the Pomeron which in turn governs the high-energy asymptotics of the cross sections: $\sigma \sim s^{\alpha_{I\!P}-1} = s^{\omega^{max}}$. The BFKL Pomeron intercept in the leading order (LO) turns out to be rather large: $\alpha_{I\!P} - 1 = \omega_{LO}^{max} = 12 \ln 2\,(\alpha_S/\pi) \simeq 0.55$ for $\alpha_S = 0.2$; hence, it is very important to know the next-to-leading order (NLO) corrections. In addition, the LO BFKL calculations have restricted phenomenological applications because, *e.g.*, the running of the QCD coupling constant α_S is not included, and the kinematic range of validity of LO BFKL is not known.

Recently the NLO corrections to the BFKL resummation of energy logarithms were calculated; see Refs. [4,5] and references therein. The NLO corrections [4,5] to the highest eigenvalue of the BFKL equation turn out to be negative and even larger than the LO contribution for $\alpha_S > 0.16$.

It should be stressed that the NLO calculations, as any finite-order perturbative results, contain both renormalization scheme and renormalization scale ambiguities. The NLO BFKL calculations[4,5] were performed by employing the modified minimal subtraction scheme ($\overline{\text{MS}}$) to regulate the ultraviolet divergences with arbitrary scale setting.

In the recent work[6] it was found that the renormalization scheme depen-

[a] *presented by V. T. Kim*

dence of the NLO BFKL resummation of energy logarithms [4,5] is not strong, *i.e.*, value of the NLO BFKL term is practically the same in the known renormalization schemes. To resolve the renormalization scale ambiguity due to the large NLO BFKL term [4,5] the Brodsky-Lepage-Mackenzie (BLM) optimal scale setting [7] has been utilized. The BLM optimal scale setting effectively resums the conformal-violating β_0-terms into the running coupling in all orders of the perturbation theory.

It was shown[6] that the reliability of QCD predictions for the effective intercept of the BFKL Pomeron at NLO evaluated at the BLM scale setting within the non-Abelian physical schemes, such as the momentum space subtraction (MOM) scheme or the Υ-scheme based on $\Upsilon \to ggg$ decay, is significantly improved compared to the $\overline{\text{MS}}$-scheme result [9,10].

One of the striking features of the analysis [6] is that the NLO value for the intercept of the BFKL Pomeron, improved by the BLM procedure, has a very weak dependence on the gluon virtuality Q^2: $\alpha_{IP} - 1 = \omega_{NLO}^{max} =\simeq 0.13 - 0.18$ at $Q^2 = 1 - 100$ GeV^2. It arises as a result of fine-tuned compensation between the LO and NLO contributions. The minor Q^2-dependence obtained leads to approximate conformal invariance.

As a phenomenological application of the NLO BFKL improved by BLM procedure one can consider the gamma-gamma scattering [11]. This process is attractive because it is theoretically more under control than the hadron-hadron and lepton-hadron collisions, where non-perturbative hadronic structure functions are involved. In addition, in the gamma-gamma scattering the unitarization (screening) corrections due to multiple Pomeron exchange would be less important than in hadron collisions.

The gamma-gamma cross sections with the BFKL resummation in the LO was considered in [2,12,13]. In the NLO BFKL case one should obtain a formula analogous to LO BFKL [11]. While exact NLO impact factor of gamma is not known yet, one can use the LO impact factor of [2,12] assuming that the main energy-dependent NLO corrections come from the NLO BFKL subprocess rather than from the impact factors [11].

In Fig. 1 the comparison of BFKL predictions for LO and NLO BFKL improved by the BLM procedure with L3 data [14] from CERN LEP2 is shown. The different curves reflect uncertainty with the choice of the Regge scale parameter which defines the beginning of the asymptotic regime. For present calculations two variants have been choosen $s_0 = Q^2/2$ and $s_0 = 2Q^2$, where Q^2 is virtuality of photons. One can see from Fig. 1 that the agreement of the NLO BFKL improved by the BLM procedure is reasonably good at LEP2 energy $\sqrt{s_{e^+e^-}} = 183$ GeV. One can notice also that sensitivity of the NLO BFKL results to the Regge parameter s_0 is much smaller than in the case of

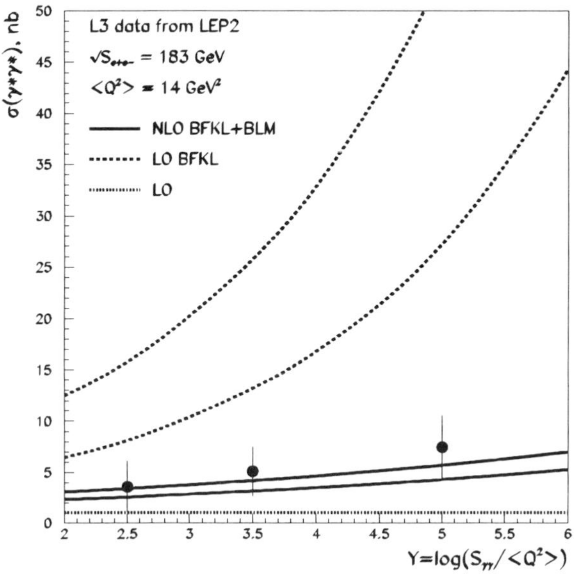

Figure 1: Virtual gamma-gamma total cross section by the NLO BFKL Pomeron within BLM approach vs L3 Collaboration data at energy 183 GeV of e^+e^- collisions. Solid curves correspond to NLO BFKL in BLM; dashed: LO BFKL and dotted: LO contribution. Two different curves are for two different choices of the Regge scale: $s_0 = Q^2/2$, $s_0 = 2Q^2$

the LO BFKL.

It was shown in Refs. [15,16] that the unitarization corrections in hadron collisions can lead to higher value of the (bare) Pomeron intercept than the effective intercept value. Since the hadronic data fit yields about 1.1 for the effective intercept value [16], then the bare Pomeron intercept value should be above this value. Therefore, assuming small unitarization corrections in the gamma-gamma scattering at large Q^2 one can accomodate the NLO BFKL Pomeron intercept value 1.13 - 1.18 [6] in the BLM optimal scale setting along with larger unitarization corrections in hadronic scattering [15], where they can lead to a smaller effective Pomeron intercept value about 1.1 for hadronic collisions.

Another possible application of the BFKL approach can be the collision energy dependence of the inclusive single jet production [17].

There have been a number of recent papers which analyze the NLO BFKL predictions in terms of rapidity correlations [18], angle-ordering [19], double transverse momentum logarithms [20,21,22], an additional $\log(1/x)$ enhancement [23] and BLM scale setting for deep inelastic structure functions [24]. Obviously, a lot of work should be done to clarify the proper expansion parameter for BFKL regime and, also the relation between those papers and the result of the present BLM approach. To confirm the result of [4,5] the independent NLO calculations (see [25] and references therein) for BFKL resummation are desirable.

To conclude, we have shown that the NLO corrections to the BFKL equation for the QCD Pomeron become controllable and meaningful provided one uses physical renormalization scales and schemes relevant to non-Abelian gauge theory. BLM optimal scale setting automatically sets the appropriate physical renormalization scale by absorbing the non-conformal β-dependent coefficients. The strong renormalization scheme and scale dependence of the NLO corrections to BFKL resummation then largely disappears. A striking feature of the NLO BFKL Pomeron intercept in the BLM approach is its very weak Q^2-dependence, which provides approximate conformal invariance. The NLO BFKL application to the total gamma-gamma cross section shows a good agreement with the L3 Collaboration data at CERN LEP2 energies.

VTK thanks the Organizing Committee of the VIIIth Blois Workshop at Institute for High Energy Physics, Protvino for their warm hospitality. This work was supported in part by INTAS, Grant No. INTAS-97-31696.

References

1. E. A. Kuraev, L. N. Lipatov and V. S. Fadin, ZhETF **71**, 840 (1976) [Sov. JETP **44**, 443 (1976)]; *ibid.* **72**, 377 (1977) [**45**, 199 (1977)]
2. I. I. Balitsky and L. N. Lipatov, Yad. Fiz. **28**, 1597 (1978) [Sov. J. Nucl. Phys. **28**, 822 (1978)]
3. L. N. Lipatov, Phys. Rept. **C286**, 131 (1997)
4. V. S. Fadin and L. N. Lipatov, Phys. Lett. **429B**, 127 (1998)
5. G. Camici and M. Ciafaloni, Phys. Lett. **430B**, 349 (1998)
6. S. J. Brodsky, V. S. Fadin, V. T. Kim, L. N. Lipatov and G. B. Pivovarov, Pisma ZhETF **70**,161 (1999) [JETP Lett.**70**,155(1999)],hep-ph/9901229
7. S. J. Brodsky, G. P. Lepage and P. B. Mackenzie, Phys. Rev. **D28**, 228 (1983)
8. L. N. Lipatov, ZhETF **90**, 1536 (1986) [Sov. JETP **63**, 904 (1986)]; in *Perturbative QCD*, ed. A.H. Mueller (World Scientific, Singapore, 1989) p. 411; R. Kirschner and L. Lipatov, Zeit. Phys. **C45**, 477 (1990)
9. D. A. Ross, Phys. Lett. **431B**, 161 (1998); Yu. V. Kovchegov and A. H. Mueller, *ibid.* **439B**, 428 (1998); N. Armesto, J. Bartels and M. A. Braun, *ibid.* **442B**, 459 (1998); E. M. Levin, Nucl. Phys. **B545**, 481 (1999)
10. J. Blümlein, V. Ravindran, W. L. van Neerven and A. Vogt, Proc. DIS

and QCD (DIS98), Brussels, Belgium, April 4-8, 1998, eds. Gh. Coremans and R. Roosen (World Scientific, Singapore, 1998) p. 211, hep-ph/9806368; R. D. Ball and S. Forte, *ibid.*, p. 770, hep-ph/9805315
11. S. J. Brodsky, V. S. Fadin, V. T. Kim, L. N. Lipatov and G. B. Pivovarov, in preparation
12. S. J. Brodsky, F. Hautmann, D. E. Soper, Phys. Rev. **D56**, 6957 (1997); Phys. Rev. Lett. **78**, 803 (1997), (E) **79**, 3544 (1997)
13. J. Bartels, A. De Roeck and H. Lotter, Phys. Lett. **389B**, 742 (1996); A. Białas, W. Czyż and W. Florkowski, Eur. Phys. J. **C2**, 683 (1998); M. Boonekamp, A. De Roeck, C. Royon and S. Wallon, Nucl. Phys. **B555**, 540 (1999); J. Kwieciński and L. Motyka, Acta Phys. Pol. **B30**, 1817 (1999)
14. L3 Coll., M. Acciari *et al.*, Phys. Lett. **453B**, 333 (1999)
15. A. B. Kaidalov, L. A. Ponomarev and K. A. Ter-Martirosyan, Yad. Fiz. **44**, 722 (1986) [Sov. J. Nucl. Phys. **44**, 468 (1986)]
16. J. R. Cudell, V. Ezhela, K. Kang, S. Lugovsky and N. Tkachenko, BROWN-HET-1184, 1999, hep-ph/9908218; J. R. Cudell, K. Kang and S. K. Kim, Phys. Lett. **395B**, 311 (1997); J. R. Cudell, A. Donnachie and P. V. Landshoff, Phys. Lett. **448B**, 281 (1999)
17. V. T. Kim and G. B. Pivovarov, Phys. Rev. **D57**, 1341 (1998)
18. V. S. Fadin and L. N. Lipatov, Proc. Theory Institute on Deep Inelastic Diffraction, ANL, Argonne, September 14-16, 1998; C. R. Schmidt, Phys. Rev. **D60**, 074003 (1999); J. R. Forshaw, D. A. Ross and A. Sabio Vera, Phys. Lett. **455B**, 273 (1999)
19. S. Catani, F. Fiorani and G. Marchesini, Nucl. Phys. **B336**, 18 (1990); M. Ciafaloni, Nucl. Phys. **B296**, 49 (1988)
20. B. Andersson, G. Gustafson and J. Samuelsson, Nucl. Phys. **B467**, 443 (1996); B. Andersson, G. Gustafson and H. Kharraziha, Phys. Rev. **D57**, 5543 (1998)
21. G. Salam, JHEP **9807**, 019 (1998)
22. M. Ciafaloni, D. Colferai and G. P. Salam, DFF-338-5-99, 1999, hep-ph/9905566; M. Ciafaloni and D. Colferai, Phys. Lett. **452B**, 372 (1999)
23. R. D. Ball and S. Forte, Edinburgh-99/5, 1999, hep-ph/9906222
24. R. S. Thorne, Phys. Rev. **D60**, 054031 (1999)
25. Z. Bern, V. Del Duca, W. B. Kilgore and C. R. Schmidt, BNL-HET-99-6, 1999, hep-ph/9903516

E. A. Kuraev

THE BFKL STATUS

E.A.KURAEV
Bogoliubov Laboratory of Theoretical Physics,
Joint Institute for Nuclear Research, Russia

The BFKL Pomeron describe the high-energy asymptotic in a perturbative QCD approach of the cross sections of hadron-hadron scattering. A lot of attention was paid to this subject especially last decades. The inters was enhanced by the recent experimental data from CERN and DESY,which need for their explanation the main features of BFKL Pomeron. The radiative correction to the intercept of Pomeron turns out to be too large,requiring some additional investigation: the scale and scheme independence results in reducing of intercept to value 0.2 which seems to be more realistic. The relation of the regimes of DGLAP and BFKL in application to DESY data also discussed.

In papers 1976-1977 of V.S.Fadin, L.N.Lipatov and E.A.Kuraev [1] the asymptotics behavior of amplitudes $2 \to 2+n$ was investigated in frames QCD in so called multiperipherical kinematics: when all the partial final particles invariant mass squares s_i much smaller compared to center of mass total energy squared s. Using the t-channel cut method of L.Lipatov it was found that effective ladder Feynman diagram (FD) description in leading logarithmical approximation (LLA) is valid:

$$A^0_{2\to 2+n} = s\frac{\Gamma^{i_1}_{AD_0}\gamma^{D_1}_{i_1 i_2}(q_1,q_2)\gamma^{D_2}_{i_2 i_3}(q_2,q_3)...\gamma^{D_n}_{i_n i_{n+1}}(q_n,q_{n+1})\Gamma^{i_{n+1}}_{AD_{n+1}}}{(q_1^2-m^2)(q_2^2-m^2)...(q_{n+1}^2-m^2)} \quad (1)$$

with some universal vertices $\gamma^D_{ij}(q1,q_2)$. The indeces i,j describe internal quantum state of created particles such as isotopic spin (group of symmetry $SU(2)$ or color $SU(3)$). The radiative correction taken into account in LLA

$$g^2 << 1, g^2\ln(s/m^2) \sim 1$$

, where g-is the coupling constant modify this Born expression by the factor

$$A_{2\to 2+n} \doteq A^0_{2\to 2+n}\Pi^{i=n+1}_{i=1}(s_i/m^2)^{\alpha(q_i^2)}$$

, which corresponds to the reggeization of the exchanged gluon with momentum q_k and the partial invariant mass squared $s_k = (P_{D_k} + P_{D_{k-1}})^2$. In the multiperipherical kinematics (providing the maximal contribution to the total cross section) they obey the inequality $m^2 << s_k << s$. The trajectory of gluon pole have a form:

$$\alpha(t) = \frac{g^2(t-m^2)}{(2\pi)^3} \int \frac{d^2k}{(k^2-m^2)((q-k)^2-m^2)}, t = q^2.$$

Using this expression we had calculate the contribution to the elastic scattering amplitude with all possible isotopic spins $T = 0, 1, 2$ in the scattering channel:

$$A(s,t) = s\Sigma_T \int \frac{d\omega}{i} (\frac{s}{m^2})^\omega \frac{(-1)^T - e^{-i\pi\omega}}{\sin\pi\omega} F_\omega^T(t),$$
$$F_\omega^T(q^2) = \frac{A^T g^2}{\omega(2\pi)^3} \int \frac{f_\omega^T(k, q-k) d^2k}{(k^2-m^2)((q-k)^2-m^2)} \qquad (2)$$

and the quantity f_ω obey the BFKL equation:

$$(\omega - \alpha(k^2) - \alpha((q-k)^2)) f_\omega^T(k, q-k) =$$
$$\frac{g^2}{A^T}\omega + \frac{g^2}{(2\pi)^3} \int \frac{d^2k' f_\omega^T(k', q-k')}{(k'^2-m^2)((q-k')^2-m^2)}$$
$$[a^T(q^2) - (2 - \frac{T(T+1)}{2}) \cdot$$
$$\frac{((k^2-m^2)((q-k')^2-m^2) + (k'^2-m^2)((q-k)^2-m^2))}{(k-k')^2-m^2}],$$
$$A^T(q^2) = m^2/2 + (q^2 - 3m^2/2)(2 - T(T+1)/2). \qquad (3)$$

It's solution for $T = 1$ agree with reggeized gluon:

$$f_\omega^1(k, q-k) = \frac{1}{q^2-m^2} \frac{\omega g^2}{\omega + 1 - \alpha(q^2)}.$$

For $T = 0$ the square root singularity is situated at $\omega = (2g^2 \ln 2)/\pi^2$. (We consider $SU(2)$ color group). For the $SU(N)$ the cross section will grow with the energy as

$$\sigma_{tot}(s) \sim (s/s_0)^\Delta, \Delta = (4N\alpha_s \ln 2)/\pi$$

and,thus violating the unitarity.

Note that the similar phenomenon was found in papers of Gribov, Frolov, Lipatov, Cheng, Wu in 1969, 1970[12], namely the cross section of arbitrary number of electron- positron pairs creation at charged particles collisions as well violate the unitarity (grows faster than any power of ln s/m^2):

$$\sigma(s) \sim (s/m^2)^{11\pi\alpha^2/32}.$$

Next steps was done in the paper of Y.Balitsky and L.Lipatov [2] and in paper of L.Lipatov [13] where BFKL equation was recovered in the lowest order of perturbation theory (PT) and the impact paramerets representation was used for the case of photon-photon scattering:

$$A(s,t) = is\int_{\sigma-i\infty}^{\sigma+i\infty} \frac{d\omega}{2\pi i}(s/s_0)^{\omega} f_{\omega}(q^2), \quad q^2 = t = -\vec{q}^2,$$

$$f_{\omega}(q^2) = \int d^2k d^2k' \Phi^A(k,q)\Phi^B(k,q) f_{\omega}(k,k',q) \tag{4}$$

with $\Phi(k,q)$-impact factors of photons and:

$$\delta^2(q-q')f_{\omega}(q^2) = \int \Pi_{r=1}^{r=2} \frac{d^2\rho_r d^2\rho_{r'}}{(2\pi)^4}\Phi^A(\rho_1,\rho_2,q)\Phi^B(\rho_{1'},\rho_{2'},q)f_{\omega}(\rho_1,\rho_2,\rho_{1'},\rho_{2'},q)$$

with gauge conditions for $\Phi^{A,B}$:

$$\Phi^i(\rho_1,\rho_2,q) = \int d^2k\Phi^i(k,q)e^{ik\rho_1}e^{i(k-q)\rho_2},$$

$$\int d^2\rho_1 \Phi^i(\rho_1,\rho_2,q) = \int d^2\rho_2 \Phi^i(\rho_1,\rho_2,q) = 0. \tag{5}$$

L.N.Lipatov show that the equation for $f_{\omega}(\rho,\rho',q)$ (being written in complex form: $\rho = (\rho_x,\rho_y)$-two dimentional vectors are written as $\rho = \rho_x + i\rho_y$ and ρ^*) is invariant regarding transformation of the conformal group and build the eigenfunctions $E^{n,\nu}$,which depend on two parameters:conformal spin-integer number n and continuous parameter ν:

$$E^{n,\nu}(\rho_{10},\rho_{20}) = e^n(z)|z|^{+2i\nu}, \quad e(z) = -z^*/|z|,$$

$$z = \frac{\rho_{12}}{\rho_{10}\rho_{20}}, \quad \rho_{ik} = \rho_i - \rho_k. \tag{6}$$

The relevant eigenfunctions have the form:

$$\omega(\nu,n) = \frac{Ng^2}{2\pi^2}\int_0^1 \frac{dx}{1-x}[x^{(|n|-1)/2}\cos(\nu\ln x) - 1].$$

The solution of BFKL equation have a form:

$$f_{\omega}(\rho_1,\rho_2,\rho_{1'},\rho_{2'},q) = \Sigma_{n=-\infty}^{n=+\infty}\int d\nu \int d^2\rho \frac{\nu^2+(n^2/4)}{[\nu^2+((n-1)^2/4)][\nu^2+((n+1)^2/4)]}$$
$$E^{n,\nu *}(\rho_{1'0},\rho_{2'0})E^{n,\nu}(\rho_{10},\rho_{20}). \tag{7}$$

The singularities in ω plane appears as a result of pinch of a contour at $\nu = 0$ at $n = 0, 1, 2, \ldots$:$\omega_0 = g^2 N \ln 2/\pi^2; \omega_1 = 0; \omega_2 = -g^2(1 - \ln 2)N/\pi^2; \ldots$

The contribution of the leading singularity is:

$$f_\omega(\rho, \rho')_{\omega \to \omega_0} = -\pi \phi_q(\rho)\phi_q(\rho')(\omega - \omega_0)^{1/2}(7Ng^2\xi(3)/2\pi^2)^{-3/2}$$

corresponds to the branch point in the plane of orbital momentum. The effect of running coupling constant,as was shown by Lipatov turns this branch point singularity to the infinite series of poles,approximately situated at $\omega_k = \omega_0/(k+1)$.The corresponding wave functions may be found solving the quasi-classical Schrodinger equation.

This approach was successfully applied in papers of N.Nikolaev and B.Zakharov [10] to explain the subtle effects in charm production by nuclei and the color transparency.

In papers of 90-th years it was recognized the relation of the BFKL equation,being written in the Hamiltonian form with the exactly solvable models of statistics such as so called XXX model (see for details the review of L.Lipatov[9]).

In papers of A.H.Mueller [11] the question of multiple emission of soft gluons by dipole of quark and antiquark was inwestgated. It was found that the generating functional describing the emission of arbitrary number of gluons (onium) obeys some nonlinear integral equation which by essentity coinsides with the Smolukhowsky equation which describes well the gases properties at sufficiently high density. It was discovered that after repairing this equation by means of formfactors insertion which provides the consevation of the normalization this equation coinsides with the BFKL equation.

This arguments show the relation of BFKL dynamics with statistical many-body problems.

The linked dipole chain (LDC) model of deepinelastic scattering (DIS) developed by B.Andersson,G.Gustafson aand J.Samuelson [7] may shed light on the connection of DIS processes at small x with BFKL regime. This model takes into account the leading as well as nonleading contributions in all orders of PT, the Sudakov formfactors,the soft gluon emission and the conditions of jets creation. It results to the modification of simplified version of BFKL equation so called Lipatov equation for the structure functions:

$$\frac{dF(x, k_\perp/k_bot^2}{d\ln(1/x)} = \bar{\alpha}\int \frac{d^2k'_\perp}{\pi q^2}[\frac{F(x, k'^2_\perp)}{k'^2_\perp} - \frac{F(x, k^2_\perp)}{k^2_\perp}\theta(k_\perp - q_\perp], q_\perp = k'_\perp - k_\perp$$

by the equation

$$\frac{dF(x, k_\perp/k^2_\perp}{d\ln(1/x)} = \bar{\alpha}\int \frac{d^2k'_\perp}{\pi q^2}[\frac{F(x', k'^2_\perp)}{k'^2_\perp} - \frac{F(x, k^2_\perp)}{k^2_\perp}\theta(k_\perp - q_\perp) = k'_\perp - k_\perp$$

with $x' = max(x, x\frac{k_\perp'^2}{k_\perp^2})$ and $x = Q^2/s$-is the Bjorken parameter. It is inportant to note that the latest equation takes into account the contributions of leading as well as nonleading sort.

At small x the behavior of the functions $F \sim x^{-\lambda}$. The values of λ is quite different: for Lipatov equation it is $\lambda = 12\alpha_s \ln 2/\pi \approx 0.55$ at $\alpha_s = 0.2$ whereas for LDC equation it is $\lambda = 0.31$. It is interesting to note that in LLA the equations coinsides. This result in agreement with the explicite calculation performed recently by V.Fadin and L.Lipatov and others[4] concerning the correction to the Pomeron intercept where the effective value of it was obtained $\lambda = 0.25$.

Another support comes from A.Kaidalov analysis of multiplicities on hadron's colliders[14]. Namely the bulk of data including the cosmic statistics may be self-consistently described in pomeron exchange model with the intercept 0.2. In such an analysis the double and many pomeron exchange in the scattering channel must be taken into account.

I underline that in the frames QCD there is no place for say double pomeron exchange-we may speak only about next to leading corrections wheras if one consider in frames of phemenology pomeron as a field theoretical object the multy pomeron exchang must be taken into account.

The result of about 10 years work on calculation of the radiative corrections (RC)to the Pomeron intercept may be written in the form:

$$\omega_{\bar{MS}}(Q_1^2, \nu) = \int d^2 Q_2 K_{\bar{MS}}(Q_1, Q_2)(Q_2^2/Q_1^2)^{-1/2+i\nu} =$$
$$N_c \kappa_L(\nu) \frac{\alpha_{\bar{MS}}(Q_1^2)}{\pi}[1 + r_{\bar{MS}}(\nu)\frac{\alpha_{\bar{MS}}(Q_1^2)}{\pi},$$
$$\kappa_L(\nu) = 2\psi(1) - \psi(1/2 + i\nu) - \psi(1/2 - i\nu), \quad \psi(z) = \Gamma'(z)/\Gamma(z); \quad (8)$$

and the quantity r is $r_{\bar{MS}} = r^\beta + r^{conf}$, $r^\beta = -\frac{\beta_0}{4}(\frac{1}{2}\kappa_L - \frac{5}{3})$, $\beta_0 = \frac{11}{3}N_c - \frac{2}{3}n_f$, The quantity r^{conf} is some complicate function of ν,which at small values of ν equal to

$$r(0) = -20.12 - 0.102 N_f + 0.067\beta_0 = -19.99, N_f = 4. \quad (9)$$

We remind here the attractive phenomenological moments of the BFKL Pomeron. First it is the possibility to match the perturbative QCD with nonperturbative and the bridge to "soft" pomeron,when taking the NLO corrections to BFKL pomeron. Second the BFKL Pomeron is only the theoretical way to prove that the density of partons become large at sufficiently high energy. HERA data definitely show that the gluon density is high. Third,the new HERA data shows that the BFKL Pomeron is needed to describe the forward jet production.

Neverseless the large value of NLO correction could change the entire understanding of the structure of the pomeron phenomenology (I follow here E.Levin talk in DESY May,99 meeting[5]). Really,in some realistic approximation the BFKL prediction for the Q^2, s dependence of the total cross section may be put in form:

$$\sigma_{tot}(BFKL) \sim e^{(y-y_0)\omega_L} e^{-\frac{(r-r_0)^2}{4D(y-y_0)}}, y - y_0 = \ln\frac{s}{s_0}, r - r_0 = \ln\frac{Q^2}{Q_0^2}, \quad (10)$$

where the effective separation of the longitudinal and transverse degrees of freedom is explicitely done. ω_L with RC taken into account turns out to be negative at $Q_0^2 < 30GeV^2$. It means that in large kinematic region the BFKL emission is suppressed and only the Born term contributes. At higher values of $Q_0^2 \sim 10^4 GeV^2$ diffusion coefficient becomes negative,no diffusion in $\ln Q^2$ take place,the oscillating character of the total cross section. It means that NLO BFKL has a serious pathalogy. In literature the several ways-out from this situation are discussed. One of them was mentioned above-the linked dipole chain model-may be considered as an alternative to NLO BFKL.

NLO calculations contain both renormalization scheme and renormalization scale ambiguities. It was suggested in paper of S.Brodsky et al.[6] to utilize the BLM optimal scale setting[8] to resolve the renormalization scale ambiguity. It is based on the physical choise of the energy scale of coupling constant renormalization (mass of upsilon meson). From the technical point of view it consist in the change of the arguments of "large logarithms" from s/Q_0^2 to the $s/\sqrt{Q^2 Q_0^2}$ the most "dangerous" term providing the numerically largest contribution tothe correction $\ln^2(Q^2/Q_0^2)$ will disappear and it result in weak dependence of intercept on the virtuality of the reggeized gluon,approaching the value 0.2 at larg values Q^2.

We discuss now the relation of two regimes BFKL and the DGLAP- regime of evolution of large virtuality according to Altarelly-Parisi- Lipatov evolution equations. It is the main idea of G.Salam and M.Giaffoloni [15]-to replace the NLO kernel of BFKL equation by the new one in which the subleading corrections have been resummed in such a way that the it will coinside with the DGLAP kernel.

I belive (see also the talk of L.Enkovsky)that it is impossibile to modify the Pomeron regime to the regime of evolution. Really the evolution imply severe hierarhy in the virtualities of subsequent hard subprocesses whereas in BFKL the real diffusion if the log of the transversal momenta take place. If one put the external condition forbidding the random walking in ,say the direction of decreasing of transversal momenta it probably will turn to DGLAP regime.

In DIS set-up I belive both regimes take place: in the first stage the decreasing of large virtuality of photon is realized through DGLAP mechanism up to values of typical parton's momenta in the targed and than in the second stage the BFKL pomeron regime start. I remind here the classical model of P.and T.Ehrenfests[3], build in 1907 year. It demonstrate the decreasing of the temperature of say a glas of hot waterin the cold environement. The fluctuation absent in the first stage of cooling and they appears after toe moment when the thermal equilibrium take place. So in first stage we are to sum the contributions of the form $(\alpha \ln(Q^2/Q_0^2))^n$ and in the kinematics when $Q^2 \sim Q_0^2$ we are to sum the contributions of the form $(\alpha \ln(1/x))^n$. In conclusion I want to mention the different topics of physics where the equation essentially equivalent to BFKL appears: solvable models of statistical mechanics, peripherical interactions of hadrons, many-body problems,described by kinetic equations.

I am grateful to L.Enkovsky,A.Gazizov,I.Mandzavidze for valuable discussions and to Igor Anikin for help.

References

1. L.N.Lipatov, Sov.J.Nucl.Phys. 23 (1976) 642; V.S.Fadin,E.A.Kuraev,L.N.Lipatov, Phys.Lett. B60 (1975) 50; E.A.Kuraev,L.N.Lipatov,V.S.Fadin, Sov.Phys.JETP 44 (1976),45; Sov.Phys. JETP 45(1977),199.
2. Ya.Ya.Balitsky and L.N.Lipatov, Sov.J.Nucl. phys.28(1978)822.
3. P. Ehrenfest,T. Ehrenfest, Uber zwei bekannte Einwande gegen das Boltzmanishsche H-Theorem,Phys.Z.,8(1907),311-314.
4. V. Fadin and L. Lipatov, Phys. Lett. B429(1998),127; M.Giafaloni,G.Camici, Phys. Lett. B430(1998),349.
5. Levin hep-ph/9905381.
6. S. Brodsky et al., SLAC-PUB-8037,IITAP-98-010,hep-ph/9901229.
7. B. Andersson et al., Nucl.Phys. B467(1996),443.
8. S.Brodsky,G.Lepage and P.Mackenzie, Phys. Rev. D 28(1983),228.
9. L.Lipatov Phys.Reports,286(1997),131.
10. N.Nikolaev,B.Zakharov, Z.Phys. C-Particles and Fields 53(1992),331.
11. A.H.Mueller et al. Nucl.Phys. B415(1994),373; B425(1994),471;B437(1995),107.
12. H. Cheng and T.T. Wu, Phys. Rev. 182 (1969), 1852-1899; V.Gribov,L.Lipatov and G.Frolov, Yad. Fiz. 12 (1970), 994.
13. L.N.Lipatov Sov.J.Phys.JETP 90 (1986),1536.
14. A.B.Kaidalov,Yad.Fiz.61(1998),911.
15. G.P.Salam, J. High Energy Phys. 9807, 019, (1998).

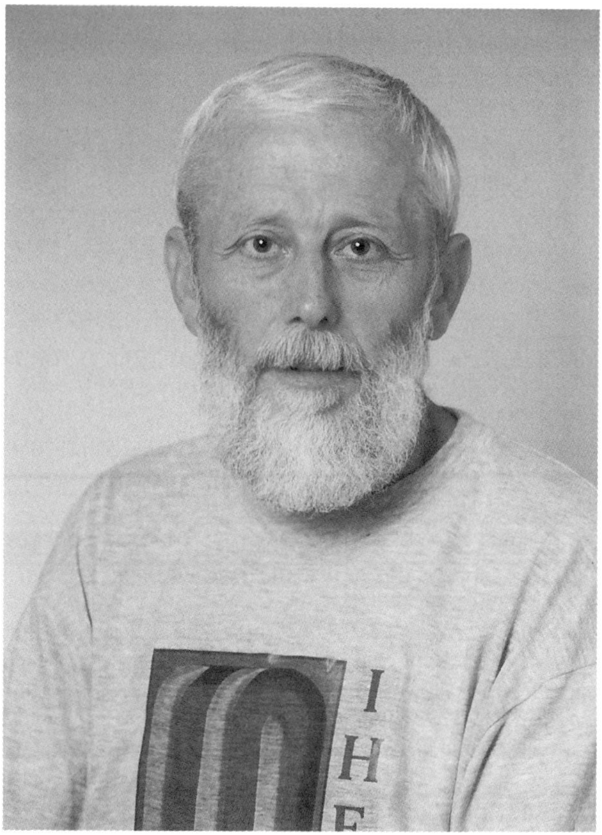

V. V. Ezhela

COMPARATIVE STUDY OF THE COMPETITIVE MODELS FOR THE FORWARD SCATTERING DATA AT HIGH ENERGIES

V. V. EZHELA, S. B. LUGOVSKY, N. P. TKACHENKO
COMPAS Group, IHEP, Protvino, Russia

J. R. CUDELL
Inst. de Physique, Université de Liège, Bât. B-5, Sart Tilman, B4000 Liège, Belgium

K. KANG
Department of Physics, Brown University, Providence RI 02912 USA and School of Physics, Korea Institute for Advanced Study, Seoul 130-012, Korea

This presentation is based on our comparative study[1] of the three popular analytic parametrizations of the total cross sections and ρ parameter on all available data accumulated in the PPDS databases[2]. The main goal of this presentation is to try to find answers to the following questions: where the parametrization with simple pole pomeron asymptitics start to be noticeably different from the others based on the unitarized pomeron asymptotics; where the asymptopia starts; what are the best model for extrapolations to higher energies.

1. Analytic Models. The proposals to use of the simple analytic models for the forward scattering amplitudes traced back to the end of 1960's (V.N.Gribov, A.A.Migdal, A. Martin, A. A. Logunov) and firstly tested on the available at that time experimental data on pp and $p\bar{p}$ by Bourelli and Fisher[3]. We perform analogous scan-fits of each model to the $p^{\pm}p, \pi^{\pm}p, K^{\pm}p, \gamma p$, and $\gamma\gamma$ total hadronic cross section data and ρ parameter for $p^{\pm}p, \pi^{\pm}p, K^{\pm}p$. All data fitted simultaneously to the following parametrizations.

RRP

$$\sigma_{\mp} = X \cdot s^{\epsilon} + Y_1 \cdot s^{-\eta_1} \pm Y_2 \cdot s^{-\eta_2}$$

$$\rho_{\mp} = \frac{1}{\sigma_{\mp}} \cdot \left[-\frac{X \cdot s^{\epsilon}}{\tan\left(\frac{(\epsilon+1)\pi}{2}\right)} - \frac{Y_1 \cdot s^{-\eta_1}}{\tan\left(\frac{(1-\eta_1)\pi}{2}\right)} \pm \frac{Y_2 \cdot s^{-\eta_2}}{\cot\left(\frac{(1-\eta_2)\pi}{2}\right)} \right]$$

RRL2

$$\sigma_{\mp} = \lambda\left(A + B \cdot \ln^2 s\right) + Y_1 \cdot s^{-\eta_1} \pm Y_2 \cdot s^{-\eta_2}$$

$$\rho_{\mp} = \frac{1}{\sigma_{\mp}} \cdot \left[\lambda \cdot B \cdot \pi \cdot \ln s - \frac{Y_1 \cdot s^{-\eta_1}}{\tan\left(\frac{(1-\eta_1)\pi}{2}\right)} \pm \frac{Y_2 \cdot s^{-\eta_2}}{\cot\left(\frac{(1-\eta_2)\pi}{2}\right)} \right]$$

RRL1

$$\sigma_{\mp} = \lambda \cdot (A + B \cdot \ln s) + Y_1 \cdot s^{-\eta_1} \pm Y_2 \cdot s^{-\eta_2}$$

$$\rho_{\mp} = \frac{1}{\sigma_{\mp}} \cdot \left[\frac{\lambda \cdot B \cdot \pi}{2} - \frac{Y_1 \cdot s^{-\eta_1}}{\tan\left(\frac{(1-\eta_1)\pi}{2}\right)} \pm \frac{Y_2 \cdot s^{-\eta_2}}{\cot\left(\frac{(1-\eta_2)\pi}{2}\right)} \right]$$

In constructing these parametrizations we use the "derivative dispersion relations" of K. Kang and B.Nicolescu [4] (see also prescriptions of Ref.[5]). To make number of parameters as low as possible we require the universality of the rising part of the cross sections and Pomeranchuk theorem in all cases as it is in the simple regge-pole motivated parametrizations as well as universality of the effective secondary regge-pole contributions. All models has the same total number of adjustable parameters, namely 16 parameters to fit 8 cross sections and 6 ρ data samples.

2. Experimental Data. Experimental data used were extracted from the Particle Physics Data System (PPDS)[2] maintained by the COMPAS Group at IHEP. Originally PPDS databases were created (1975-1980) by merging CERN-HERA cross sections data compilation, old LRL and LBL bibliographic and numerical data compilations. New data now gathered by collaborative efforts of COMPAS(IHEP), BPDG(LBNL), PDG(Durham,UK), ITEP(Moscow), and JINR(Dubna).
Data used for these studies were repeatedly checked with original publications, all preliminary data were removed, all known published errata fixed in the databases. Computer readable data files posted on the WEB [6]. It is our hope that this dataset will eventually become the standard reference when studying the validity of models for the forward quantities.

3. Scan-fits. We produce scan fits of each model to the data in the intervals from E^{cm}_{min} to the maximal evailable cm energies for each collisions and vary E^{cm}_{min} starting from 3-GeV in 1-GeV steps up to $E^{cm}_{min} = \mathbf{13}$-GeV, to be able to establish the area of applicability of each model (demanding $\chi^2/DoF \leq 1.$) and to see if the model parameters have weak dependence upon E^{cm}_{min}. The stability pattern presented on the plots in the Professor Kang report, here we give only figures for comparison of different models and to show their areas of applicability.

4. Summary. From the Figs 1.-3. we see that our three models reproduce the data sufficiently well and give practically the same fit qualities starting from $E^{cm}_{min} = 9$-GeV. However, the factorization of the coupligs in RRP model and reproduction of the quark counting rules[1] (discussed in details in the Professor

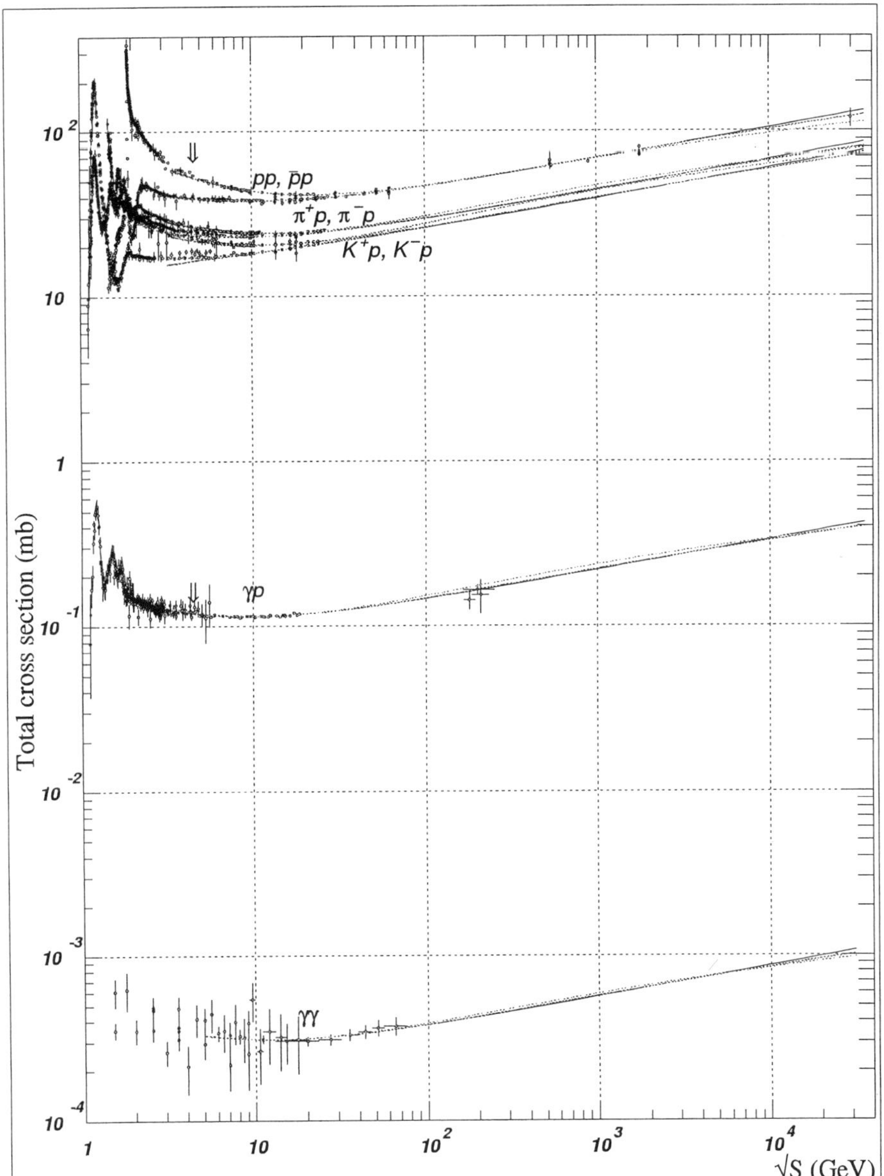

Figure 1: Three models global comparison. Curves are drawn with parameters obtained at the beginning of the parameters stability regions: for the RRP and RRL2 it is $E_{min}^{cm} = 9$-GeV and for the RRL1 it is $E_{min}^{cm} = 5$-GeV.

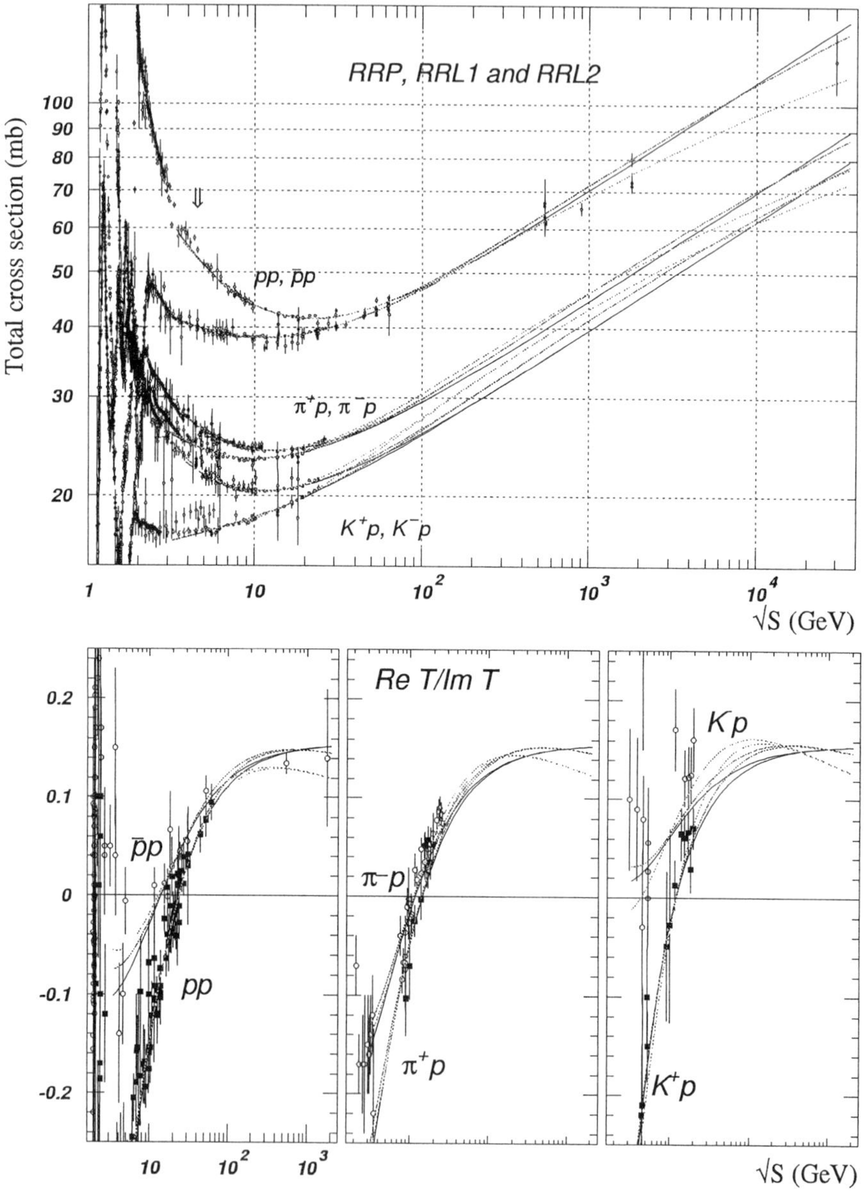

Figure 2: Hadronic collisions only. Curves are drawn with parameters obtained at the beginning of the parameters stability regions: for the RRP and RRL2 it is E^{cm}_{min} = 9-GeV and for the RRL1 it is E^{cm}_{min} = 5-GeV.

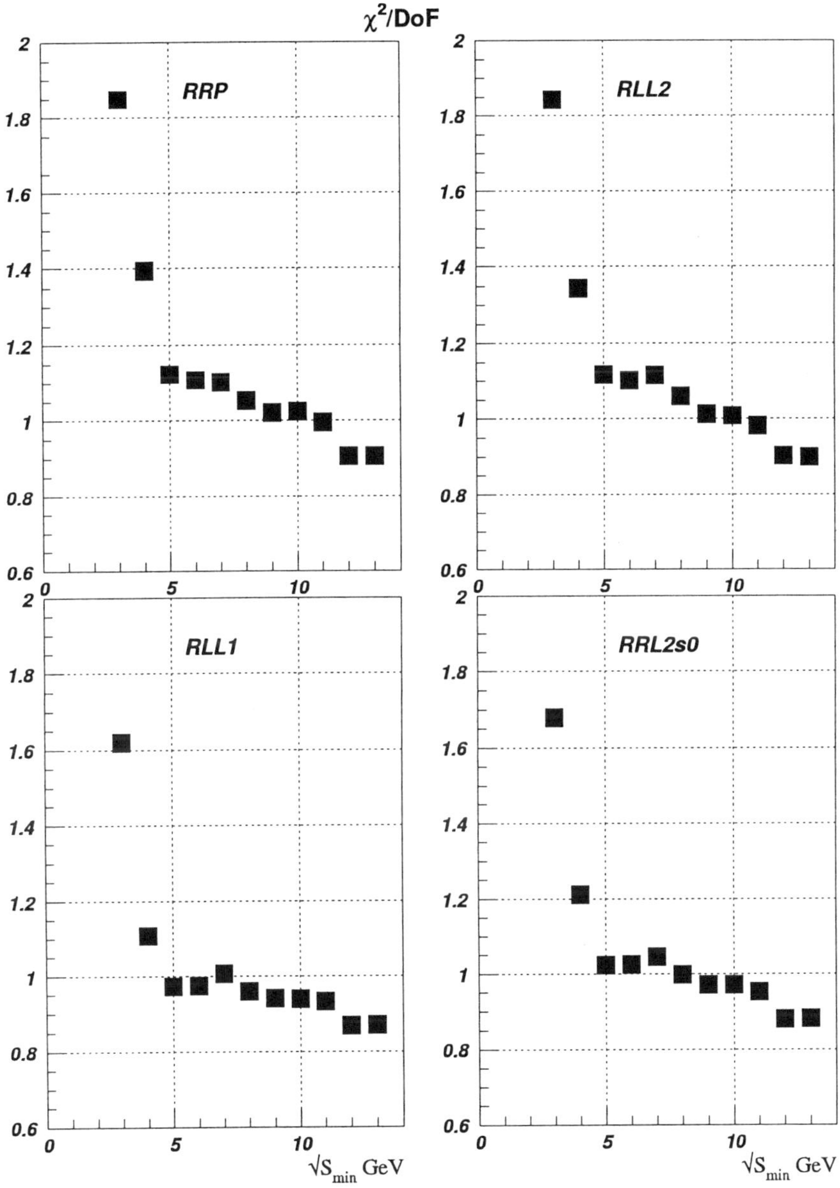

Figure 3: χ^2/Dof dependencies upon E^{cm}_{min}.

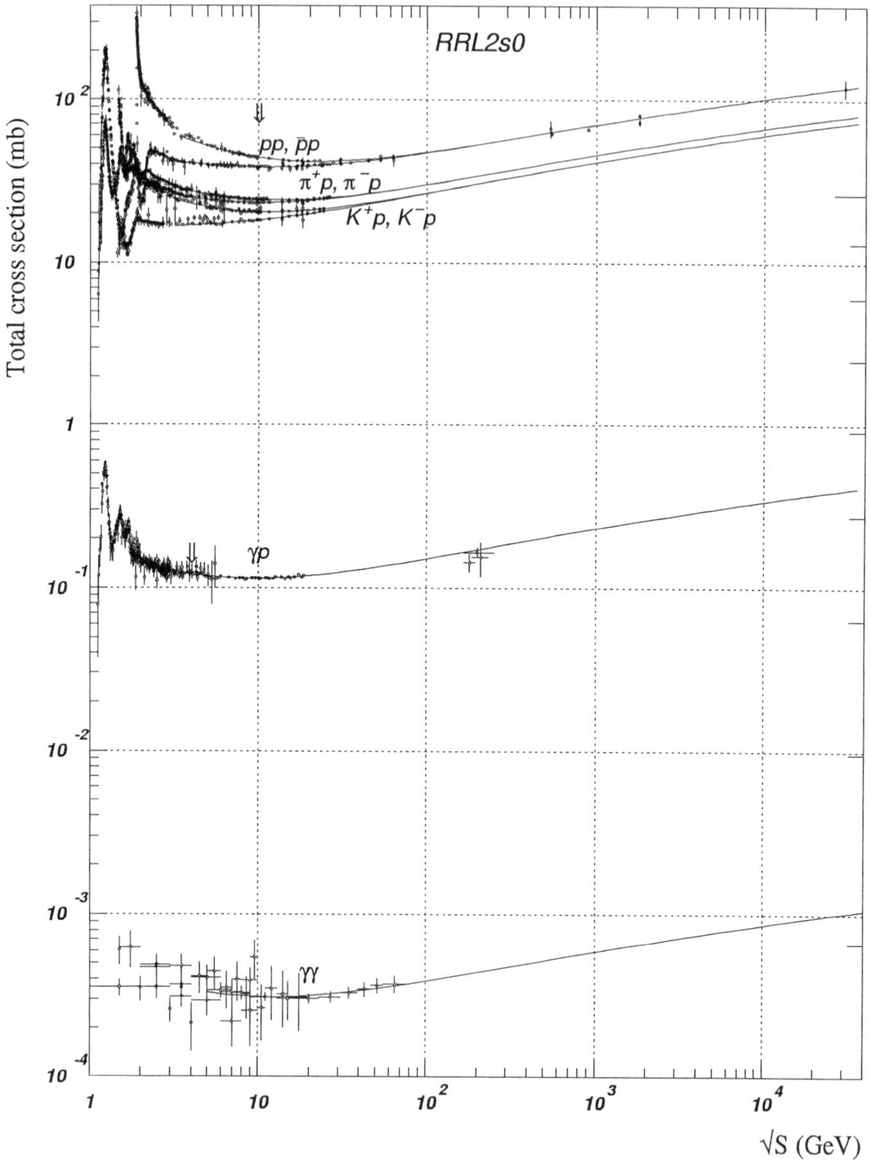

Figure 4: **Data description by RRL2s0 model**

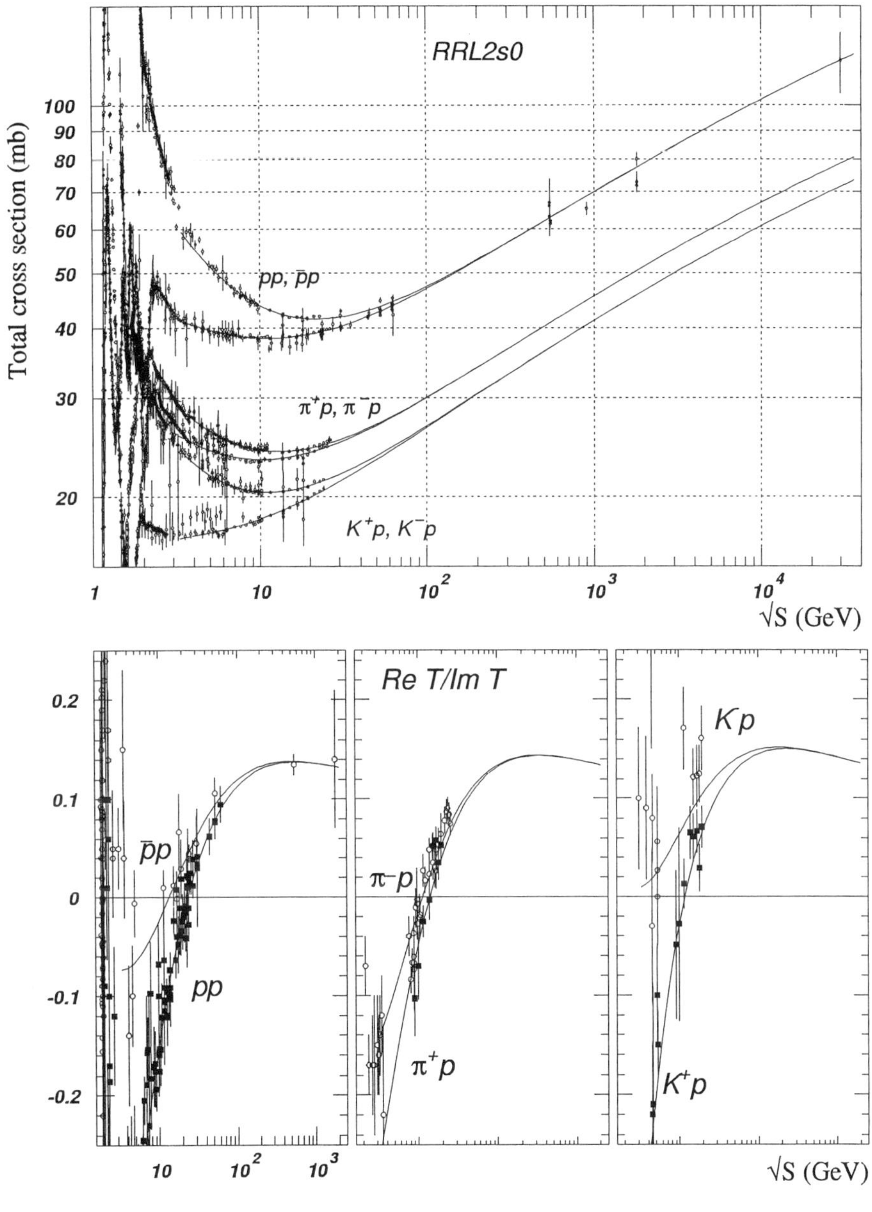

Figure 5: **Data description by RRL2s0 model. Hadronic part only**

Kang talk) make it more attractive for quick calculations. From the other side some kind of "ranking tendency" is revealed: models with unitarized (logarithmic) pomeron are preferable: RRL2 and RRL1 give slightly better $\chi 2/Dof$ values, more extended intervals of applicability, and more slow variation of parameters with E^{cm}_{min}.

5. Afterwords. After discussions at the conference around the scale parameter s_0, that was fixed in our fits at $s_0 = \mathbf{1\text{-}GeV^2}$ for the sake of reduction the number of parameters and correlations, we found a possibility to modify RRL2 parametrization in such a way that it still have 16 adjustable parameters. We remove parameters A and λ_{ab}, and introduced parameters B_{ab} and s_0.

$$\boxed{\textbf{RRL2s0}} \qquad \sigma_{\mp} = \lambda\left(B \cdot \ln^2(s/s_0)\right) + Y_1 \cdot (s/s_0)^{-\eta_1} \pm Y_2 \cdot (s/s_0)^{-\eta_2}$$

$$\rho_{\mp} = \frac{1}{\sigma_{\mp}} \cdot \left[B \cdot \pi \cdot \ln(s/s_0) - \frac{Y_1 \cdot (s/s_0)^{-\eta_1}}{\tan\left(\frac{(1-\eta_1)\pi}{2}\right)} \pm \frac{Y_2 \cdot (s/s_0)^{-\eta_2}}{\cot\left(\frac{(1-\eta_2)\pi}{2}\right)} \right]$$

The obtained picture of the scan-fit is much better than in the RRL2 case and is quite similar to that of RRL1 (see Figs 3-5).

As it was expected[1] the value of $\sqrt{s_0}$ turned to be very small, approximately 20-MeV.

References

1. J.R. Cudell, V. Ezhela, K. Kang, S. Lugovsky, N. Tkachenko, *High-Energy Forward Scattering and the Pomeron: Simple Pole versus Unitarized Models*, eprint **hep-ph/9908218**
2. COMPAS Group, *Particle Physics Data System*, **http://wwwppds.ihep.su:8001/ppds.html**
3. C. Bourrely and J. Fisher, *Analytic Parametrization of the High-Energy Forward Scattering Amplitude. I. **pp** and $\bar{p}p$ Scattering. Nucl. Phys.* B **61**, 513 (1973).
4. K.Kang and B.Nicolescu, *Models for hadron-hadron scattering at high energies and rising total cross sections.*, Phys. Rev. D11 (1975) 2461.
5. M.M. Block and R.N. Cahn, *High-energy $p\bar{p}$ and **pp** forward elastic scattering and total cross sections*, Rev. Mod. Phys. **57** (1985) 563.
6. Particle Data Group, *Review of Particle Physics*, The European Physical Journal **C3** (1998) 1
http://pdg.lbl.gov/xsect/contents.html

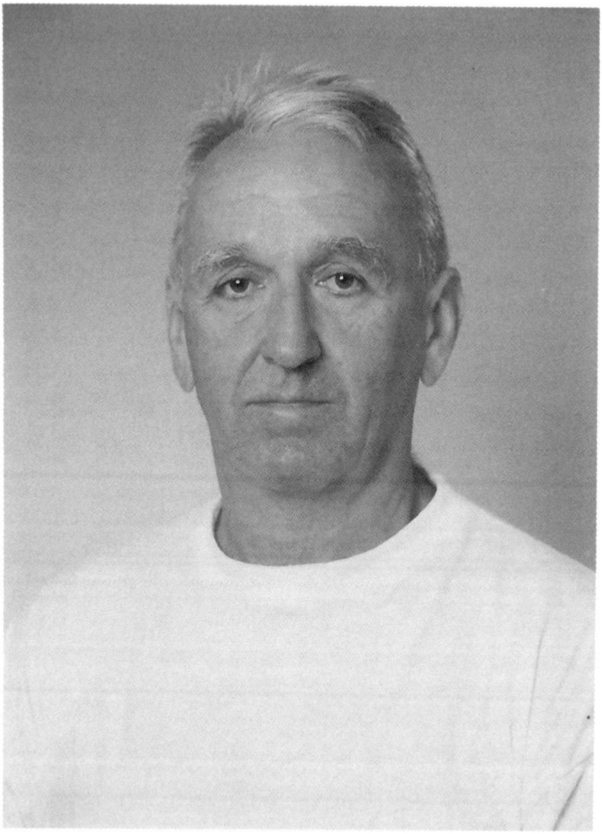

L. Jenkovszky

RELATING ELASTIC AND DEEP INELASTIC SCATTERING

R. Fiore ([a]) L. Jenkovszky ([b])
([a]) Universita degli Studi della Calabria ([b]) Bogolyubov Institute for Theoretical Physics, Kiev-143, Ukraine.

We relate the elastic hadron-hadron scattering amplitude to deep inelastic structure functions by identifying the integrand of the four-point dual amplitude with Mandelstam analyticity (DAMA) with the deeply virtual Compton scattering (DVCS) amplitude, whose imaginary part at $t = 0$ is the hadron structure function (SF). We also identify the integration variable x in DAMA with the partons' momentum fraction (Bjorken variable).

As is well known [1], the cross section for deep inelatic scattering

$$\frac{d\sigma}{dxdQ^2} = \frac{2\pi\alpha^2}{Q^4 x}[2(1-y)+y^2)F_2(x,Q^2) - y^2 F_L(x,Q^2)] = kF_2, \qquad (1)$$

where $y = Q^2/xs$, x is the the Bjorken variable and s the c.m.s. energy, can be related via unitarity to the imaginary part of the virtual Compton scattering amplitude with

$$\sigma_L + \sigma_T = \frac{4\pi^2\alpha(1+4m^2x^2/Q^2)}{Q^2(1-x)} F_2(x,Q^2), \qquad (2)$$

and further, by assuming vector meson dominance – to hadronic total cross sections.

Below we suggest an alternative way of relating (on shell) elastic hadron scattering amplitudes to the structure functions of deep inelastic scattering, based on the integral representation of dual amplitudes with Mandelstam analyticity (DAMA) [2].

Before doing so, let us remind that in the dual, crossing-symmetric Veneziano model [3],

$$V(s,t) = \int_0^1 dx(1-x)^{-\alpha(s)}x^{-\alpha(t)}, \qquad (3)$$

where s and t are the Mandelstam variables, like in any other narrow-resonance dual model, the Regge trajectories are constrained to be linear and consequently the model describes extended objects (hadrons, strings etc) only, whose partonic content (manifest in the wide angle scaling behaviour, typical of the

[a]E-mail: fiore@fis.unical.it
[b]E-mail: jenk@bitp.kiev.ua

interaction of point-like particles) remains beyond the range of the narrow-resonance dual models.

In DAMA, that may be considered as a generalization of narrow-resonance dual models incorporating broad resonances, physical thresholds required by unitarity and Mandelstam analyticity,

$$D(s,t) = \int_0^1 dx \left(\frac{1-x}{g}\right)^{-\alpha(sx)} \left(\frac{x}{g}\right)^{-\alpha(t(1-x))}, \tag{4}$$

where g is a constant, the Regge trajectories are essentially nonlinear. Their form is constrained by the threshold and asymptotic behavior of the amplitude. In fact, by requiring Regge asymptotic and correct threshold behaviour of the amplitude to hold, one gets a Regge trajectory typical of DAMA [2]:

$$\alpha(t) = \alpha_0 - \beta\sqrt{t_0 - t}$$

or

$$\alpha(t) = \alpha_0 - \sum_i \beta_i \sqrt{t_i - t}, \tag{5}$$

where t_i are hadronic thresholds and β is a free parameter with the meaning of the coupling strength. Moreover, wide angle scaling behavior in DAMA imposes an even stronger upper bound on the asymptotic behaviour of the trajectories, namely [4] $\alpha(t) \geq \ln|t|$. A representative trajectory that combines the abovementioned properties (linear at small, square-root at intermediate and logarithmic at large $|t|$) is [4]

$$\alpha(t) = \alpha_0 - \gamma \ln(1 - \beta\sqrt{t_0 - t}), \tag{6}$$

(and its generalization with multiple thresholds included, as in (5)) γ being another free parameter.

Note also that DAMA, contrary to the narrow resonance dual models, is equally applicable both for the reggeon (non-singlet) and Pomeron (singlet) components. Relevant trajectories have the same functional form, differing only by the values of their parameters (for a recent discussion and applications see e.g. [5] and [6]).

First attempts to give DAMA a dynamical partonic interpretation, based on its wide angle scaling behaviour are due to M.G.Schmidt [7], who associated the integration variable x in DAMA with the fraction of the partons' momenta in the infinite momentum frame and wrote the hadronic scattering amplitude in terms of the quark bound states. As shown in the subsequent papers [4], wide angle scaling scattering corresponds to values of x near 1/2 (90^0 scattering corresponds exactly to $x = 1/2$), i.e. the colliding hadron shares its momentum equally between the constituents. This behaviour in DAMA corresponds to the perturbative regime in QCD [6]. The resonance- and Regge behaviour, both essentially nonperturbative, are generated near the integration end points $x = 0$ and 1 [2]. Although they cannot be treated in perturbative QCD, we tentatively assume that in these processes the colliding hadrons interact inelastically via the exchange of a virtual parton producing a bunch of partons that hadronize producing again two final hadrons.

We identify the integrand in (4) with the deeply virtual Compton scattering (DVCS) amplitude,

$$f(x,s,t) = c\left(\frac{x}{g}\right)^{-\alpha(t(1-x)}\left(\frac{1-x}{g}\right)^{-\alpha(tx)}, \tag{7}$$

where

$$s = Q^2(1/x - 1) + m^2,$$

m being the hadron (proton) mass. By definition, the nucleon structure function is

$$F_2(x, Q^2) = 4\pi^2\alpha x \mathrm{Im} f(x, Q^2, t = 0), \tag{8}$$

By inserting trajectory (6) into this expression, one gets the following expression for the SF:

$$F_2(x, Q^2) = k\left(\frac{x}{g}\right)^{-\alpha_t+1} \mathrm{Im}\left(\frac{1-x}{g}\right)^{-\alpha_s+\gamma\ln[1+\beta\sqrt{m^2-(Q^2-m^2)(1-x)}]}, \tag{9}$$

where α_t and α_s are constants and k absorbs all the constants in front of the SF.

The first factor is typical of the small-x Regge behavior of any SF. It scales in x, so evolution in Q^2 is to be introduced. The behaviour of the second, large-x factor, that may produce oscillations in F_2, is of interest. It has a strong Q^2 dependence. SF (8) is nonzero above the threshold $Q^2 = m^2$. The low-Q^2 limit of (9) requires special treatment.

As usual, one writes the nucleon structure function F_2 as a sum of a singlet F_2^S and non-singlet F_2^{NS} components, $F_2 = F_2^S + F_2^{NS}$. In our case they have similar forms, differing only in the values of the parameters: e.g. $\alpha_t(0)$ is 1.08 for the singlet part (Pomeron) and 0.5 for the non-singlet etc. These parameters as well as the rest ones listed in the text may be fitted to the data.

The relation between the hadronic scattering amplitude and the SF presented in this contribution may be used either to gain information on the former from the latter or vice versa.

We thank F. Paccanoni A. Papa and E. Predazzi for discussions and criticism.

References

1. P. D. B. Collins: An Introduction to Regge Theory and High Energy Physics, Cambridge University Press, 1977.
2. A. Bugrij et al., Fortschritte der Physik **21** (1973) 427.
3. G. Veneziano, Nuovo Cim. **57A** (1968) 190.
4. L. L. Jenkovszky and Z. E. Chikovani, Yad. Fiz. **30** (1979) 531; A. I. Bugrij, L. L. Jenkovszky and Z. E. Chikovani, Z.Phys. C**4** (1979) 45.
5. R. Fiore, L. L. Jenkovszky, V. Magas and F. Paccanoni, hep-ph/9904202, Phys. Rev. D, in press;
6. R. Fiore, L. L. Jenkovszky, V. Magas, A. Papa and F. Paccanoni, to be published.
7. M. G. Schmidt, Phys.Letters **43B** (1973) 417.

E. Martynov

SOFT DIPOLE POMERON IN HADRONIC ELASTIC AND IN DEEP INELASTIC SCATTERING

E. MARTYNOV
Bogoliubov Institute for Theoretical Physics, 03143, Kiev-143, Ukraine
E-mail: martynov@bitp.kiev.ua

A brief review on the Dipole Pomeron model is given. The model not only describes data on hadron-hadron interactions, but also allows to describe data on the proton structure function with a Q^2 independent intercept. Moreover the chosen Dipole Pomeron has an intercept equal to one and does not violate unitarity limit on the total elastic cross-section.

The Regge approach [1] is one of the most powerful method to investigate hadronic interactions at high energies. In spite of many unsolved problems, its actual application is slowed down because of lacking new experimental data. Now, the main efforts of the experimentalists and theoreticians are concentrated on the deep inelastic scattering (DIS) and related processes. Again the Regge method shows its effectiveness in a wide kinematical domain.

It follows quite evidently from the observed growth of the hadronic total cross-sections and of the structure functions that the true Pomeron singularity is more complicated than that a simple j-pole. Even if a simple pole with an intercept larger than unity is used as an input, it should be unitarized in order to restore the Froissart-Martin bound [2], $\sigma_t(s) < C\ln^2 s$. Unfortunately a well defined strict procedure of unitarization is still absent, while approximate methods are used; as consequence, we do not know what is really the true Pomeron.

The well known and quite popular example of a model with an input Pomeron violating the unitarity bound (because it has $\alpha_{\mathcal{P}}(0) \approx 1.08 > 1$)) is the Donnachie-Landshoff Pomeron (D-L)[3]. Moreover, and regrettably, in some papers this model is identified with the Regge theory and its inability to describe some experimental data is declared as a problem of Regge theory. However, and happily, the D-L Pomeron is not the only phenomenological possibility to describe the above mentioned growth of observed quantities. Another way to satisfy this property is to suppose that the Pomeron is harder than a simple pole at $j = 1$. The simplest realization of such an hypothesis is the Dipole Pomeron model [4].

The Dipole Pomeron (DP) is a combination of the double and simple poles in the partial j-plane amplitude

$$\Phi(j,t) \propto \frac{\phi(j,t)}{[j-\alpha_{\mathcal{P}}(t)]^2} \approx \frac{\phi_1(t)}{[j-\alpha_{\mathcal{P}}(t)]^2} + \frac{\phi_2(t)}{j-\alpha_{\mathcal{P}}(t)}$$

with the trajectory $\alpha_{\mathcal{P}}(t) \approx 1+\alpha'_{\mathcal{P}}t$, linear at small t. The Dipole Pomeron is the unitarity limit for an isolated j-singularity with a unit intercept and with a linear trajectory at small t. If

$$\Phi(j,t) \propto (j-1-\alpha' t)^{-\nu-1} \quad \text{then} \quad \sigma_{tot} \propto \ln^{\nu} s/s_0, \quad \sigma_{el} \propto \ln^{2\nu-1} s/s_0,$$

$$\sigma_{el}/\sigma_{tot} \leq 1 \quad \Rightarrow \quad \nu \leq 1.$$

Thus in the present Dipole Pomeron model at $s \to \infty$:

$$\sigma_{tot}^{(\mathcal{P})}(s) \propto \sigma_{el}^{(\mathcal{P})}(s) \propto \sigma_{inel}^{(\mathcal{P})}(s) \propto \ln(s/s_0).$$

The Dipole Pomeron in hadronic processes. At preasymptotic energies the contribution of other reggeons (f, ω *etc...*) should be added. If only pp and $\bar{p}p$ amplitudes are considered one can write [a]

$$A_{(pp)}^{(\bar{p}p)}(s,t) = \mathcal{P}(s,t) + f(s,t) \pm \omega(s,t) \tag{1}$$

with

$$\mathcal{P}(s,t) = i[g_1(t)\ln(-is/s_0) + g_2(t)](-is/s_0)^{\alpha_{\mathcal{P}}(t)-1}, \quad s_0 = const, \tag{2}$$

$$r(s,t) = \eta_r\, g_r(t)(-is/s_0)^{\alpha_r(t)-1}, \quad r = f,\omega, \quad \eta_f = i, \quad \eta_\omega = 1. \tag{3}$$

The DP model was applied [5] to the analysis of available data on the total cross-sections and on the real to imaginary part ratios (at $t=0$ and $\sqrt{s} \geq 5$ GeV) for meson-nucleon and nucleon-nucleon interactions. It leads to the best description ($\chi^2/d.o.f. \approx 1.12$) comparing with other models. A new very detailed and circumstantial analysis of these data confirming our conclusions is presented at this conference in the talk of V. Ezhela [6]

Furthermore, in the framework of a modified additive quark model (where corrections for the coupling of Pomeron with two quark lines as well as new counting rules for the secondary reggeons were taken into account) [7] the Dipole Pomeron was used to describe simultaneously cross-sections of $p^{\pm}p, \pi^{\pm}p, \gamma p, \gamma\gamma$ interactions, related by factorization conditions.

An important property of the Dipole Pomeron model is that all fits give a high value of the f-reggeon intercept, $\alpha_f(0) \approx 0.8 \div 0.82$. Does such an intercept contradict to the data on the f-trajectory known from a resonance

[a] We do not consider here Odderon because it is out of the subject of talk.

region? The answer is "yes" if trajectory is assumed to be linear. However besides of general theoretical arguments in favor of nonlinear trajectories, the experimental data on resonances lying on f-trajectory indicate its nonlinearity (see Fig.1). A more realistic trajectory than those shown in the Fig.1, $\alpha_f(t) = \alpha_f(0)+\gamma_1(\sqrt{4m_\pi^2}-\sqrt{4m_\pi^2-t})+\gamma_2(\sqrt{t_1}-\sqrt{t_1-t})$, gives $0.77 < \alpha_f(0) < 0.87$.

In the Fig.2, we show how $\alpha_f(0)$ is correlated with a power of $\ln s$ in behaviour of the total cross-section if the amplitudes are parameterized in the form (1-3), but with the replacement $\ln(-is/s_0) \to \ln^\gamma(-is/s_0)$ (see [5] for details).

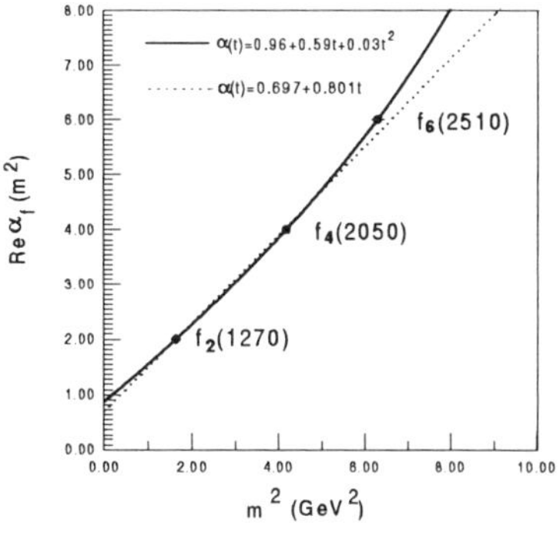

Figure 1. Real part of f-trajectory.

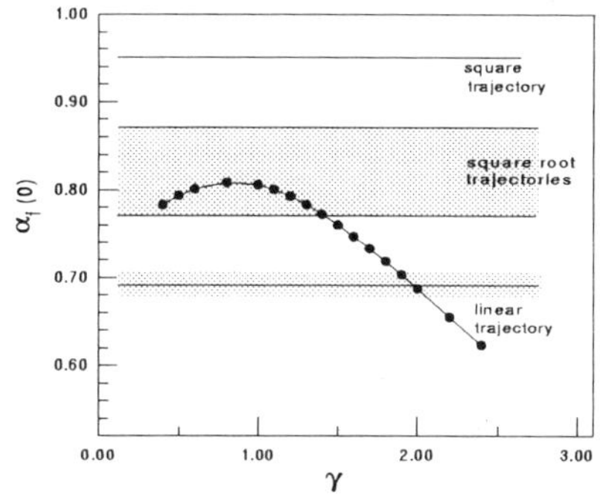

Figure 2. Intercept of f-trajectory vs. γ.

The Dipole Pomeron in deep inelastic scattering. We support and investigate the point of view that there is only one Pomeron (two Pomerons, "soft" and "hard", are considered in [9,10]). This unique Pomeron is universal and factorizable (at least the main term should satisfy factorization at c.m.s. energy $W \gg m_p$), *i.e.* it is the same in all processes, only vertex functions depend on interacting particles. Thus we think that Pomeron has a Q^2 independent trajectory. As in the pure hadron case we use a Pomeron trajectory with a unit intecept ($\alpha_{\mathcal{P}}(0) = 1$).

Defining the Dipole Pomeron model for DIS, we start from the expression connecting the transverse cross-section of $\gamma^* p$ interaction to the proton structure function F_2^p and the optical theorem for forward scattering amplitude

$$\sigma_T^{\gamma^* p} = \Im m A(W^2, Q^2; t=0) = \frac{4\pi^2\alpha}{Q^2(1-x)}(1 + 4m_p^2 x^2/Q^2) F_2^p(x, Q^2), \quad (4)$$

where $\sigma_L^{\gamma^* p} = 0$ is assumed. The forward scattering at $W^2 = Q^2(1/x-1)+m_p^2$

far from the threshold $W_{th} = m_p$ is dominated by the Pomeron and the f-reggeon [b]

$$A(W^2, Q^2; t=0) = P(W, Q^2) + f(W, Q^2), \qquad P(W^2, Q^2) = P_1 + P_2,$$
$$P_1 = iG_1(Q^2)\ln(-iW^2/m_p^2)(1-x)^{B_1}, \qquad P_2 = iG_2(Q^2)(1-x)^{B_2}.$$
$$f(W^2, Q^2) = iG_f(Q^2)(-iW^2/m_p^2)^{\alpha_f(0)-1}(1-x)^{B_f}.$$

It evidently follows from the experimental data (Fig.3) that $Q^2\sigma^{\gamma^* p}$ decreases with Q^2 at least at high Q^2. Therefore we choose

$$G_i(Q^2) = g_i/(1+Q^2/Q_i^2)^{D_i},$$

expecting $D_i > 1$ at high Q^2. Here we only note that as it follows from the fit, D_i and B_i should be functions of Q^2. By hypothesis each of them varies between two constants when Q^2 goes from 0 to ∞. The details of the parameterization of the real functions $D_i(Q^2)$, $B_i(Q^2)$ (as well as the references for the data) are given in the paper[8].

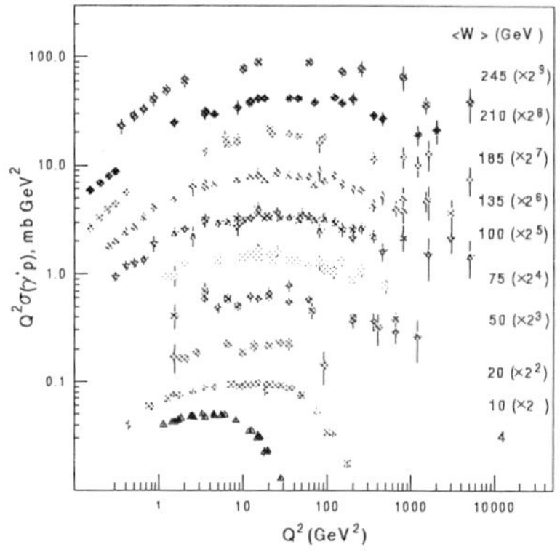

Figure 3. Data on $Q^2\sigma^{\gamma^* p}$ as function of Q^2 at fixed W.

A fit was performed to the 1209 experimental points in the region

$$W \geq 3GeV^2, \quad 0 \leq Q^2 \leq 5000GeV^2, \quad 0 \leq x \leq 0.85,$$

and a quite good description of data was obtained: $\chi^2/d.o.f. = 1.11$ and $\chi^2/d.o.f. = 0.77$ if $x \leq 0.1$ and any dependence on x is canceled ($B_i = 0$).

A detailed analysis[8] of the fit leads to the following conclusions:

• The possibility to describe available data with the Pomeron that does not violate the Froissart-Martin limit is shown.

• The preasymptotic contributions, P_2 and f, play an important role in the considered kinematical domain.

• The Dipole Pomeron has a Q^2 independent intercept. There is no contradictions with the "experimental data" on $\Delta(Q^2) = \alpha_{\mathcal{P}} - 1$. Indeed, the effective Pomeron intercept is extracted from the data on $F_2^p(x, Q^2)$ making use of the simplified parameterization of the structure function $F_2^p = a + bx^{-\Delta}$ at fixed Q^2.

---

[b]We ignored an a_2-reggeon considering the f-term as an effective one at $W > 3$ GeV.

On the second hand, one can write

$$F_2^p(x, Q^2) = G(Q^2)(1/x)^{\Delta_{eff}(x,Q^2)}$$

and consequently

$$\frac{\partial \ln F_2^p}{\partial \ln(1/x)} = \Delta_{eff} + x\frac{\partial \Delta_{eff}}{\partial x}\ln(1/x)$$

If $\Delta_{eff}(x, Q^2) \approx$ const at $x \ll 1$ then

$$\frac{\partial \ln F_2^p}{\partial \ln(1/x)} \approx \Delta_{eff}.$$

Comparison of this derivative in our model with the experimental data on Δ_{eff} (or λ_{eff} in the notation of the experimentalists) is shown in Fig.4. The open triangles and the curve correspond to the $\frac{\partial \ln F_2^p}{\partial \ln(1/x)}$ calculated in the given model.

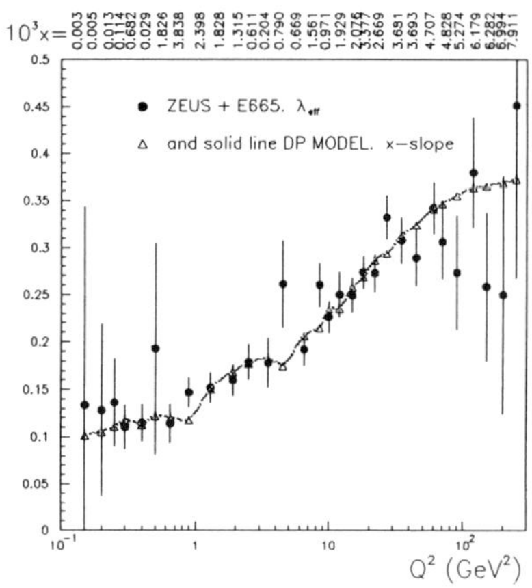

Figure 4. $\Delta_{eff}(Q^2)$, extracted from experimental data and compared to the x-slope calculated in the Dipole Pomeron Model.

There is an important prediction of the model concerning $\frac{\partial \ln F_2^p}{\partial \ln(1/x)}$ at fixed x. This x-slope begins to decrease when $Q^2 > \sim 100 GeV^2$. It would be interesting to compare the model with new high Q^2 data when they are published. A comparison of this quantity in our model [8], in a QCD motivated ALLM model [11] and in the interpolating model [12] (between a Regge-type behaviour and a solution of the evolution equation) is shown in Fig.5

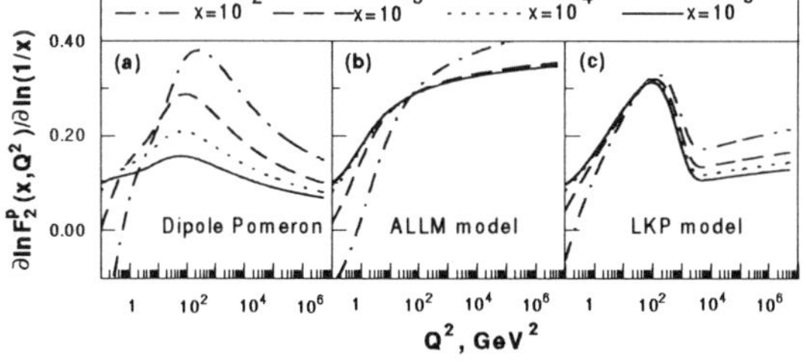

Figure 5. A behaviour of x-slope in three models (see text for details).

The Soft Dipole Pomeron model well describes also the Q-slope or

$\frac{\partial F_2^p(x,Q^2)}{\partial \ln Q^2}$. A more detailed discussion of the properties of the Dipole Pomeron Model (comparison with the data on $\sigma^{\gamma^* p}$, F_2^p and so on) as well as some predictions of the model are given in the paper [8].

Conclusion. The available data on hadron-hadron and photon-hadron cross-sections can be described in a traditional Regge approach with a Soft Pomeron. It is not necessary to use an intercept depending on Q^2 and higher than one. A scale in Q^2, where a nonperturbative Regge behaviour is transformed to a QCD perturbative one, is still unknown in the presented approach and requires an additional consideration.

I thank very much my coworkers P.Desgrolard, M.Giffon, A.Lengyel and E.Predazzi for many fruitful discussions, for interesting and useful collaboration.

References

1. P. D. B. Collins, *An introduction to Regge theory & high energy physics*, Cambridge University press (Cambridge) 1977.
2. M. Froissart, *Phys. Rev.* **123**, 1053 (1961); A. Martin, *Phys. Rev.* **129**, 432 (1963) and **29**, 993 (1963).
3. A. Donnachie and P. Landshoff, *Phys. Lett.* B **296**, 227 (1992);
4. T. Sawada, *Nuovo Cimento* A **48**, 534 (1967);R. J. N. Phillips, Preprint of Rutheford Lab. RL-73-04, 1973; A. I. Bugrij, L. L. Jenkovszky, N. A. Kobylinsky and V. P. Shelest, *Lett. Nuovo Cimento* **6**, 577 (1973); L. L. Jenkovszky, E. S. Martynov and B. V. Struminsky, *Phys. Lett.* B **249**, 535 (1990).
5. P. Desgrolard, M. Giffon, A. Lengyel and E. Martynov, *Nuovo Cimento* A **107**, 637 (1994); P. Desgrolard, M. Giffon and E. Martynov, *Nuovo Cimento* A **110**, 537 (1997).
6. V. Ezhela *et al.*, talk at this Conference, hep-ph/9908218.
7. P. Desgrolard, M. Giffon, E. Martynov and E. Predazzi, Eur. Phys. Journ. C **9**, 623 (1999).
8. M. Bertini, *et al.*, *Rivista Nuovo Cimento* **19**, 1 (1996).
9. A. Donnachie and P. Landshoff, *Phys. Lett.* B **437**, 408 (1998).
10. P. Desgrolard, A. Lengyel and E.Martynov, Eur. Phys. Journ. C **7**, 655 (1999); hep-ph/9811380.
11. H. Abramovicz, A. Levy, hep-ph/9712415.
12. P. Desgrolard, L. Jenkovszky, F. Paccanoni, Eur. Phys. Journ. C **7**, 263 (1999).

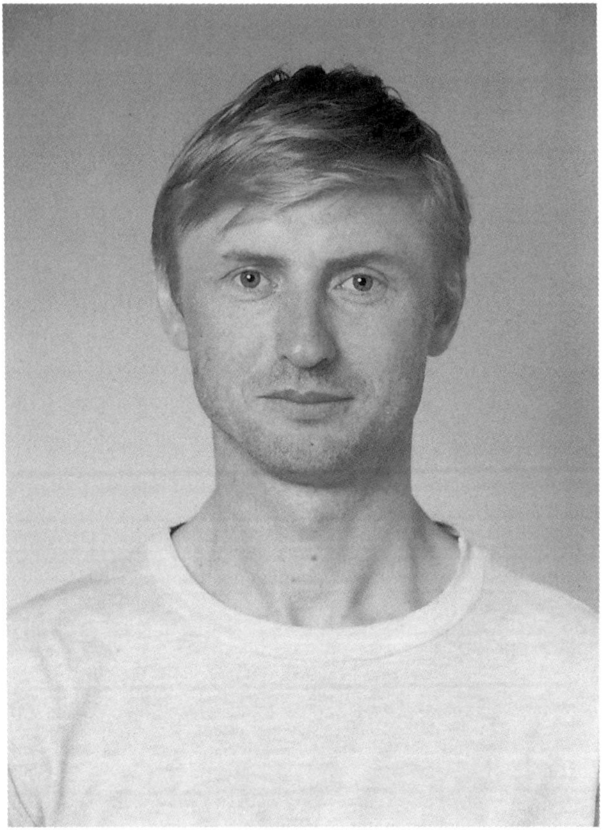

A. V. Kotikov

SMALL X BEHAVIOUR OF PARTON DISTRIBUTIONS HAVING SOFT INITIAL CONDITIONS

A. V. KOTIKOV
Particle Physics Laboratory
Joint Institute for Nuclear Research
141980 Dubna, Russia
E-mail: kotikov@sunse.jinr.ru

G. PARENTE
Departamento de Física de Partículas
Universidade de Santiago de Compostela
15706 Santiago de Compostela, Spain
E-mail: gonzalo@gaes.usc.es

We investigate the Q^2 evolution of parton distributions at small x values, recently obtained in the case of soft initial conditions. The results are in excellent agreement with deep inelastic scattering experimental data from HERA.

The measurements of the deep-inelastic scattering structure function F_2 in HERA [1] have permitted the access to a very interesting kinematical range for testing the theoretical ideas on the behavior of quarks and gluons carrying a very low fraction of momentum of the proton, the so-called small x region. In this limit one expects that non-perturbative effects may give essential contributions. However, the resonable agreement between HERA data and the NLO approximation of perturbative QCD that has been observed for $Q^2 > 1\text{GeV}^2$ (see the recent review in [2]) indicates that perturbative QCD could describe the evolution of structure functions up to very low Q^2 values, traditionally explained by soft processes. It is of fundamental importance to find out the kinematical region where the well-established perturbative QCD formalism can be safely applied at small x.

The standard program to study the small x behavior of quarks and gluons is carried out by comparison of data with the numerical solution of the DGLAP equations [a] by fitting the parameters of the x profile of partons at some initial Q_0^2 and the QCD energy scale Λ (see, for example, [3]). However, if one is interested in analyzing exclusively the small x region, there is the alternative of doing a simpler analysis by using some of the existing analytical solutions of DGLAP in the small x limit (see [2] for review). This was done so in Ref. [4,5,6]

[a] At small x there is another approach based on the BFKL equation, whose application is out of the scope of this work.

where it was pointed out that the HERA small x data can be interpreted in terms of the so called doubled asymptotic scaling phenomenon related to the asymptotic behaviour of the DGLAP evolution discovered [7] many years ago. Here we will illustrate the results obtained recently in [6].

1. Thus, our purpose is to find out the small x asymptotic form of parton distributions in the framework of the DGLAP equation starting at some Q_0^2 with the flat function:

$$f_a(Q_0^2) = A_a \quad (a = q, g), \tag{1}$$

where f_a are the parton distributions multiplied by x and A_a are unknown parameters that have to be determined from data. Through this work at small x we neglect the non-singlet quark component.

The effective method to reproduce the x-dependence of parton distributions at low x is based on the separation of the singular and regular parts of the exact solution for the moments [6]. The method allows in simple way to reproduce the leading order (LO) results [4] and to construct the x-dependence of parton distributions at next-to-leading order (NLO):

$$\begin{aligned}
f_a(x,Q^2) &= f_a^+(x,Q^2) + f_a^-(x,Q^2) \qquad \text{and} \\
f_a^-(x,Q^2) &= A_a^-(Q^2,Q_0^2)exp(-d_-(1)s - D_-(1)p) \ + \ O(x) \\
f_g^+(x,Q^2) &= A_g^+(Q^2,Q_0^2)I_0(\sigma)exp(-\overline{d}_+(1)s - \overline{D}_+(1)p) \ + \ O(\rho) \\
f_q^+(x,Q^2) &= A_q^+(Q^2,Q_0^2)\Big[(1-\overline{d}_{\pm}^{q}(1)\alpha(Q^2))\rho I_1(\sigma) + 20\alpha(Q^2)I_0(\sigma)\Big] \\
&\quad \cdot exp(-\overline{d}_+(1)s - \overline{D}_+(1)p) + O(\rho) \\
F_2(x,Q^2) &= e \cdot \left(f_q(x,Q^2) + \frac{2}{3} f\alpha(Q^2) f_g(x,Q^2) \right)
\end{aligned} \tag{2}$$

where $I_\nu(\sigma)$ is the modified Bessel function, s and p are given by $s = ln(\alpha(Q_0^2)/\alpha(Q^2))$, $p = \alpha(Q_0^2) - \alpha(Q^2)$, and

$$\sigma = 2\sqrt{(\hat{d}_+ s + \hat{D}_+ p) ln x} \ , \qquad \rho = \sqrt{\frac{(\hat{d}_+ s + \hat{D}_+ p)}{ln x}} = \frac{\sigma}{2ln(1/x)}$$

are NLO generalizations of the corresponding Ball-Forte variables and

$$\begin{aligned}
A_g^+ &= \Big[1 - \frac{80}{81} f\alpha(Q^2)\Big] A_g + \frac{4}{9}\Big[1 + 3(1+\frac{1}{81}f)\alpha(Q_0^2) - \frac{80}{81} f\alpha(Q^2)\Big] A_q \ , \\
A_g^- &= A_g - A_g^+, \ A_q^+ = \frac{f}{9}\left(A_g + \frac{4}{9}A_q\right), \ A_q^- = A_q - 20\alpha(Q_0^2)A_q^+
\end{aligned} \tag{3}$$

Wherever in this work we use the notation $\alpha = \alpha_s/(4\pi)$.

The nonzero components of the singular and regular parts of the terms $d_{\pm}$ and $D_{\pm}$ have the form:

$$\begin{aligned}
\hat{d}_{+} &= -\frac{12}{\beta_0} \; , \quad \overline{d}_{+}(1) = 1 + \frac{4}{3\beta_0} f \; , \quad d_{-}(1) = \frac{16}{27\beta_0} f \; , \\
\hat{d}_{++} &= \frac{412}{27\beta_0} f \; , \quad \hat{d}^{q}_{+-} = -20 \; , \quad \overline{d}^{g}_{+-}(1) = \frac{80}{81} f \; , \\
\overline{d}_{++}(1) &= \frac{8}{\beta_0}\left(36\zeta_3 + 33\zeta_2 - \frac{1643}{12} + \frac{2}{9} f \left[\frac{68}{9} - 4\zeta_2 - \frac{13}{243} f\right]\right) \; , \\
\overline{d}^{q}_{+-}(1) &= \frac{134}{3} - 12\zeta_2 - \frac{13}{81} f \; , \quad \overline{d}^{g}_{-+}(1) = -3\left(1 + \frac{f}{81}\right), \\
d_{--}(1) &= \frac{16}{9\beta_0}\left(2\zeta_3 - 3\zeta_2 + \frac{13}{4} + f\left[4\zeta_2 - \frac{23}{18} + \frac{13}{243} f\right]\right) \; ,
\end{aligned} \tag{4}$$

where β_0 is the first coefficient of the QCD β-function, ζ_n are the Euler ζ-functions and f is the number of active quarks.

Our NLO results given in Eq. (2) coincide with the corresponding of Ball and Forte [5] if one neglects the "$-$" component, expands our NLO singular terms $(\rho)^k I_{k+1}(\sigma)$ in the vicinity of the point $\sigma = \sigma_{LO}$ and ignores the NLO regular terms (i.e. put $exp(-\overline{D}_{+}(1)p) = exp(-D_{-}(1)p) = 1$ and cancel the terms proportional to $\alpha(Q_0^2)$ into the normalization factors $A_g^{\pm}$ and A_q^{-}). We think, however, that this expansion is not so correct because it generates NLO corrections of the order of the LO terms.

2. With the help of the results obtained in the previous section we have analyzed F_2 HERA data at small x from the H1 collaboration [1]. In order to keep the analysis as simple as possible we have fixed $\Lambda_{\overline{\rm MS}}(n_f = 4) = 250$ MeV which is a reasonable value extracted from the traditional (higher x) experiments. The initial scale of the parton distributions was also fixed into the fits to $Q_0^2 = 1\ GeV^2$, although later it was released to study the sensitivity of the fit to the variation of this parameter. The analyzed data region was restricted to $x < 0.01$ to remain within the kinematical range where our results are accurate. Finally, the number of active flavors was fixed to f=4.

Fig. 1 shows F_2 calculated from the fit with $Q^2 > 1$ GeV^2 in comparison with H1 data. Only the lower Q^2 bins are shown. One can observe that the NLO result (dot-dashed line) lies closer to the data than the LO curve (dashed

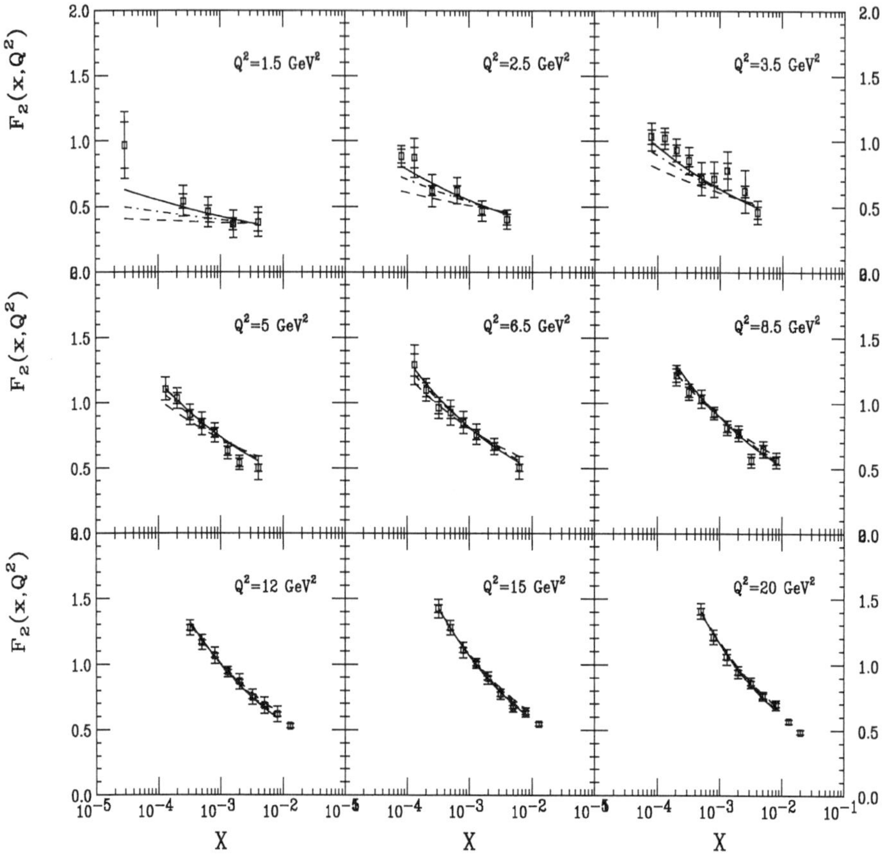

Figure 1: The structure function F_2 as a function of x for different Q^2 bins. The experimental points are from H1. The inner error bars are statistic while the outer bars represent statistic and systimatic errors added in quadrature. The dashen and dot-dashed curves are obtained from fits at LO and NLO respectively with fixed $Q_0^2 = 1$ GeV^2. The solid line is from the fit at NLO giving $Q_0^2 = 0.55$ GeV^2.

line). The lack of agreement between data and lines observed at the lowest x and Q^2 bins suggests that the initial flat behavior should occur at Q^2 lower than 1 GeV^2. In order to study this point we have done the analysis considering Q_0^2 as a free parameter. Comparing the results of the fits (see [6]) one can notice the better agreement with the experiment of the NLO curve at fitted $Q_0^2 = 0.55GeV^2$ (solid curve) is apparent at the lowest kinematical bins.

3. *Resume.* We have shown that the results developed recently in [6] have quite simple form and reproduce many properties of parton distributions at small x, that have been known from global fits.

We found very good agreement between our approach based on QCD at NLO approximation and HERA data, as it has been observed earlier with other approaches (see the review [2]). Thus, the nonperturbative contributions as shadowing effects, higher twist effects and others seems to be quite small or seems to be canceled between them and/or with $ln(1/x)$ terms contained by higher orders of perturbation theory. To clear up the correct contributions of nonperturbative dynamics and higher orders containing strong $ln(1/x)$ terms, it is necessary more precise data and further efforts in developing of theoretical approaches.

Acknowledgments

One of the authors (G.P.) was supported in part by Xunta de Galicia (XUGA-20602B98) and CICYT (AEN96-1673).

References

1. H1 Collab.: S. Aid et al., *Nucl.Phys.* B **470**, 3 (1996); ZEUS Collab.: M. Derrick et al., *Zeit.Phys.* C **72**, 399 (1996); C **69**, 607 (1996).
2. A.M. Cooper-Sarkar et al., *Int.J.Mod..Phys.* A **13**, 3385 (1998).
3. A.D. Martin, W.S. Stirling and R.G. Roberts, *Phys.Lett.* B **387**, 419 (1996); M. Gluck, E. Reya and A. Vogt, *Eur.Phys.J.* C **387**, 461 (1998).
4. R.D. Ball and S. Forte, *Phys.Lett.* B **336**, 77 (1994); L. Mankiewicz, A. Saalfeld and T. Weigl, *Phys.Lett.* B **393** (1997) 175.
5. R.D. Ball and S. Forte, *Acta Phys.Polon.* B **26**, 2097 (1995); *Nucl.Phys.Proc.Suppl.* A **54**, 163 (1997).
6. A.V. Kotikov and G. Parente, *Nucl.Phys.* B **549**, 242 (1999).
7. A. De Rújula et al., *Phys.Rev.* **D10** (1974) 1649.

HADRONIC FINAL STATES IN DIFFRACTION AT HERA

A. BUNYATYAN
Max-Planck-Institut für Kernphysik,
Saupfercheckweg 1, 69117 Heidelberg, Germany
and Yerevan Physics Institute, Armenia
E-mail: bunar@mail.desy.de

Representing the H1 and ZEUS Collaborations

Measurements are presented of global event shapes, energy flow, charged particle multiplicity, high p_T jets and open charm production in diffractive ep scattering, performed by the H1 and ZEUS Collaborations at HERA. The results are in agreement with models based on the parton densities extracted from inclusive ep diffractive scattering at HERA and support the hypothesis that the momentum of the diffractive exchange is carried largely by hard gluons.

1 Introduction

There are considerable attempts to understand the underlying dynamics of diffractive interactions in terms of quantum chromodynamics. At HERA, the contribution to the proton structure function from diffractive processes has been measured, for example, in terms of a diffractive structure function $F_2^{D(3)}$ [1,2]. Within the DGLAP ansatz, diffractive parton distributions have been extracted that are dominated at low Q^2 by gluons carrying large fractions of the exchanged momentum. With the assumption of diffractive factorization, i.e. a flux of colourless objects and partonic coupling, the parton distributions for the Pomeron extracted from $F_2^{D(3)}$ are expected to universally describe hard diffractive interactions once perturbation theory may be applied.

The analysis of hadronic final states in diffractive interactions provides an independent test of this picture. In this report some recent results of the H1 and ZEUS experiments are reviewed.

2 Selection of diffractive events

The generic diffractive process $ep \rightarrow eXY$ is illustrated in Fig.1. The systems X and Y are defined by the largest gap in rapidity of final state hadrons and Y is the system closest to the outgoing proton direction. The diffractive process can be described in terms of:

$$M_X^2 = p_X^2 \quad M_Y^2 = p_Y^2 \quad t = (p_p - p_Y)^2 \quad x_{I\!P} = \frac{q \cdot (p_p - p_Y)}{q \cdot p_p}.$$

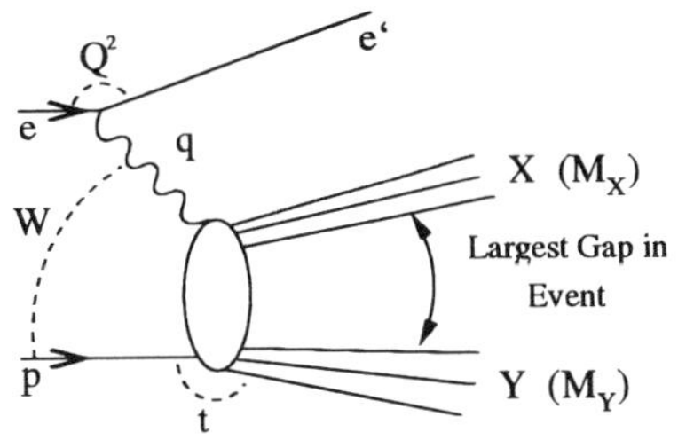

$$s = (p_e + p_p)^2 \quad Q^2 = -q^2$$

$$W^2 = (q + p_p)^2$$

$$x = \frac{Q^2}{2q \cdot p_p} \qquad y = \frac{q \cdot p_p}{p_e \cdot p_p}$$

Figure 1: Illustration of the generic diffractive process $ep \to eXY$

t is the squared four-momentum transferred between the photon and the incoming proton, and $x_{I\!P}$ is the fraction of the proton beam momentum transferred to the diffractive system X. Experimentally, in analyses of hadronic final states at HERA, two different methods are used to select diffractive dissociation:

• Detection of the leading proton (LPS method): both experiments have dedicated proton spectrometers to detect and measure directly protons which are produced at very small scattering angles ($\theta \lesssim 1\ mrad$). The data sample selected with this method corresponds to single photon dissociation ($M_Y \equiv M_p$).

• Large Rapidity Gap selection (LRG method): diffractive events are characterized by a rapidity gap between the two hadronic subsystems X and Y of the size $\Delta\eta \simeq ln(W^2/M_X^2)$. The gap is identified by an absence of activity in the forward part of detector. Since the system Y is not detected, the selected data sample also contains events where the proton dissociates to low mass states (e.g. $M_Y \lesssim 1.6\ GeV$ in H1 analyses).

3 Models for diffractive hard scattering

The models for diffractive hard scattering which are used in analyses for comparison with data can be in general subdivided into three classes:

• Factorizible models (e.g. Monte–Carlo programs RAPGAP[3] and POMPYT[4]): In these models the photon–proton diffractive scattering is factorized as a flux of spacelike Pomerons from the proton and subsequent photon–Pomeron interaction. The parton densities in the Pomeron have been analyzed in the form of QCD fits to $F_2^{D(3)}$ data [1,15]. The expected topology of the hadronic final state depends strongly on the partonic structure of the Pomeron. For a quark dominated Pomeron the final state in lowest order is defined by a $q\bar{q}$ configuration with low transverse momentum relative to the photon direction in the $\gamma^* I\!P$ center of mass (CM) frame. A gluon dominated Pomeron interacts largely by the boson–gluon fusion (BGF) mechanism with a $q\bar{q}g$ system in the final state: the transverse momenta of the outgoing partons and the amount of gluon radiation are therefore expected to be larger than in the case of a quark dominated Pomeron.

• Two-Gluon models (e.g. Monte–Carlo program RIDI [5]):
These models describe the diffractive $\gamma^* p$ interaction in terms of the elastic scattering of partonic fluctuations of the photon ($q\bar{q}$, $q\bar{q}g$,...) from the proton via the exchange of two gluons in a net colour–singlet configuration. The simplest $q\bar{q}$ state is relevant only at small transverse momenta $\hat{p}_T$ of the outgoing partons due to colour transparency effect. At large $\hat{p}_T$ the cross–section is therefore dominated by the scattering of $q\bar{q}g$ and higher multiplicity components.

• Soft Colour Interactions (SCI) (e.g. Monte–Carlo program LEPTO [6]):
In this model LRG events are the result of normal ep DIS scattering with subsequent soft interactions between the outgoing partons. These interactions alter the final–state colour configurations, and hence may give rise to large rapidity gaps.

4 Global Event Shapes

The topological features of the hadronic final state in γ^* dissociation were studied at HERA by analyzing the event–shape variables thrust $\langle T \rangle$ and sphericity $\langle S \rangle$ [7,8]. These variables have been used extensively in e^+e^- experiments at various CM energies [9], and thus are interesting for a comparison with diffractive DIS data. Figures 2a,b show the average values of thrust and sphericity as a function of M_X for ZEUS DIS events, selected with the leading proton (LPS) method. For comparison, the data from e^+e^- annihilation events are plotted at $M_X = \sqrt{s_{e^+e^-}}$. The following features are observed:

• $\langle T \rangle$ increases ($\langle S \rangle$ decreases) with M_X. This signifies an increasing back-to-back correlation of the momentum flow as the final state mass increases, and the thrust direction defines the topological axis of the event.

• There is a striking similarity between the energy dependence of $\langle T \rangle$ and $\langle S \rangle$ in e^+e^- annihilation and in virtual photon dissociation, both in shape and absolute values. This observation suggests that the diffractive DIS hadronic system can be predominantly viewed as a simple $q\bar{q}$ system. On the other hand, the RAPGAP Monte–Carlo, which implements a gluon dominated Pomeron and uses the ARIADNE fragmentation scheme can reproduce the event shape variables well. The RIDI Monte–Carlo fails to describe ZEUS data, a deficiency that might be remedied by tuning the relative contributions of $q\bar{q}$ and $q\bar{q}g$ processes.

Also plotted in Fig.2a are the average thrust values measured by H1 where the diffractive DIS sample is tagged with large rapidity gaps (LRG). The $\langle T \rangle$ of H1 LRG events is systematically lower than in the ZEUS LPS data and e^+e^- annihilation events at low M_X. This result $\langle T \rangle_{H1} < \langle T \rangle_{e^+e^-}$ means that the parton multiplicities higher than a simple $q\bar{q}$ state are more important in H1 LRG DIS events than in e^+e^- annihilation. The discrepancy between the H1 and ZEUS results is still under investigation, but it may be due to different methods of diffractive event tagging.

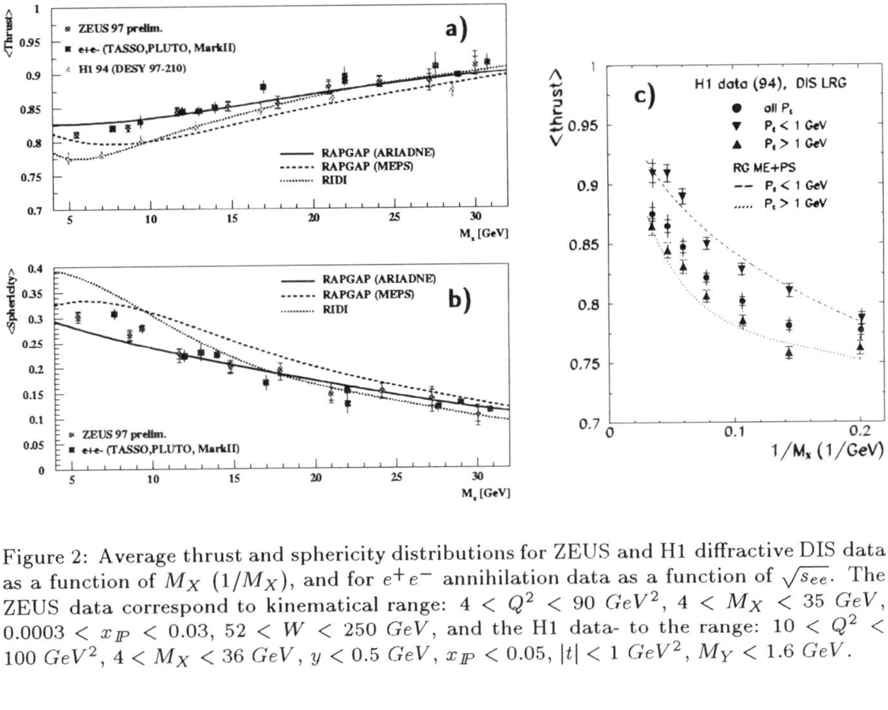

Figure 2: Average thrust and sphericity distributions for ZEUS and H1 diffractive DIS data as a function of M_X $(1/M_X)$, and for e^+e^- annihilation data as a function of $\sqrt{s_{ee}}$. The ZEUS data correspond to kinematical range: $4 < Q^2 < 90\ GeV^2$, $4 < M_X < 35\ GeV$, $0.0003 < x_{I\!P} < 0.03$, $52 < W < 250\ GeV$, and the H1 data- to the range: $10 < Q^2 < 100\ GeV^2$, $4 < M_X < 36\ GeV$, $y < 0.5\ GeV$, $x_{I\!P} < 0.05$, $|t| < 1\ GeV^2$, $M_Y < 1.6\ GeV$.

Given the thrust axis, two thrust jets can be defined as groups of particles with positive or negative momentum components along the thrust axis. In Fig.2c the average thrust is shown as a function of $1/M_X$ for events with differing transverse thrust jet momentum relative to the proton beam direction. At each M_X point the average thrust for large P_t is lower than for small P_t, indicating the presence of configurations with at least 3 non-collinear partons. RAPGAP predictions for a mixture of $q\bar{q}$ and $q\bar{q}g$ configurations confirm this conclusion.

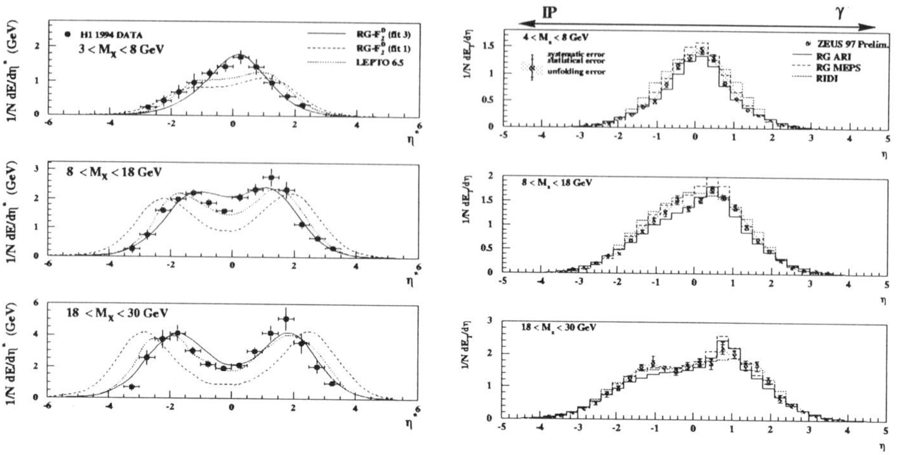

Figure 3: The energy flow from H1 (a) and the transverse energy flow from ZEUS (b) for the diffractive DIS data as a function of pseudorapidity for 3 different M_X intervals.

5 Hadron Production

Figure 3 (left) shows the event-normalized energy flow against pseudorapidity $\eta^* = -\ln(\tan\ \theta/2)$ in three regions of M_X, where θ is the polar angle of energy deposit with respect to the a $\gamma^* I\!P$ CM axis. The distribution is approximately symmetrical about $\eta^* = 0$, with similar levels of energy flow in the two hemispheres. At higher masses, $M_X > 8\ GeV$, a two–peaked structure is seen, indicating that the major topological property of these events is of a 2-jet nature (current jet and remnant jet). The data are well described by the RAPGAP model with the gluon dominated Pomeron parton densities ($RG - F_2^D$-fit 3), and for $M_X > 8\ GeV$ by LEPTO, whereas the the RAPGAP model with the quark dominated Pomeron parton densities ($RG - F_2^D$-fit 1) model predicts too much energy flow at the largest accessible pseudorapidity and fails to account for the observed energy flow in the central region. A similar behavior has been seen by the ZEUS experiment using the LPS selection method [8]. Figure 3 (right) shows the distribution of transverse energy as a function of η^* measured by ZEUS. A slight asymmetry is visible at high M_X with less energy going in the Pomeron direction, which indicates a suppression of particle production associated with the Pomeron remnant. This asymmetry is predicted from the Monte–Carlo.

The observed features can be related to the gluon content of the Pomeron and to the role of BGF. The transverse momentum generated by this process reduces the accessible pseudorapidity range along the $\gamma^* I\!P$ axis at fixed M_X and induces additional energy flow in more central regions.

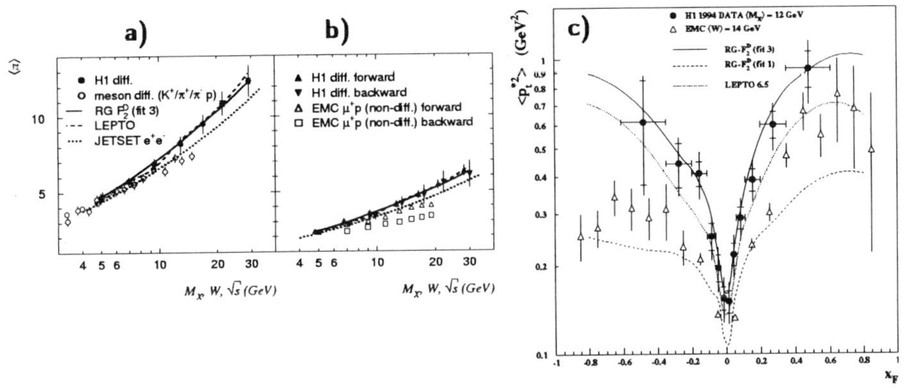

Figure 4: Average charged particle multiplicity in full phase space (a) and in single hemispheres (b) as a function of M_X. Seagull plot (c) showing H1 diffractive DIS data together with EMC μp DIS data.

Complementary to the energy flow analyses we studied the multiplicity of charged particles [11]. Figure 4a shows the dependence of the mean charged particle multiplicity $\langle n \rangle$ on M_X for the H1 diffractive DIS data compared with results from meson diffraction [12] and e^+e^- annihilation (modeled by JETSET). The multiplicities for all these processes are similar at low masses, but for $M_X \gtrsim 10\ GeV$ the multiplicity grows faster with mass in diffractive DIS than in other processes. It may be concluded from this result that low-mass diffraction in hadron collisions and photon dissociation in DIS at HERA can be interpreted as the hadronization of a single string. High-mass diffraction involves additional mechanisms besides $q\bar{q}$ fragmentation and gluon bremsstrahlung from the quark. Figure 4b shows the mean multiplicity in the forward and backward hemispheres separately. There is no evidence for an asymmetry between the two hemispheres, in contrast to what is observed in fixed-target μp DIS.

Confirmation of the general trends discussed above is obtained from the "seagull plot" distribution, which was studied both by H1 [10] and ZEUS [8]. The H1 result is shown in Fig.4c, in which the mean transverse momentum squared, $\langle p_T^{*2} \rangle$, is plotted as a function of the Feynman-x variable, $x_F = 2p_{||}^*/M_X$, where $p_{||}^*$ is the component of the particle's momentum in the direction of the incoming proton. The plot shows a greater degree of symmetry than is present in the EMC $\gamma^* p$ data [13]. This is consistent with having a more point-like partonic system in the target fragmentation hemisphere in the diffractive data than in $\gamma^* p$ data, where the extended nature of the proton remnant leads to a restricted phase space for parton radiation. $\langle p_T^{*2} \rangle$ is significantly higher in the H1 diffractive data than in $\gamma^* p$ data which can be attributed to the transverse momentum generated by the BGF process. This behavior is well described by the gluon dominated Pomeron model (RAPGAP-fit 3) whereas the $\langle p_T^{*2} \rangle$ values predicted by the quark dominated Pomeron model (RAPGAP-fit 1) model are too low at all x_F.

In conclusion, the studies of global event shapes, energy flow and charged particle multiplicity support the conclusion reached in the analysis of the diffractive structure function $F_2^{D(3)}$ that, at low Q^2, the momentum of the diffractive exchange is carried largely by hard gluons.

6 High p_T Jet production

Jet production has been investigated both in photoproduction ($Q^2 \sim 0$) and DIS ($Q^2 \gg 0$) [14,15]. Events were selected which contained two jets with transverse energies above 5 GeV (H1) and 6 GeV (ZEUS). Figure 5a shows the photoproduction cross-section differential in the jet energy E_T^{jet} transverse to the $\gamma^* I\!P$ interaction axis. The result is compared to the model predictions, which are based on the combined QCD analysis of the diffractive structure function in DIS and of the di-jet cross-sections in photoproduction, and which require a substantial hard momentum component of gluons in the Pomeron [15]. The models describe the data distributions well (also in other variables), confirming the results of other HERA diffractive analyses that a large gluon component is required in the Pomeron parton distributions.

The jet cross–section was measured also differentially in the variables x_γ and $z_{I\!P}$ which can be interpreted as a fraction of photon and Pomeron momenta respectively carried by the partons involved in the hard interactions. The cross-section $d\sigma/dx_\gamma^{obs}$ is shown in Fig.5b. The distribution peaks at high values of x_γ, with a pronounced tail that extends to low x_γ values. This result shows the presence of both a direct ($x_\gamma = 1$) and a resolved ($x_\gamma < 1$) component in diffractive di-jet photoproduction and is in good agreement with Monte–Carlo expectations.

The cross-section $d\sigma/dz_{I\!P}^{jet}$ in H1 photoproduction data is shown in Fig.6a. It indicates the sizeable contribution to di-jet production from events in which a large fraction of the Pomeron momentum participates in the hard scattering.

The model in which the diffractive parton distributions consist solely of quarks at the starting scale underestimates the cross-section by factor 5 (not shown in plots). The POMPYT Monte–Carlo with hard gluon simulation

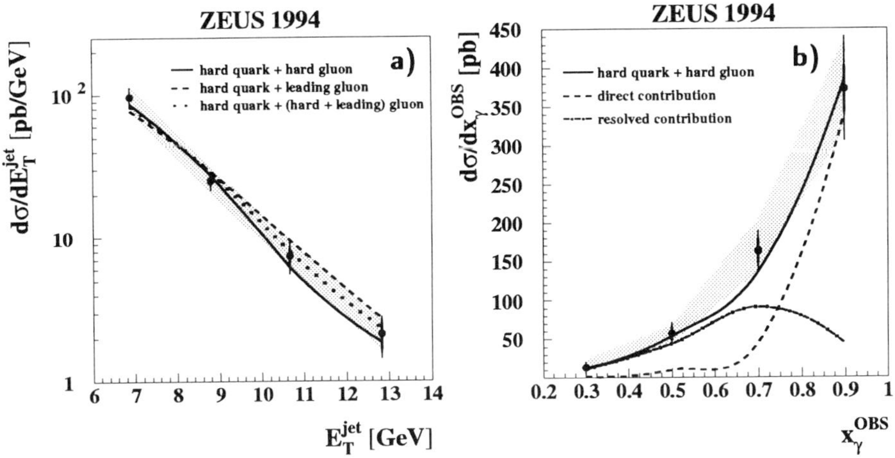

Figure 5: Differential cross-section for photoproduction of di-jets: (a) as function of E_T^{jet} and (b) as function of x_γ^{obs} (ZEUS). The shaded band displays the uncertainty due to the absolute energy scale of the jets.

reproduces the shape of distribution but lies slightly above the data. The difference in normalization between the data and Monte–Carlo predictions can be explained by following argument. In resolved photon interactions multiple interactions between spectator partons might destroy rapidity gaps, an effect which would represent a breaking of diffractive factorization. This effect is not expected to be present in direct photon interactions. Since no full calculations of spectator interactions in diffractive photoproduction exist, we compare the data to the POMPYT models after applying constant weighting factors, so called "survival probabilities", $< S >= 0.4 \div 0.6$ to all events generated with $x_\gamma < 0.8$. The introduction of this factor improves the distributions at all $z_{I\!P}^{jet}$. At the Tevatron, the comparison of diffractive di-jet production data with models suggest that the rapidity gap survival probability may be as small as 0.1 for $p\bar{p}$ interactions at $\sqrt{s} = 1800\ GeV$ [17]. Within the context of this very crude model, a possible interpretation is therefore that the rapidity gap survival probability for resolved photoproduction at HERA is larger that for $p\bar{p}$ interactions at higher energies at Tevatron.

The DIS di-jet production cross-section differential in $z_{I\!P}^{jets}$ is presented and compared to Monte–Carlo prediction in Fig.6b. There is a significant contribution at the largest $z_{I\!P}^{jets}$ and the differential cross section increases as $z_{I\!P}^{jets}$ decreases. The calculation for the $q\bar{q}$ Fock state [16] is compared via RAPGAP simulation. This model falls short of the data. Higher parton multiplicities are

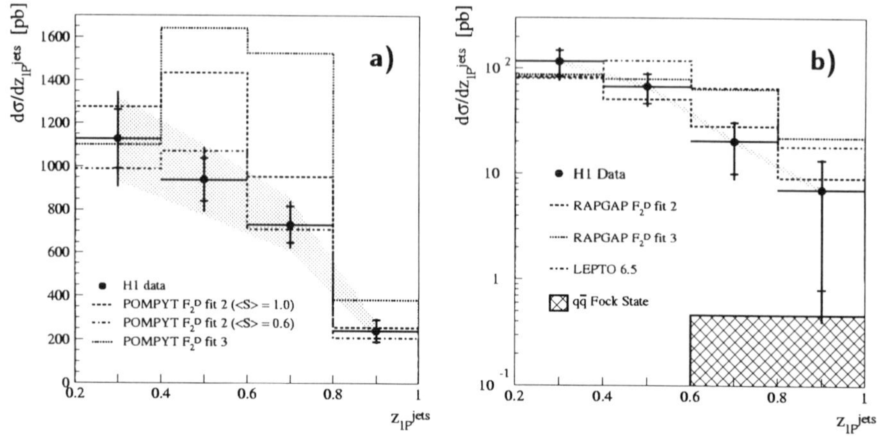

Figure 6: Differential cross-section as a function of $z_{I\!P}^{jet}$ in (a) photoproduction and (b) DIS regimes. The shaded band shows the overall normalization uncertainties.

presumably dominant in the kinematic regime studied here.

7 Open charm production

Open charm production in diffractive DIS is a sensitive probe of the underlying partonic nature of the production mechanism. Both the H1 and ZEUS Collaborations have observed a signal of $D^{*\pm}(2010)$ mesons in diffractive DIS interactions [18,19]. The D^* mesons are reconstructed using the decay channels

$$
\begin{aligned}
D^{*\pm} \to D^0\pi^\pm &\to (K\pi)\pi^\pm \\
&\to (K\pi\pi\pi)\pi^\pm
\end{aligned}
$$

H1 uses only the $K\pi\pi$ mode whereas ZEUS uses both.

The fraction of diffractively produced D^* to all D^* produced in a given kinematic region was measured by ZEUS. Within the errors the result is rather independent of Q^2 and W and is equal to $7.0 \pm 1.3^{+1.7}_{-1.8}$ % ($D^* \to K\pi\pi$), $8.9 \pm 2.4^{+1.7}_{-1.6}$ % ($D^* \to K4\pi$). These numbers are consistent with the fraction of the total ep cross-section attributed to diffractive scattering [2].

The differential D^* cross-sections measured by ZEUS in $D^* \to K4\pi$ decay channel are shown in Fig.7. Results are compared with different model predictions [5,20,21]. Given the large statistical errors, all distributions are reasonably well reproduced by the different predictions. However, it was found that the data favour predictions made with a gluon–dominated compared to a quark–dominated Pomeron in the resolved Pomeron model, and favour predictions of a two–gluon model which includes higher order corrections. Similar measurements were done by ZEUS using the $D^* \to K\pi\pi$ channel. Good agreement between the two decay channels was obtained.

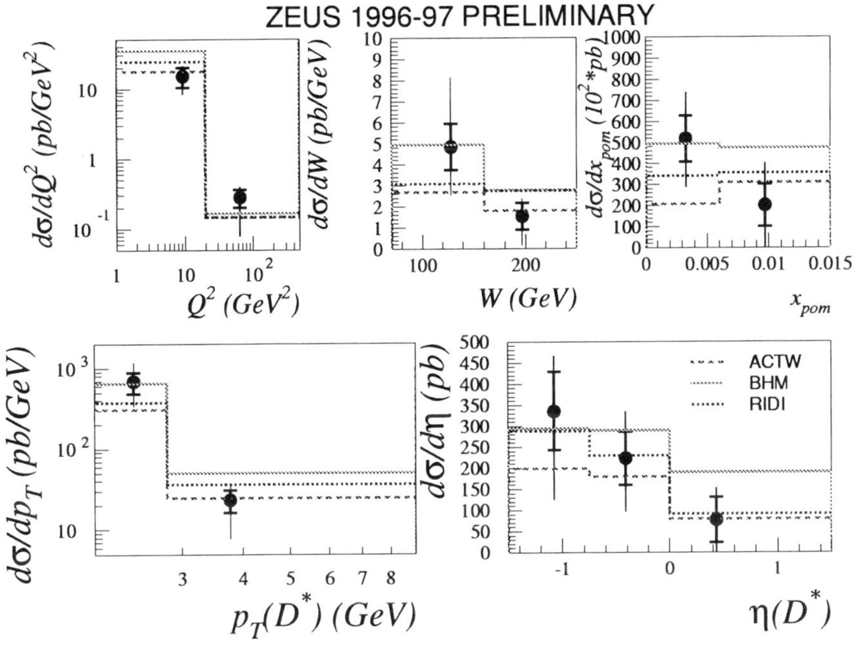

Figure 7: The differential diffractive D^* cross-sections as a function of Q^2, W, $x_{I\!P}$, $p_T^*(D^*)$, $\eta(D^*)$ (ZEUS).

The differential cross-sections measured by H1 are shown in Fig.8 as functions of p_T^* and the hadronic observable $z_{I\!P}^{obs}$, defined by the equation:

$$z_{I\!P} = \beta \cdot (1 + \frac{\hat{s}}{Q^2}) \approx \frac{(M_{c\bar{c}}^2 + Q^2)}{(M_X^2 + Q^2)} = z_{I\!P}^{obs}$$

The resolved Pomeron approach reproduces well the shape of the spectra and in particular the fact that the data show a sizeable fraction of charm production in the low $z_{I\!P}$ region. However, while the model agrees with the ZEUS data, it falls

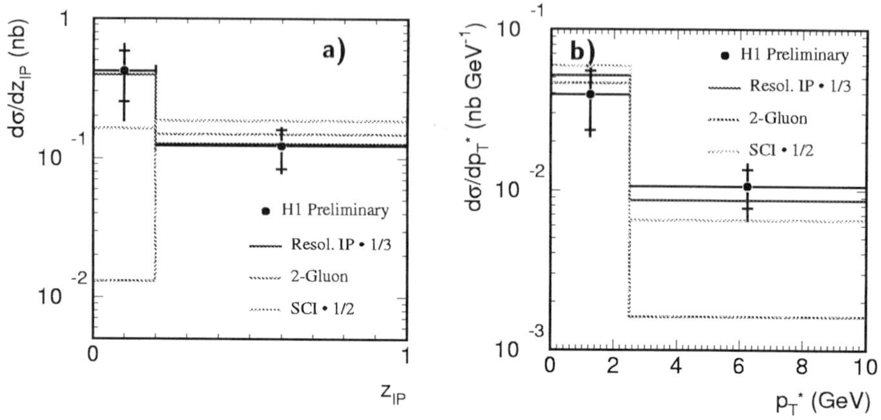

Figure 8: The differential diffractive D^* cross-sections as a function of $z_{I\!P}$ and p_T^* (H1).

a factor of 2 to 3 below the H1 data. Reasons for this discrepancy are still under investigation. The two gluon model for exclusive charm production $\gamma^* + p \to c\bar{c} + p$, in contrast to the resolved Pomeron model, does not describe the region of low $z_{I\!P}$ and high p_T^*. It is apparent that the higher order contributions such as $\gamma^* + p \to c\bar{c}g + p$ are needed.

8 Summary

Some properties of hadronic final states in diffractive ep interactions at HERA have been presented.

Measurements of event shape evolution with M_X, the energy flow, charged particle multiplicities are in general consistently described by resolved Pomeron models with gluon dominated parton densities evolving with DGLAP. These studies confirm the results of the inclusive $F_2^{D(3)}$ measurements, however there are some puzzles to be understood in the rate of di-jet events and D^* production.

References

1. H1 Collab., *Z. Phys.* C **76**, 613 (1997)
2. ZEUS Collab., *Z. Phys.* C **68**, 569 (1995);
   ZEUS Collab., *Eur. Phys.* *J* C **6**, 43 (1999)
3. H.Jung, *Comp. Phys. Comm.* **86** (1995) 147

4. P.Bruni, G.Ingelman, Proc. of the Europhysics Conference, Marseilles, July 1993, 595
5. M.G.Ryskin, Sov. Journ. Nucl. Phys. **52** (1990) 529;
M.G.Ryskin and A.Solano, Proc. of Workshop *'Monte-Carlo Generators for HERA Physics'*, Ed. G.Grindhammer, G.Ingelman, H.Jung and T.Doyle (1999), Hamburg, Germany
6. G.Ingelman, A.Edin, J.Rathsman, *Comp. Phys. Comm.* **101** (1997) 108
7. H1 Collab., *Eur. Phys. J* C **1**, 495 (1998)
8. ZEUS Collab., Contributed paper N505 to Intern. Europhys. Conference HEP99, Tampere, Finland, Jul.1999
9. PLUTO Collab., *Z. Phys.* C **12**, 297 (1982);
MARK II Collab., *Phys. Rev.* D **37**, 1 (1988);
TASSO Collab., *Z. Phys.* C **47**, 187 (1990);
AMY Collab., *Phys. Rev.* D **41**, 2675 (1990);
DELPHI Collab., *Z. Phys.* C **73**, 229 (1997)
10. H1 Collab., *Phys. Lett.* B **428**, 206 (1998)
11. H1 Collab., *Eur. Phys. J* C **5**, 439 (1998)
12. NA22 Collab., *Z. Phys.* C **55**, 531 (1992);
F.C.Winkelman et al.,*Phys. Rev. Lett.* **32**, 121 (1974);
R.L.Cool et al.,*Phys. Rev. Lett.* **48**, 1451 (1982)
13. EMC Collab., *Phys. Lett.* B **149**, 415 (1984);
EMC Collab., *Z. Phys.* C **35**, 417 (1987)
14. H1 Collab., *Eur. Phys. J* C **6**, 421 (1999)
15. H1 Collab., *Eur. Phys. J* C **5**, 41 (1998)
16. J.Bartels, H.Lotter, M.Wüsthoff, *Phys. Lett.* B **379**, 239 (1996);
J.Bartels et al., *Phys. Lett.* B **386**, 389 (1996)
17. CDF Collab., *Phys. Rev. Lett.* **79**, 2636 (1997);
K. Borras (CDF), P.Rubinov (D0), In Proc. of the 6th Intern. Workshop on DIS and QCD, Brussels, Belgium, April 1998, pp. 422-429
18. ZEUS Collab., Contributed paper N527 to Intern. Europhys. Conference HEP99, Tampere, Finland, Jul.1999
19. H1 Collab., Contributed paper N157ag to Intern. Europhys. Conference HEP99, Tampere, Finland, Jul.1999
20. L.Alvero et al., *Phys. Rev.* D **59**, 74022 (1999)
21. W.Buchmüller, A.Hebecker and M.F.McDermott, *Phys. Lett.* B **404**, 353 (1997)

DIFFRACTION AND EXCLUSIVE PROCESSES AT HERA

J. OLSSON (on behalf of the H1 and ZEUS collaborations)
DESY, Notkestraße 85, 22603 Hamburg, Germany
E-mail: jan.olsson @ desy.de

A review is given of selected recent results from the H1 and ZEUS experiments at the HERA ep collider. The results cover exclusive elastic and proton-dissociative diffractive vector meson production, and are obtained in wide ranges of $W_{\gamma p}$, Q^2, $|t|$ and vector meson mass M_V. A comparison is made with recent theoretical calculations and it is shown that perturbative QCD models are successful in describing the processes when the involved scales have large values.

1 Introduction

The large data samples collected during the years 1994-97 by the H1 and ZEUS experiments at the ep collider HERA at DESY are the basis for the results, which are presented in this report. The beam energies in these years were 27.5 GeV(e^+) and 820 GeV(p), yielding a center of mass (CM) energy of $\sqrt{s} = 300$ GeV.

The simplest exclusive production process in diffractive ep scattering is the process

$$ep \to eVp, \quad \text{where } V \text{ is a vector meson } \rho^\circ, \omega, \phi, \rho', J/\Psi, \Psi', \Upsilon.$$

These processes have been studied since a long time at lower energies, in fixed target experiments. Although a simple reaction, the exclusive vector meson production has served as a rich source of information on the nature of the diffractive process. The HERA colliding beam experiments greatly extend the accessible range both of the center of mass energy and of the various scales which influence the reaction. Thus, with the data from HERA detailed studies of the diffractive vector meson production are made possible, both in the soft, low-energy regime already explored by the fixed target experiments, and in the new regimes, where the involved scales reach large values and perturbative QCD (pQCD) can be applied.

Although the bulk of the results cover the elastic vector meson production, i.e. those reactions where the outgoing proton stays intact, some results will also be described, in which the proton dissociates into a low-mass hadronic system.

The diagram of Fig.1 illustrates that HERA in fact can be viewed as a collider of virtual photons γ^* with protons. The following quantities will be used in the discussion of the results:

- Q^2: The photon virtuality, i.e. the absolute value of the squared 4-momentum transfer at the electron (the generic term 'electron' is used for both e^+ and e^-) vertex. In the results presented here Q^2 ranges from ~ 0 to ~ 80 GeV^2.
- $W_{\gamma p}$: the CM energy of the $\gamma^* p$ system, here in the range $20 \lesssim W_{\gamma p} \lesssim 285$ GeV.
- t: The squared 4-momentum transfer at the proton vertex, with $0 \lesssim |t| \lesssim 10$ GeV^2 in the present results. For small values of Q^2, $|t|$ is well approximated by the squared transverse momentum of the produced vector meson V, $t \approx -p_t^2$.

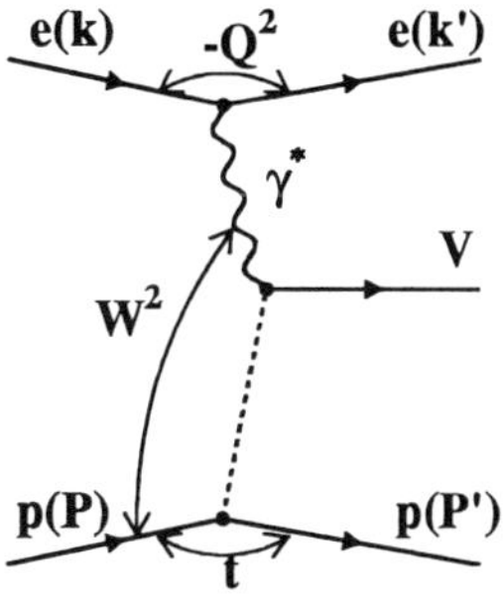

Figure 1: Schematic diagram of elastic vector meson production in ep scattering.

The γ^* energies available at HERA make it possible, within a single experiment, to study the process of exclusive vector meson production over a large $W_{\gamma p}$ interval and with a very wide range of the scales: Q^2, t and M_V, where M_V is the mass of the vector meson V.

The detectors of the H1 and ZEUS experiments are both all-purpose detectors, excellently suited for detection and reconstruction of the vector meson decays, for measuring the Q^2 and t values of the reactions, and for distinguishing the pure elastic processes from those where the proton dissociates. Also in such p-dissociative reactions the few-body decay products of the vector meson are clearly separated, by several units of rapidity, from the low-mass multi-particle system resulting from the proton dissociation.

The following presentation is structured into a photoproduction section, where results in which Q^2 is small are discussed, and an electroproduction section, in which Q^2 takes larger values. In each of these regimes results will be presented for both low and high V masses. In the last section photoproduction results, in which the variable $|t|$ takes on large values, are presented.

2 Photoproduction

Fig. 2 shows a compilation of the cross section measurements $\sigma_{\gamma p \to Vp}$ vs. $W_{\gamma p}$, for exclusive photoproduction of vector mesons at HERA and at lower (fixed target) CM energies. Also the total photoproduction cross section is shown. The measurements show one of the typical features of soft diffractive processes, namely the slow rise of the cross section with $W_{\gamma p}$, for $W_{\gamma p} \gtrsim 10$ GeV. Soft diffractive processes are well described within the framework of Regge phenomenology and the Vector Meson Dominance (VMD) model. In this approach the exclusive vector meson production at high energies is assumed to proceed via the exchange of a Pomeron trajectory; the $W_{\gamma p}$ and t dependences of the cross section are coupled:

$$\frac{d\sigma_{\gamma p}}{d|t|} \propto e^{-b_0|t|} \left(\frac{W^2_{\gamma p}}{W^2_0} \right)^{2(\alpha(t)-1)} \qquad (1)$$

Here $\alpha(t) = \alpha(0) + \alpha' t$; the constants b_0 and W_0 are process dependent.

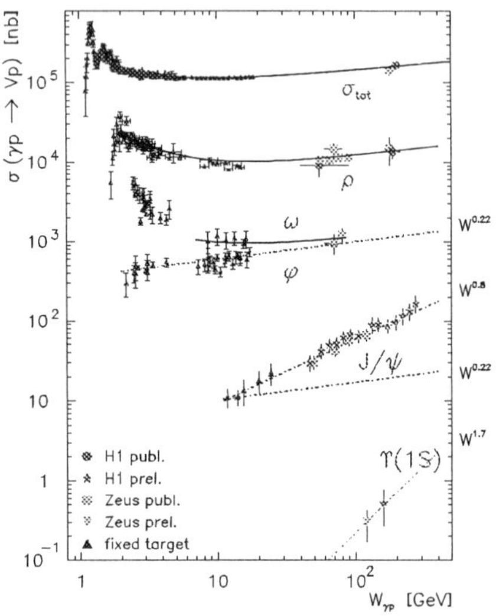

Figure 2: Compilation of $\sigma^{tot}_{\gamma p}$ and $\sigma_{\gamma p}$ for exclusive photoproduction of the vector mesons $\rho^\circ, \omega, \phi, J/\Psi$ and Υ, vs. $W_{\gamma p}$. Full curves represent fits using Regge parametrizations and assuming single pomeron exchange. The dashed lines indicate $W_{\gamma p}$ dependences as written at the right hand side.

The slope b of the t-distribution is then $b = b_0 + 2\alpha' \ln(W^2_{\gamma p}/W^2_0)$. After integration over t, the effective power of a $W^\delta_{\gamma p}$ dependence of the cross section is $\delta \simeq 4(\alpha(0) - 1 - \alpha'/b)$. Fits[1] to hadron-hadron scattering and photoproduction data give $\alpha(0) = 1.08$ and $\alpha' = 0.25$ GeV^{-2}. For the light vector mesons slope values of typically $b \simeq 10$ GeV^{-2} are found. As can be seen in Fig. 2, the predicted δ values of ~ 0.22 are in good agreement with the measurements for the light vector mesons ρ°, ω and ϕ.

No such agreement is seen for the photoproduction of heavy vector mesons. The increase of the cross section of J/Ψ is much steeper than the soft prediction, with a power $\delta \approx 0.8$. Moreover, the measured slope of the $|t|$-distribution for J/Ψ production is smaller than for the light vector mesons,

$b_{J/\Psi} \simeq 5\ \mathrm{GeV}^{-2}$, giving the expectation $\delta \simeq 0.14$.
The exclusive J/Ψ photoproduction data, including brand new data from H1[2], are shown again in Fig.3 together with two different theoretical approaches.

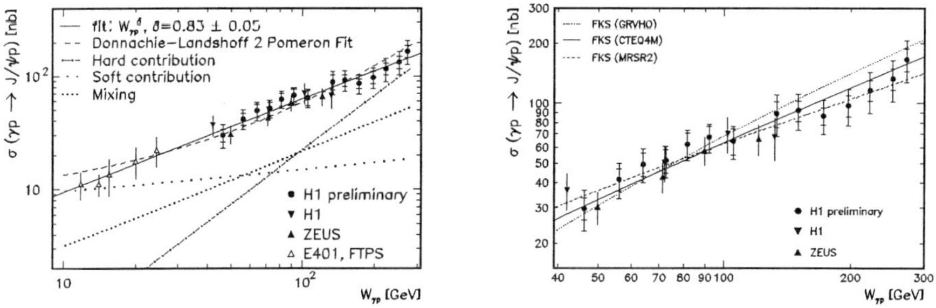

Figure 3: Total elastic J/ψ cross section $\sigma_{\gamma p}$ as a function of $W_{\gamma p}$ compared to different models.

The Regge approach: In an effort to save the approach within the framework of Regge phenomenology, Donnachie and Landshoff[3,4] added a second, "hard" pomeron trajectory to the exchange. Fig.3 shows that a fit with 2 pomerons indeed provides a reasonable description of the data. A good theoretical motivation for the "hard" pomeron is however lacking, although the authors discuss the possibility of glueball exchange.

The pQCD approach: The presence of a hard scale, provided through the heavy c quark mass, suggests an approach within perturbative QCD. Motivated by the very different time scales involved, factorization is assumed (and later proven) in such models, i.e. the $\gamma^* p \to Vp$ scattering is viewed as a sequence of events[5] well separated in time. This is illustrated in Fig.4. The steps are:

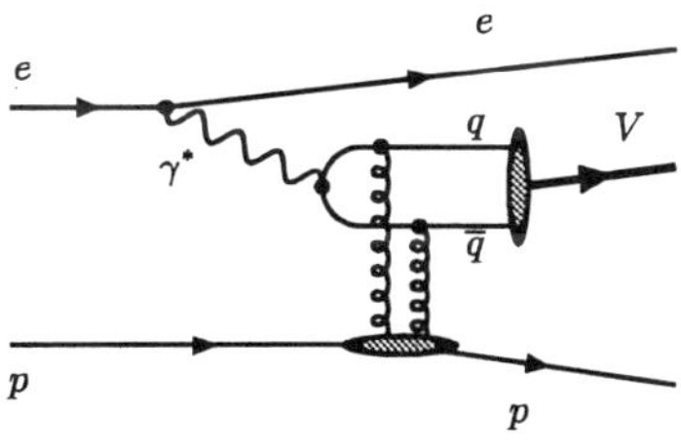

Figure 4: Schematic diagram of elastic vector meson production in ep scattering, via t-channel exchange of a gluon pair.

- The virtual photon fluctuates into a $q\bar{q}$ pair.
- The $q\bar{q}$ pair scatters off the proton.
- The scattered $q\bar{q}$ pair forms a bound state.

In leading order the scattering of the $q\bar{q}$ pair off the proton is mediated by the t-channel exchange of two gluons, in a colour singlet state. The cross section becomes proportional to the square of the gluon density $g(x, Q^2)$ in the proton,

$$\sigma \propto \alpha_s^2(Q^2)/Q^6 \; [xg(x, Q^2)]^2, \tag{2}$$

where x is the Bjorken x. Fits of the model of Frankfurt, Koepf and Strikman[6] (FKS) are shown in Fig.3, for several parameterisations of the gluon density in the proton. It is seen that the cross section is very sensitive to the gluon density parameterisation; while the CTEQ4M and MRSR2 parametrisations can describe the data, the GRVHO parameterisation has a tendency which is steeper than the data.

The steep rise of the $\gamma p \to J/\Psi p$ cross section with $W_{\gamma p}$ has a simple explanation in the pQCD models: it is a consequence of the increasing gluon density of the proton at small x. Using Ryskin's[7] proposed scale $\mu^2 = (Q^2 + M_{J/\Psi}^2 + |t|)/4$ for photoproduction at small $|t|$, a value $\mu^2 = 2.4$ GeV^2 is obtained. At this scale, and at small x, the gluon density rises[8] as $xg(x) \propto x^{-0.2}$. Using the approximate relation $xW_{\gamma p}^2 \approx Q^2$, valid at small x and at fixed Q^2, one finds that $\sigma_{\gamma p \to J/\Psi p} \propto W_{\gamma p}^{0.8}$, in good agreement with the observation.

Shrinkage: The logarithmic $W_{\gamma p}$ dependence of the elastic slope parameter b, known as shrinkage, is a prediction of the soft pomeron model, and has been observed for the light vector mesons. pQCD, in which there is no coupled $W_{\gamma p}$ and t dependence, in contrast predicts no shrinkage of the diffractive peak. It is therefore interesting to determine the value of α' in J/Ψ photoproduction. The new H1 data have been used also to this effect.

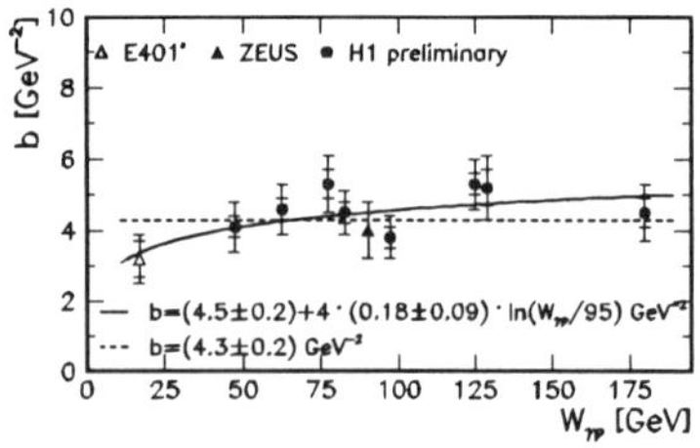

Figure 5: Slope b vs. $W_{\gamma p}$.

Fig.5 shows the slope parameter b, obtained in 5 different $W_{\gamma p}$ bins, together with data from ZEUS and from the fixed target experiment E401[9]. Fits with and without (i.e. α' set to 0) the logarithmic $W_{\gamma p}$ dependence indicate that shrinkage cannot be excluded; the effect is however only 2 s.d., with $\alpha' = 0.18 \pm 0.09$ GeV^{-2}.

A more direct determination of the Regge trajectory has been made by Levy[10] and by H1. In their analysis, H1 determined $\alpha(t) = \alpha(0) + \alpha' t$ in 5 bins of t, fitting in each t-bin i the $W_{\gamma p}$ dependence of eq.(1). Fig. 6 shows the result of the linear fit of the resulting 5 values of $\alpha(t_i)$, with the central value and its

1 s.d. band. Again no clear conclusion can be drawn, the data are compatible both with no shrinkage and with the presence of shrinkage, weaker than in the soft pomeron model.

Fig. 6 also shows the trajectories expected for the soft and hard pomeron (for the latter $\alpha' = 0.1$ GeV^{-2} is estimated[4]); the H1 measurement lies between these predictions. The BFKL (Balitsky, Fadin, Kuraev, Lipatov) pomeron[11] trajectory is in rough agreement with the measurement (the value $\alpha' = 0$ was assumed by H1, since only the intercept is predicted).

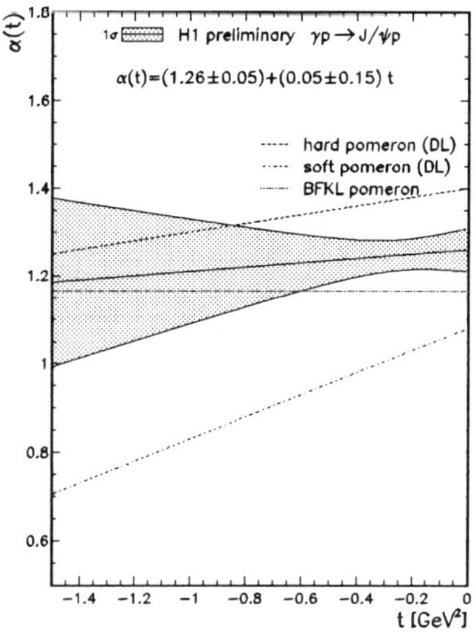

Figure 6: The measured Regge trajectory $\alpha(t) = \alpha(0) + \alpha' t$ for elastic J/Ψ photoproduction.

Elastic photoproduction of Υ: The cross section measurements[12,13] of Υ photoproduction are shown in Fig. 7. The published ZEUS measurement, based on the $\mu^+\mu^-$ decay of Υ, has now been combined with a new, preliminary measurement based on detection in the e^+e^- decay channel[14]. A number of theoretical predictions of the cross section as function of $W_{\gamma p}$ are also shown in Fig. 7. The early prediction by Frankfurt, Koepf and Strikman[6] is about 2 s.d. too low. The measurements of H1 and ZEUS have inspired recent improvements of theory. Taking into account the fact that at large vector meson masses the off-diagonal gluon distributions in the exchange are important, and including also the real part of the amplitude, Frankfurt, McDermott and Strikman[15] (FSM), and Martin, Ryskin and Teubner[16] (MRT), can now give a good description of the measurements. The importance of including the real part of the amplitude in the calculations is shown by the 2 curves labelled FSM and FSM(only Im).

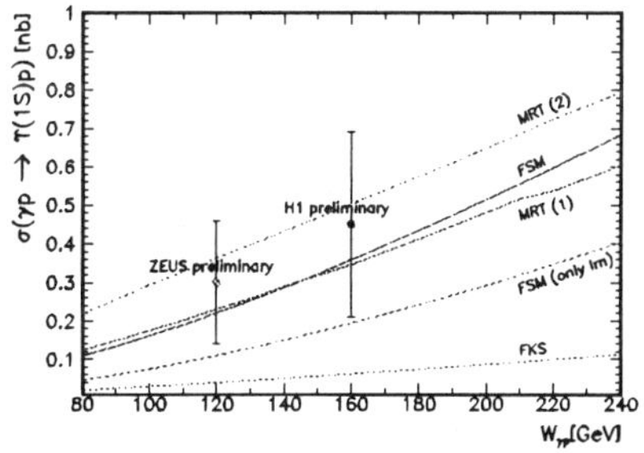

Figure 7: Υ photoproduction cross section vs. $W_{\gamma p}$, together with theoretical predictions.

3 Electroproduction

Elastic electroproduction of vector mesons at HERA offers new rich possibilities for tests of QCD, since here the scale Q^2 can be varied over a large range. Not only can the Q^2 dependence of the cross section be studied and compared to that of eq.(2), but the measurement of the Q^2 dependence of other observables will provide insight into the transition from the soft, non-perturbative to the hard, perturbative regime. In particular the light vector meson electroproduction is of great interest. New, high statistics data from H1 and ZEUS on the electroproction of ρ^o [17,18], ω [19], ϕ [20] and J/Ψ [17,21] are described in the following.

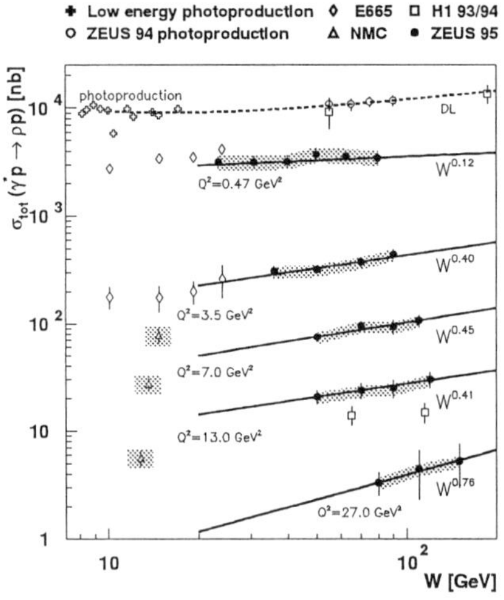

Figure 7: Cross section for electroproduction of ρ^o, vs. $W_{\gamma p}$. Error bands are indicated. The curves represent fits of the dependence $\sigma \sim W_{\gamma p}^{\delta}$

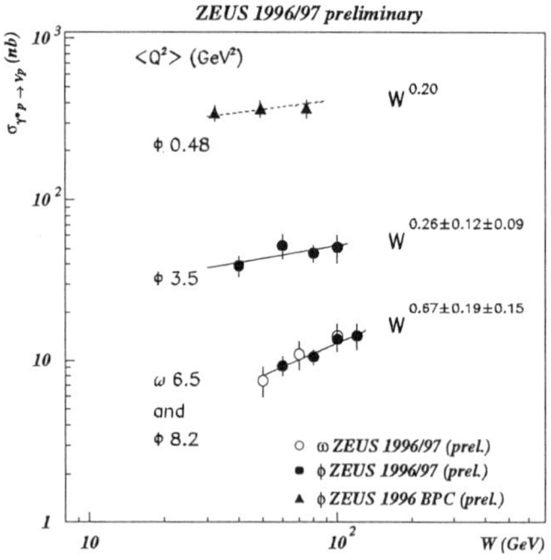

Figure 8: Cross sections for electroproduction of ω and ϕ, as function of $W_{\gamma p}$. Solid curves show fits to the ϕ cross section. The dotted line shows the $W_{\gamma p}^{0.20}$ dependence for comparison.

$W_{\gamma p}$ dependence: The new ZEUS cross section measurements of elastic ρ^o, ϕ and ω electroproduction are presented in Figs. 7 and 8, as function of $W_{\gamma p}$ and for several values of the average Q^2. The cross sections rise with increasing $W_{\gamma p}$, and the higher the average Q^2, the faster is the rise. This observation can be quantified with fits to the form $W_{\gamma p}^{\delta}$ in each Q^2 bin. As already discussed, the parameter δ is in Regge phenomenology related to the exchange trajectory, $\delta \simeq 4(\alpha(\langle t \rangle) - 1)$, with $\alpha(t) = \alpha(0) + \alpha' t$. Using $\langle t \rangle = 1/b$

(see below) and $\alpha' = 0.25$ GeV^{-2}, H1 obtained with their new ρ° electroproduction data the Q^2 dependence of the intercept $\alpha(0)$, Fig.9. The data suggest that with increasing Q^2 the intercept of the effective trajectory governing ρ° electroproduction is larger than 1.08, the "soft pomeron" intercept.

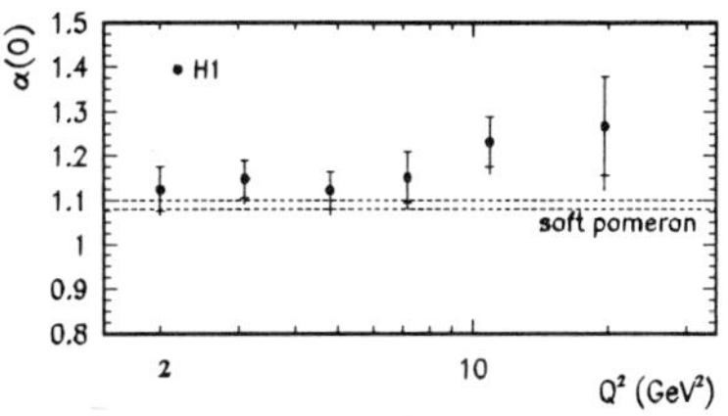

Figure 9: Q^2 dependence of the intercept $\alpha(0)$ for elastic ρ° production. The dashed lines represent the range of values obtained for the "soft pomeron" intercept[3,22].

Q^2 dependence of $\sigma(\gamma^* p \rightarrow Vp)$: The cross section eq.(2) contains the factor $1/Q^6$, characteristic for "soft" diffraction. In pQCD this steep Q^2 dependence is however modified by the Q^2 dependence of α_s and of the gluon density. The steepness of the Q^2 dependence of the cross section is also modified by e.g. the Fermi motion of the quarks[5,23], which suggests that, beyond the pQCD tests, precise measurements of the Q^2 dependence of the cross section for elastic vector meson electroproduction can reveal information also about the wave functions of the vector mesons.

Fig.10 shows the Q^2 dependence of the H1 and ZEUS data on exclusive elastic ρ° and ϕ electroproduction, at $\langle W_{\gamma p} \rangle$ values of 75 and 80 GeV, respectively. The curves show fits of the form $(Q^2 + M_\rho^2)^{-n}$ (ρ°) and $(Q^2)^{-n}$ (ϕ). The values obtained, $n = 2.24 \pm 0.09$ and $n = 2.18 \pm 0.10(stat.) \pm 0.15(syst.)$ respectively, confirm the less steep Q^2 dependence expected from pQCD. Similar agreement with the form $(Q^2 + M_\omega^2)^{-2}$ is obtained for the ω electroproduction data (not shown). Note that in the case of the ϕ data the fit includes only data points at $Q^2 \geq 5$ GeV^2, and the dashed curve extrapolates the fit to lower Q^2 values. Here the exclusion of the M_ϕ^2 term lets the fit lie far above the data.

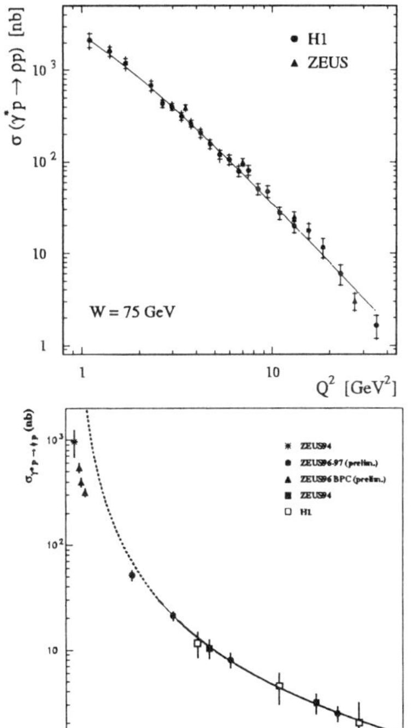

Figure 10: ρ° and ϕ electroproduction cross sections vs. Q^2. The curves are explained in the text.

In Fig.11 the Q^2 dependence of the $\sigma(\gamma^* p \to \rho^\circ p)$ data is directly compared to several model calculations: The Generalized VMD model (GVDM) of Schildknecht, Schuler and Surrow[24] falls above the data at larger Q^2 values, while the pQCD inspired model of Cudell and Royen[25] gives a good description of the data, down to the photoproduction measurements. The model of Martin, Ryskin and Teubner[26], also pQCD inspired, cannot describe the data.
It is interesting to note that the Q^2 dependence of the cross section is also reasonably described in a Regge-eikonal approach, reported by Prokudin elsewhere in these proceedings[27].

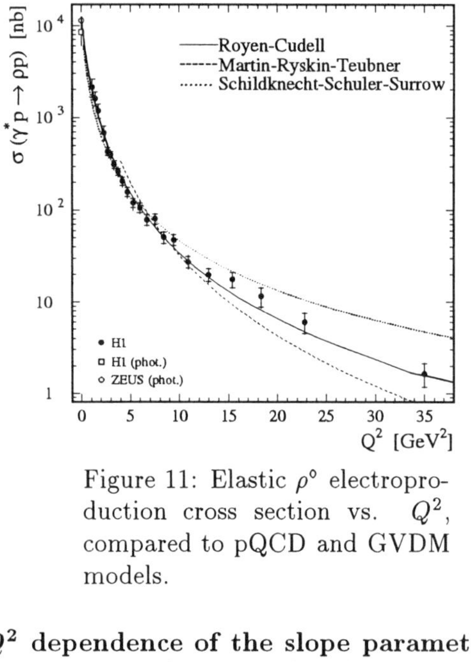

Figure 11: Elastic ρ° electroproduction cross section vs. Q^2, compared to pQCD and GVDM models.

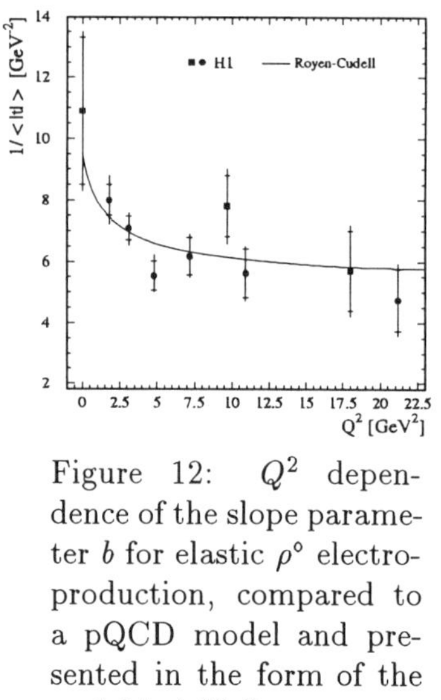

Figure 12: Q^2 dependence of the slope parameter b for elastic ρ° electroproduction, compared to a pQCD model and presented in the form of the variable $1/\langle|t|\rangle$.

Q^2 dependence of the slope parameter b: It was already mentioned in the previous discussion that the slope parameter b of the diffractive peak is considerably smaller (~ 5 GeV^{-2}) in elastic J/Ψ photoproduction, compared to the values found (~ 10 GeV^{-2}) for the light vector mesons. This can be interpreted as a reflection of the smaller transverse size of the virtual photon, fluctuating into a $q\bar{q}$ system, when the quark mass scale is increased. It is then interesting to see if increasing another scale, Q^2, leads to a similar decrease of the slope parameter b. Fig.12 shows the H1 results for the Q^2 dependence of the parameter b in elastic ρ° production, plotted here in the form of the variable $1/\langle|t|\rangle$ (identical to b for an exponential t distribution). A strong

decrease is indeed seen, going from the photoproduction point to higher Q^2 values. Relating b to the radii of the scattering particles, $b \propto R_p^2 \oplus R_{q\bar{q}}^2$, and remembering that $R_p^2 \sim 4 - 5$ GeV^{-2}, one can conclude that with increasing Q^2 the $q\bar{q}$ system becomes small compared to the proton and that it therefore starts to probe the structure of the proton.
Fig.12 also shows a comparison of the data with the model of Cudell and Royen[25]. Good agreement is seen.
At the lowest Q^2 values (Beam Pipe Calorimeter, BPC data) ZEUS find evidence for shrinkage of the diffractive peak in the elastic ρ° electroproduction data, while at larger Q^2 values no conclusive results are obtained.

$W_{\gamma p}$ dependence of $\sigma(\gamma^* p \to J/\Psi p)$: Fig.13 shows the $W_{\gamma p}$ dependence of the cross section for elastic J/Ψ electroproduction, for several values of Q^2. New data from H1 and ZEUS provide these results. As was already discussed in the previous photoproduction section, the c quark mass provides a hard scale, so that with increasing Q^2 now two hard scales come into play. The dash-dotted curves in Fig.13, which show fits of the form $W_{\gamma p}^{\delta}$, demonstrate that adding a new hard scale does not change the behaviour of the $W_{\gamma p}$ dependence of the cross section. The steep rise of the cross section, which was observed already in the J/Ψ photoproduction data, stays practically the same at larger values of Q^2.
The data are compared to the model calculations (full and dashed curves)

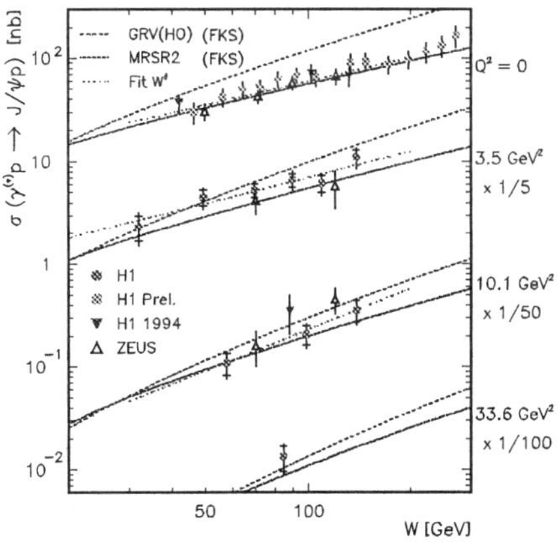

Figure 13: J/Ψ elastic electroproduction cross section vs. $W_{\gamma p}$ for several values of Q^2. The curves are explained in the text.

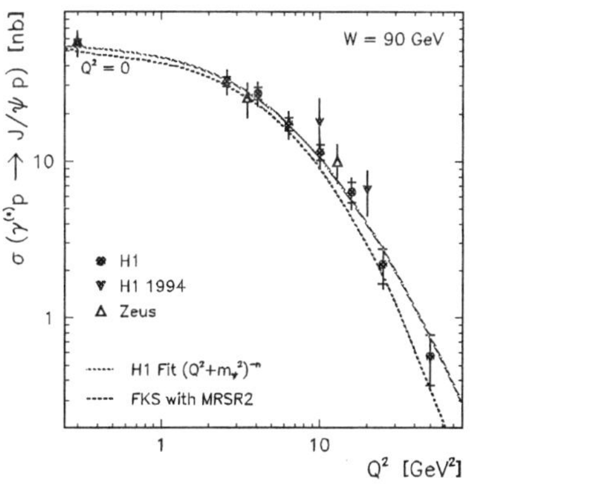

Figure 14: Q^2 dependence of elastic J/Ψ electroproduction. The curves are explained in the text.

of Frankfurt, Koepf and Strikman[6], using two different parameterisations of the proton gluon density. Data are precise enough to allow some distinction of the parameterisations.

Q^2 dependence of $\sigma(\gamma^* p \to J/\Psi p)$: The Q^2 dependence of elastic J/Ψ production is shown in Fig.14. The data have been fitted (full curve) to the form $(Q^2 + M_{J\Psi}^2)^{-n}$, with $n = 2.38 \pm 0.11$ as result. The dotted curve shows the Frankfurt, Koepf and Strikman model calculation, using the MRSR2 proton gluon parameterisation.
It is worthwhile to note that the value of the power n, 2.38, is close to the values found for similar fits to the Q^2 dependence of the low mass vector meson cross sections. A compilation of all HERA data on $\sigma(\gamma^* p \to Vp)$, using a universal scale $(Q^2 + M_V)/4$, has recently been presented and discussed[28].

Q^2 dependence of the slope b in J/Ψ electroproduction: H1 have used their elastic J/Ψ electroproduction data to determine the diffractive slope parameter b in 2 bins of Q^2, namely $2-8$ GeV^2 and $8-80$ GeV^2. The results, $b = 4.4 \pm 0.4 \pm 0.4$ and $b = 2.5 \pm 0.6 \pm 0.6$ GeV^{-2}, respectively, give a weak indication that b decreases with increasing Q^2. $b = 5.1 \pm 1.3$ GeV^{-2} for $2 < Q^2 < 40$ GeV^2 is found by ZEUS. The low Q^2 results for both experiments are compatible with the photoproduction values for the slope parameter b. None of the experiments find any significant dependence on $W_{\gamma p}$ for b.

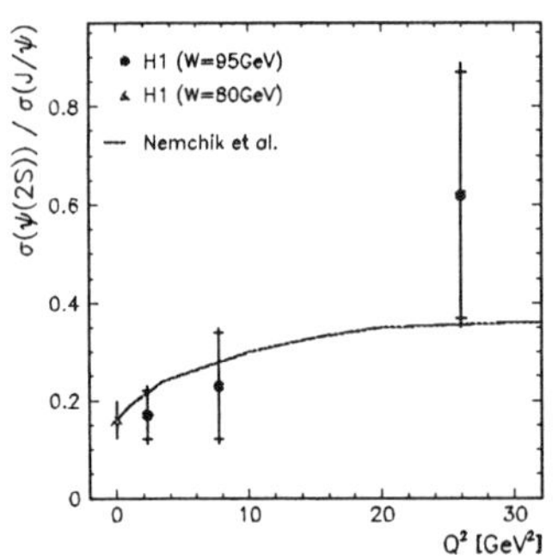

Figure 15: Ratio of the cross sections for quasi-elastic production of $\Psi(2S)$ and J/Ψ, vs. Q^2. The curve shows the calculation of Nemchik et al.[29].

Q^2 dependence of Ψ' electroproduction: H1 has recently also presented data on electroproduction of $\Psi(2S)$, using the decay $\Psi' \to J/\Psi\pi^+\pi^-, J/\Psi \to e^+e^-, \mu^+\mu^-$. The scarce data ("quasi-elastic", i.e. including both elastic and low-mass proton dissociative events) have been divided into 3 Q^2 bins and are shown in Fig.15, in the form of the ratio $\sigma(\gamma^* p \to \Psi(2S)p)/\sigma(\gamma^* p \to J/\Psi p)$ vs. Q^2. The data at low Q^2 values agree with the previous photoproduction measurement. There is an indication that the cross section ratio rises with increasing Q^2, which is also predicted in pQCD models[6,29]. An asymptotic value of 0.5 is predicted in[6]. The curve in Fig.15 shows the model calculation of Nemchik et al.[29], based on colour dipole phenomenology.

Flavour independence at large Q^2 values: At sufficiently high scales one expects that the vector meson production process becomes independent of the participating quark flavour. Simple counting of the quark charges, together with the assumption of a point-like electromagnetic coupling to the quark charges, give the ratios, well known from SU(4), of 9 : 1 : 2 : 8 for the ρ°, ω, ϕ and J/Ψ vector mesons. It is also well known that at low energies this expectation is not fulfilled by the cross section ratios $\sigma_{\gamma^* p \to Vp}/\sigma_{\gamma^* p \to \rho^\circ p}$, with $V = \phi, \omega$ and J/Ψ. Figs.16 and 17 show the HERA results for these cross section ratios as function of Q^2, with the limits expected from SU(4) indicated as dashed lines. For the light vector mesons good agreement is seen between the cross section ratios and the expectation; in the case of ω already for small Q^2, in the case of ϕ for Q^2 values $\gtrsim 8$ GeV^2. In the case of the J/Ψ meson, the data in Fig.17 indicate that the limit is reached only at still larger Q^2 values.

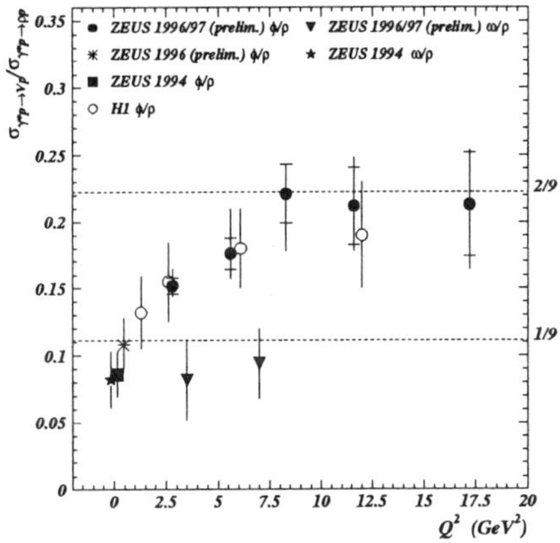

Figure 16: Cross section ratios $\sigma_{\gamma^* p \to Vp}/\sigma_{\gamma^* p \to \rho^\circ p}$ vs. Q^2, with $V = \phi$ and ω. The $SU(4)$ predictions are indicated by dashed lines.

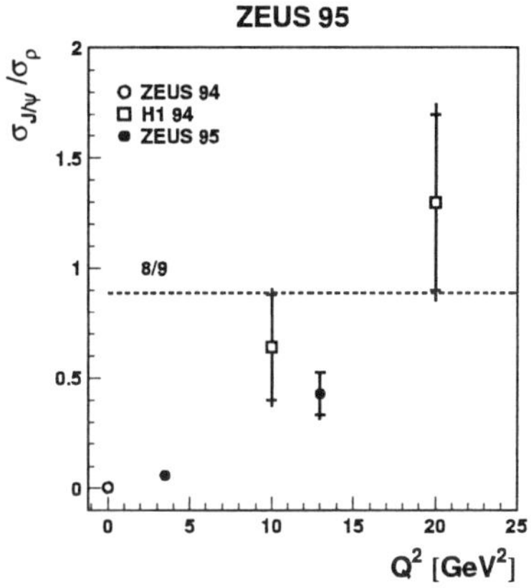

Figure 17: Cross section ratio $\sigma_{\gamma^* p \to J\Psi p}/\sigma_{\gamma^* p \to \rho^\circ p}$ vs. Q^2. The dashed line shows the SU(4) prediction.

Helicity Studies: The study of the angular distributions of the production and decay of the vector mesons V in the reaction $\gamma^* p \to Vp$ provides information about the polarization states of the photon and the vector mesons V. Such studies are particularly interesting at large values of Q^2, since many theoretical models make firm predictions about the polarization in this regime. Thus pQCD models predict dominance at high Q^2 of longitudinally polarized vector mesons from longitudinally polarized photons.
In their studies, which are so far limited to 2-body decays of vector mesons, H1 and ZEUS define three angles in the helicity system, the commonly used reference frame, as shown for $V = \rho^\circ$ in Fig.18. The angles are:

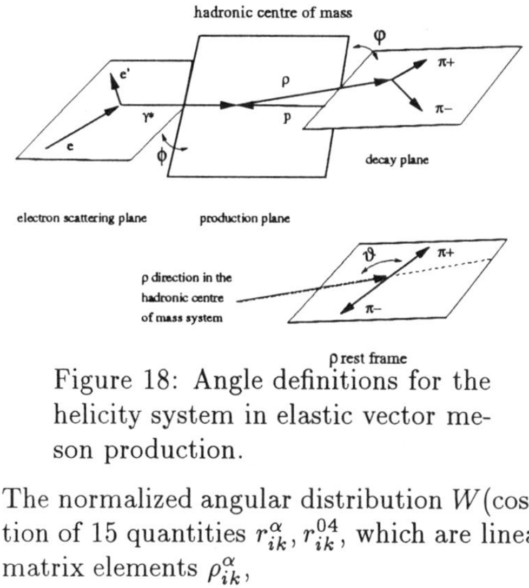

Figure 18: Angle definitions for the helicity system in elastic vector meson production.

- ϕ: in the hadronic (γp) CM the azimuthal angle between the electron scattering plane and the plane containing V and the scattered proton,
- θ: the decay angle of the 2-body decay of V, defined by the positive decay particle in the rest system of V, with the quantization axis taken as the direction of V in the γp CM,
- φ: the azimuthal angle of the positive decay particle in the rest system of V.

The normalized angular distribution $W(\cos\theta, \varphi, \phi)$ can be written[30] as a function of 15 quantities $r^\alpha_{ik}, r^{04}_{ik}$, which are linear combinations of the spin density matrix elements ρ^α_{ik},

$$r^{04}_{ik} = \frac{\rho^0_{ik} + \epsilon R \rho^4 ik}{1 + \epsilon R} \qquad r^\alpha_{ik} = \frac{\rho^\alpha_{ik}}{1+\epsilon R}, \ \alpha = 1,2 \qquad r^\alpha_{ik} = \frac{\sqrt{R}\,\rho^\alpha_{ik}}{1+\epsilon R}, \ \alpha = 5,6$$

Here ϵ is the polarization parameter of the virtual photon, being ≈ 0.99 in the present analyses. $R = \sigma_L/\sigma_T$ is the ratio of the longitudinal and transverse $\gamma^* p$ cross sections. The ρ^α_{ik} elements depend on bilinear combinations of the helicity amplitudes and the subscripts i and k take the values of the possible helicity states $-1, 0, 1$. The superscripts 0 and 4 refer respectively to unpolarized transverse photons and longitudinally polarized photons, superscripts 1 and 2 correspond to linearly polarized transverse photons, and superscripts 5 and 6 represent the interference of transverse and longitudinal amplitudes.

In the case of S-channel helicity conservation (SCHC) and natural parity exchange (NPE) the 15 independent density matrix elements ρ_{ik}^{α} are constrained, and only 5 elements are non-zero.

H1 and ZEUS have performed large statistics helicity analyses of elastic ρ° [18,31] and ϕ [20] electroproduction. The result for the ρ° analysis is shown for the two experiments in Fig.19, where all 15 quantities $r_{ik}^{\alpha}, r_{ik}^{04}$ have been determined. One notes the very good agreement between the two experiments. The prediction from SCHC and NPE is also shown as a dotted curve. The data agree with this prediction, except for the quantity r_{00}^{5}, which is clearly non-zero where the SCHC prediction is zero. The pQCD model of Ivanov and Kirschner[32], which takes into consideration the momentum distribution of the vector meson quark constituents, in contrast predicts this violation of SCHC, and is in excellent agreement with the data. Recent calculations of Royen and Nikolaev[33] also reproduce the observed SCHC violation.

In their analysis of the ϕ data (not shown) ZEUS obtained a similar, although statistically less significant, result.

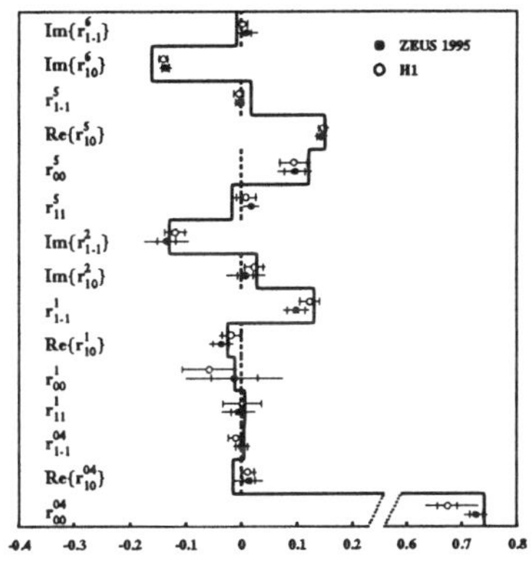

Figure 19: The 15 combinations $r_{ik}^{04}, r_{ik}^{\alpha}$ of the spin density matrix elements, obtained in the ρ° analysis. The curves are explained in the text.

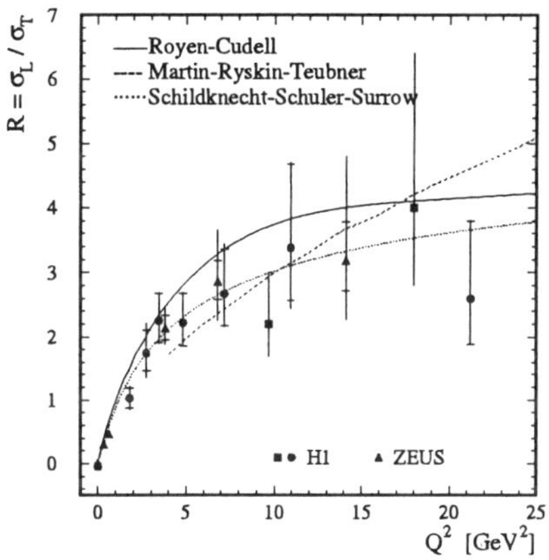

Figure 20: Ratio R of the longitudinal to transverse cross sections for elastic ρ electroproduction, vs. Q^2. The curves represent predictions from models based on pQCD and GVDM.

The element r_{00}^{5} is approximately proportional to the amplitude T_{01} for a transverse photon to produce a longitudinal vector meson. H1 have estimated the size of the T_{01} amplitude relative to the non-flip amplitudes, finding that the effect is 8 ± 3 %. This is in rough agreement with similar results, obtained at lower energy in fixed target experiments[34].

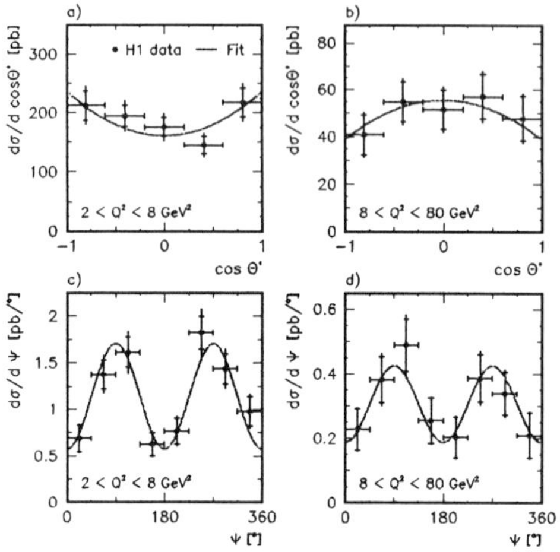

Figure 21: Angular distributions for the positive lepton in J/Ψ decays, obtained in quasi-elastic $\gamma^* p$ production of J/Ψ in 2 bins of Q^2 and for $40 < W_{\gamma p} < 160$ GeV. The curves are explained in the text.

The SCHC violation effect being small, one can further explore the angular dependence under the assumption of validity of SCHC and NPE. One prediction is then the relation $r^1_{1-1} = 1/2\,(1 - r^{04}_{00})$, valid independent of the value of the ratio R. The measurements are in agreement with this relation within 1 s.d., thereby providing an independent check of the assumption. Furthermore, if SCHC is valid for the interaction, then the ratio R can be directly obtained, via the relation

$$R = \frac{\sigma_L}{\sigma_T} = \frac{1}{\epsilon}\,\frac{r^{04}_{00}}{1 - r^{04}_{00}}.$$

r^{04}_{00} is the probability to produce longitudinally polarized vector mesons, from either transversely or longitudinally polarized photons. The R values measured under this assumption (the influence of the SCHC violating effect on R results in a 2-3 % overestimation) are shown in Fig.20, as function of Q^2. A steep increase of R is observed when Q^2 increases from the photoproduction point (R is expected to be 0 for $Q^2 = 0$), while R seems to flatten out at larger Q^2 values. A clear dominance of the longitudinal cross section is observed already at Q^2 values of a few GeV^2. The data are in good agreement with previous fixed target measurements (not shown). The magnitude of the ratio, as well as its Q^2 dependence, are well reproduced by model calculations, some of which are shown in Fig.20. This is true both for the pQCD inspired models[25,26] and for the GVDM model[24].

The value of R has been determined also for J/Ψ production, in data where also low-mass proton dissociative states are included. The statistics do not allow a full angular analysis as in the ρ° and ϕ cases. Instead, 1-dimensional distributions of $\cos\theta$ and $\Psi = \varphi - \phi$ are used to extract the spin density matrix element combinations r_{00}^{04} and r_{1-1}^{1}:

$$\frac{d\sigma}{d\cos\theta} \propto 1 + r_{00}^{04} + (1 - 3r_{00}^{04})\cos^2\theta, \qquad \frac{d\sigma}{d\Psi} \propto 1 - \epsilon r_{1-1}^{1}\cos 2\Psi.$$

The H1 data and the fits to these distributions are shown in Fig.21.

The extracted values for r_{1-1}^{1} and r_{00}^{04} are found to satisfy the relation $r_{1-1}^{1} = 1/2\,(1 - r_{00}^{04})$ within 1 s.d., thus giving support for the assumption of SCHC and NPE validity in the J/Ψ production process.

$< Q^2 >$ (GeV^2)	R	exp.
0	0.17 ± 0.14	H1
0	-0.01 ± 0.09	ZEUS
4	$0.18^{+0.18}_{-0.14}$	H1
5.9	$0.41^{+0.45}_{-0.52}$	ZEUS
16	$0.94^{+0.79}_{-0.43}$	H1

Table 1: Values for the ratio R in J/Ψ production.

The values for the ratio R, shown in Table 1, suggest that R increases with increasing Q^2, in particular when taking the earlier photoproduction results[35,36] into account.
The values of R obtained here are much smaller than the ones determined for elastic ρ and ϕ production at similar Q^2. However, when considering the measurements on the global scale Q^2/M_V^2, they are of the same order.

4 Photoproduction of vector mesons at large $|t|$

New data have been presented by ZEUS[19], for photoproduction of the vector mesons ρ°, ϕ and J/Ψ at large values of $|t|$, from 1 up to 11 GeV^2. While in the case of the scale Q^2 the vector meson p_t is balanced by the momentum transfer to the electron, it is here balanced by the momentum transfer to the proton. The question addressed is thus whether also $|t|$, like Q^2, can serve as a hard scale in perturbative QCD descriptions of the processes, and if so, at which values of $|t|$ the calculations become reliable.
At large values of $|t|$ the exclusive vector meson production process is dominated by low-mass proton dissociative events and detector components close to the proton beamline can be used to distinguish these from the elastic events. The scattered electron was tagged by a calorimeter close to the electron beamline, so that $Q^2 < 0.01$ GeV^2 and $80 < W_{\gamma p} < 120$ GeV. Two of these new ZEUS results are discussed here.

Helicity studies of ρ° production: The spin density matrix element combinations r^{04}_{00}, Re$[r^{04}_{10}]$ and r^{04}_{1-1} are shown vs. $|t|$ in Fig.22.

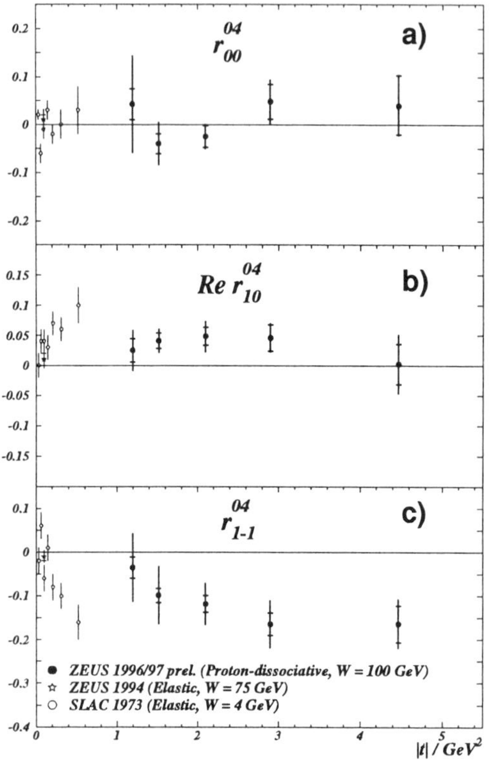

Figure 22: The spin-density matrix element combinations r^{04}_{00}, Re$[r^{04}_{10}]$ and r^{04}_{1-1}, vs. $|t|$. ZEUS photoproduction data $\gamma p \to VN$, with N being a proton or a low-mass dissociated state.

Ivanov and Ginzburg[37] and Brodsky et al.[38] predict that at large values of $|t|$ the production of longitudinally polarized ρ° will dominate. Since r^{04}_{00} is the probability to obtained longitudinal ρ°, it is thus expected to be large at large values of $|t|$. As seen in Fig.22, data up to $|t|$ values of 4-5 GeV^2 do not support this prediction, r^{04}_{00} is compatible with being zero everywhere. One can conclude that the perturbative regime with $|t|$ as a scale is not yet reached.

The quantity Re$[r^{04}_{10}]$, which corresponds to an interference between the helicity conserving amplitude and the single flip amplitude, is non-zero at higher $|t|$ values and thus indicates evidence for a single flip contribution in the production process.

The quantity r^{04}_{1-1} is significantly non-zero at larger values of $|t|$. This indicates that a double flip contribution becomes important, i.e. a transversely polarized photon produces a ρ° with the opposite helicity. This is a violation of SCHC.

The latter results confirm previous fixed target measurements at SLAC[39], also shown in Fig.22. However, the results are still preliminary. One uncertainty is a possible non-resonant di-pion background under the ρ°.

Flavour independence limit: As already discussed in the case of the scale Q^2, one expects in pQCD the interaction to be flavour independent at large enough values of the scale. ZEUS have used their data at high $|t|$ to test this prediction also in the case when the scale is provided by the momentum transfer at the proton vertex. Fig.23 shows ratios of the cross sections for ϕ and J/Ψ production, relative to the cross section for ρ°, vs $|t|$. The expected limits from a pointlike electromagnetic interaction with the quarks forming the vector mesons are indicated with dotted lines. It is seen that the ϕ data are consistent with this expectation over the whole $|t|$ range, while the J/Ψ data are significantly lower than the limit, for $|t|$ values below $\sim$ 4 GeV^2.

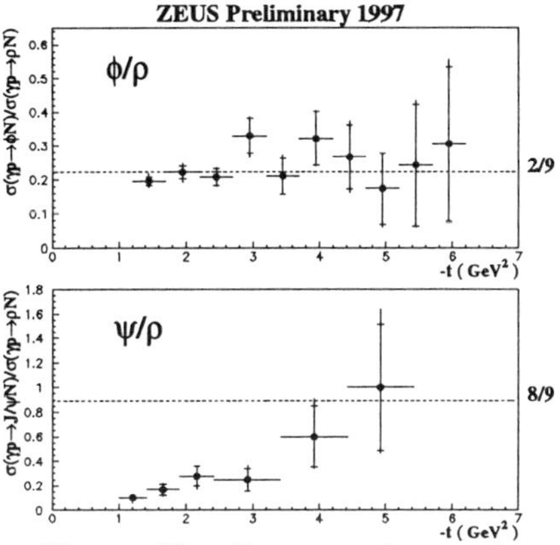

Figure 23: Cross section ratios $\sigma(\gamma^* p \to VX)/\sigma(\gamma^* p \to \rho^{\circ} X)$ vs. $|t|$, with V being J/Ψ or ϕ, and X being a proton or a proton dissociative system. ZEUS preliminary data. The dotted lines are explained in the text.

5 Conclusions

The years of high luminosity running at HERA have opened up a rich field of precise data on exclusive vector meson production, spanning large intervals of $W_{\gamma p}$, Q^2, M_V and $|t|$. This new wealth of data extends the results obtained in the earlier fixed target experiments. Whereas these early results could be successfully interpreted in a "soft", VMD-Regge approach, the new HERA data open up access to larger values of several scales, thereby entering the realm of validity of perturbative QCD. Thus the data span the very interesting region of transition between "soft" and "hard" diffractive interactions. While there is good evidence in the present data that both Q^2, M_V and $|t|$ all serve successfully as hard scales in perturbative QCD calculations for diffractive reactions, this field of research remains very active. Further fruitful interaction between theory and experiments will certainly provide new and interesting knowledge, not only about the structure of the proton and the vector mesons, but also about the complex nature of the diffractive interactions.

Acknowledgments

It is a pleasure to thank the organizers for the warm and agreeable atmosphere in a most interesting and well prepared conference. I also wish to thank all my colleagues in H1 and ZEUS, for providing the data shown in this report and for all their help given to me.

References

1. A. Donnachie and P.V. Landshoff, Phys. Lett. **B296** (1992) 227
2. H1 Collab., C. Adloff et al., contr. paper 157aj to EPS-HEP 1999, Tampere, Finland
3. A. Donnachie and P.V. Landshoff, Phys. Lett. **B348** (1995) 213
4. A. Donnachie and P.V. Landshoff, Phys. Lett. **B437** (1998) 408
5. S.J. Brodsky et al., Phys. Rev. **D50** (1994) 3134;
   J.C. Collins, L. Frankfurt and M. Strikman, Phys. Rev. **D56** (1997) 2982
6. L. Frankfurt, W. Koepf and M. Strikman, Phys. Rev. **D57** (1998) 512
7. M.G. Ryskin, Z. Phys. **C57** (1993) 89
8. M. Erdmann, in Proceedings of the 29th Intern. Conference on High Energy Physics, Vancouver 1998, Editors A. Astbury, D. Axen and J. Robinson, Vol. 1, p. 217
9. E401 Collab., M. Binkley et al., Phys. Rev. Lett. **48** (1982) 73
10. A. Levy, Phys. Lett. **B424** (1998) 191
11. S.J. Brodsky et al., hep-ph/9901229 (1999)
12. ZEUS Collab., J. Breitweg et al., Phys. Lett. **B437** (1998) 432
13. H1 Collab., C. Adloff et al., contr. paper 574 to ICHEP 1998, Vancouver
14. P. Merkel, to be published in the proceedings, EPS-HEP 1999, Tampere, Finland
15. L. Frankfurt, M.F. McDermott and M. Strikman, JHEP **9902** (1999) 2
16. A.D. Martin, M.G. Ryskin and T. Teubner, Phys. Lett. **B454** (1999) 339
17. ZEUS Collab., J. Breitweg et al., Eur. Phys. J. **C6** (1999) 603
18. H1 Collab., C. Adloff et al., DESY-99-010, to appear in Eur. Phys. J. C
19. ZEUS Collab., contr. paper 499 to EPS-HEP 1999, Tampere, Finland
20. ZEUS Collab., contr. paper 793 to ICHEP 1998, Vancouver
21. H1 Collab., C. Adloff et al., Eur.Phys.J. **C10** (1999) 373
22. J.R. Cudell, K. Kang and S. Kim, Phys. Lett. **B395** (1997) 311
23. L. Frankfurt, W. Koepf and M. Strikman, Phys. Rev. **D54** (1996) 3194
24. D. Schildknecht, G.A. Schuler and B. Surrow, Phys. Lett. **B449** (1999) 328

25. J.R. Cudell and I. Royen, Phys. Lett. **B545** (1999) 505
26. A.D. Martin, M.G. Ryskin and T. Teubner, Phys. Rev. **D55** (1997) 4329
27. V.A. Petrov and A.V. Prokudin, these proceedings and hep-ph/9912245(1999)
28. B. Clerbaux, hep-ph/9908519
29. J. Nemchik et al., Phys. Lett. **B258** (1991) 493;
J. Nemchik et al., J. Exp. Theor. PHys. **86** (1998) 1054
30. K. Schilling and G. Wolf, Nucl. Phys. **B61** (1973) 381;
for a detailed discussion, see also J.A. Crittenden, Springer Tracts in Modern Physics, Volume 140 (Springer, Berlin Heidelberg, 1997)
31. ZEUS Collab., J. Breitweg et al., DESY-99-102, to appear in Eur. Phys. J. C
32. D.Yu.Ivanov and R. Kirschner, Phys. Rev. **D58** (1998) 114026
33. I. Royen, N. Nikolaev, to be published in the proceedings, 7th Intern. Conference on Deep Inelastic Scattering and QCD, 19-23 April, 1999, Zeuthen, Germany
34. P. Joos et al., Nucl. Phys. **B113** (1976) 53;
CHIO Collab., W.D. Shambroom et al., Phys. Rev. **D25** (1982) 1
35. H1 Collab., S. Aid et al., Nucl. Phys. **B472** (1996) 3
36. ZEUS Collab., M. Derrick et al., Phys. Lett. **B377** (1996) 259
37. D. Yu. Ivanov, Phys. Rev. **D53** (1996) 3564;
I.F. Ginzburg and D.Yu. Ivanov, Phys. Rev. **D54** (1996) 5523
38. S.J. Brodsky et al., Phys. Lett. **B449** (1999) 306
39. J. Ballam et al., Phys. Rev. **D7** (1973) 3150

S. B. Nurushev

Study of the Total and Differential Cross Sections, and Polarization Effects in pp Elastic Scattering at RHIC

S.B.Nurushev
Institute for High Energy Physics, 142284 Protvino, Moscow Region, Russia
E-mail:nurushev@mx.ihep.su

We present the physics program, setup and perfomance of an experiment to measure the total cross section in proton-proton interaction, the differential cross section and polarization in elastic pp-scattering at the Relativistic Heavy Ion Collider, RHIC, at Brookhaven National Laboratory, USA.

Paper presented by Sandibek Nurushev on behalf of pp2pp(R7) Collaboration [1] *at 8-th Blois Conference, June 28-July 2, 1999, Protvino, Russia.*

1 Introduction

The Relativistic Heavy Ion Collider(RHIC) at Brookhaven National Laboratory (USA) is a dedicated machine for study of heavy ion interactions(up to gold ions) at the top beam energy of 100 GeV/amu. RHIC can be at the same time a powerful pp-collider, expanding the ISR energy domain by one order of magnitude. This collider will open a wide avenue for the measurements of the total and elastic scattering cross sections in a preasymptotic range of energy and for checking the predictions of theoretical models. Such motivation lead us to submit the Proposal R7 entitled " Experiment to Measure Total and Elastic pp Cross Sections at RHIC" which was approved in 1994, and it also included a spin physics program.

Further addition of a polarized and unpolarized jet target in R7 setup allows to extend the physics program to the fixed target domain of the RHIC. The next attractive and achievable scientific item is a study of the single diffractive dissociation process. We intend to include later both items in our research program too.

2 Physics Objectives

R7 Collaboration [1] aims to measure systematically the total pp-cross section and elastic scattering from $\sqrt{s} = 50 GeV$ to $\sqrt{s} = 500 GeV$. The measurements of the elastic scattering differential cross section will cover two kinematical regions: 1) $4 \cdot 10^{-4} <| t |< 0.12 (GeV/c)^2$: we shall measure and study the s dependences of the total and elastic cross sections, σ_{tot} and σ_{el}; the ratio of the real to the imaginary parts of the forward elastic spin non flip amplitude ρ; and the nuclear slope parameter of the pp elastic scattering, B. The possibility to reach the region of $|t| = 1 \cdot 10^{-4} (GeV/c)^2$ for searching of possible oscillations in the differential cross-section [2] will be analyzed by R7 Collaboration; 2) $0.006 <| t |< 1.5(GeV/c)^2$: we plan to study the evolution of the dip structure observed at ISR in the differential elastic cross section, $d\sigma_{el}/dt$, with s, and the s and $| t |$ dependences of B.

The measurements that can be performed with transversally and longitudinally polarized protons in RHIC are: 1) the difference of the total cross-sections as a function of pure initial spin states, $\Delta\sigma_T$ and $\Delta\sigma_L$; 2) the analyzing power, A_N; 3) the spin correlation parameters, A_{NN}, A_{LL} and A_{NL}.

In order to explore the fixed target energy region of the RHIC R7 Collaboration plans to install the jet target(unpolarized at the first stage and later polarized) and add the recoil arm detectors.

Nucleon-nucleon elastic scattering is described by the 5 helicity amplitudes [3]: $\phi_1 =< ++|++>, \phi_2 =< ++|-->, \phi_3 =< +-|+->, \phi_4 =< +-|-+>$ $, \phi_5 =< ++|+->$, where + and - refer to the nucleon helicities in the c.m. frame. Due to the helicity conservation at 0°, $\phi_4(0) = 0 = \phi_5(0)$ and only three amplitudes survive. So we see two amplitudes with spin nonflip ϕ_1 and ϕ_3, two others with double spin flip ϕ_2 and ϕ_4 and only one amplitude with single spin flip ϕ_5. In order to determine all amplitudes at any fixed s and t one must measure 9 observables. Such set of experiments is called the complete set of experiment [4]. Our strategy is to fulfill such measurements step by step using polarized protons at RHIC.

The imaginary parts of the three elastic scattering amplitudes surviving at zero degree are related to the three types of total cross sections through the generalized optical theorems in the following way [5], [6]: $\sigma_{tot} = \frac{2\pi}{k} Im[\phi_1(0) + \phi_3(0)], \Delta\sigma_L = \frac{4\pi}{k} Im[\phi_1(0) - \phi_3(0)], \Delta\sigma_T = -\frac{4\pi}{k} Im[\phi_2(0)]$, where k is the c.m. momentum. Here σ_{tot} is the unpolarized total cross-section, $\Delta\sigma_L$ is measured as a difference of the total cross-sections with longitudinally polarized beams, while $\Delta\sigma_T$ is measured with transversally polarized beams. Therefore, by measuring three total cross-sections one can reconstruct the imaginary parts of three nonzero amplitudes at zero degree, $\phi_1(0), \phi_2(0)$ and $\phi_3(0)$.

In order to get a hint for pp total cross section at RHIC we refer to the description of the total cross-secion based on the Froissart-Martin bound [7] and given in [8]. In the RHIC domain of energy according to this paper the pp total and elastic cross sections should increase essentially(from 43 to 60 mb for total and from 8 to 13 mb for elastic cross sections) and such growths will be easily detected. The difference between pp and $\bar{p}p$ total cross sections is $\Delta\sigma_{tot}$=1.18mb at the RHIC injection energy and is expected to be 40μb at $\sqrt{s}$=500 GeV. Recently an alternative approach to this problem was proposed[9].

According to this paper the failure of production of Centauro events in $\bar{p}p$ colliders might be caused by the fact that Centauro events are originated purely from pp interaction and not from the $\bar{p}p-$ events. It means that there is an extra source of interaction in pp and, therefore, at sufficiently high energy(around $\sqrt{s} = 500$ Gev) $\sigma_{tot}(pp)$ becomes equal to the $\sigma_{tot}(\bar{p}p)$ and then it grows faster. In this case one can expect that $\Delta\sigma_{tot}$ rises like $\approx \lg s$.

Another important subject is the ratio $r_{el} = \frac{\sigma_{el}}{\sigma_{tot}}$. There is an indication that σ_{el} and r_{el} continue to rise and r_{el} approaches the black body limit which is 0.5 . At the RHIC energy region it is expected that r_{el} rises from 0.174 to 0.217, which can be easily measured. This ratio for $\bar{p}p$ collision is 0.2202±0.0078 according to a recent measurement [10].

The dependence of the differential elastic cross-section $\frac{d\sigma_{el}}{dt}$ on $|t|$ can be divided into three regions: 1) the Coulomb region, 2) the CNI region, 3) the hadronic region. At small t the Coulomb term dominates, and $\frac{d\sigma_{el}}{dt}$ has a $\frac{1}{t^2}$ dependence. Since the Coulomb amplitude is precisely known, the measurement at very small $|t|$ gives direct determination of the machine's luminosity and, consequently, the absolute normalization of the hadronic amplitude. But such measurement will be hard to do.

The measurements in the CNI region allow to determine the very crucial parameter - the ratio of the real to imaginary parts of the leading spin non flip amplitudes. This parameter is related to the total cross-section, $\sigma_{tot}(pp)$, via dispersion relation and is a check of the first principles of the strong interaction dynamics like analyticity, unitarity and crossing symmetry.

Near forward direction, $|t| < 0.12(GeV/c)^2$, the elastic differential cross section is well described by the simple exponential, $\exp^{-B\cdot|t|}$. In the classical Regge model the slope parameter depends on the energy like B(s)=$B_0 + 2 \cdot \alpha' \log s$ if the leading pole is the pomeron. Such prediction is consistent with the experiments. The pp2pp data will extend a study of such dependence beyond the ISR energy region and improve the estimate of the effective slope of the Pomeron trajectory, α'. Beyond the above t region the $d\sigma_{el}(t)/dt$ does not follow a simple exponential form and the slope parameter B(t) for pp scattering may show a complicated t and s dependence.

At the ISR it was discovered that at $|t| > 0.5(GeV)^2$ the pp elastic differential cross section shows a diffraction-like picture. At $|t| = 1.5(GeV/c)^2$ and $\sqrt{s}$=23.5GeV there is a pronounced dip followed by bump. With growth of $\sqrt{s}$ this structure shifts to the smaller $|t|$ and becomes not so pronounced. Such behaviour is described by several models, like the "impact picture" model of Bourrely et al.[11], the "multiple diffraction" model of Glauber [12], the specific Regge model [13]. They show differences in predictions at the large t-region($> 2(GeV/c)^2$). The large luminosity of RHIC will make possible a detailed study of scattering at large momentum transfer.

As apparent from the previous discussion, the basic question to be addressed is: if a new diffraction-like structure will emerge in this large $|t|$-region, or will the $|t|$-distribution be smooth and energy independent, controlled by a single QCD diagram[14]? The measurements at RHIC will be useful for checking these predictions.

3 Spin Physics Program

RHIC will have the unique capability of accelerating polarized protons up to $\sqrt{s}$=500 GeV with a high average polarization, $\approx 70\%$ for each beam, and a high luminosity, $L = 2 \cdot 10^{32} cm^{-2} \cdot s^{-1}$. This will enable us to measure the spin dependent parameters of elastic pp scattering at much higher *cms* energies compared to the highest energy data to date ($\sqrt{s} = 24$ GeV).

The main polarization measurements that will be performed by R7 Collaboration are listed below. We assume that the transversly and later the longitudinally polarized proton beams will be available at the Interaction Region(IR).

By colliding two polarized protons with polarizations P_1 and P_2 we can measure the total cross section

$$\sigma_{tot}(\vec{P}_1, \vec{P}_2) = \sigma_{tot} + \sigma_1 \cdot (\vec{P}_1 \cdot \vec{P}_2) + \sigma_2 \cdot (\vec{P}_1 \cdot \vec{k})(\vec{P}_2 \cdot \vec{k}), \tag{1}$$

where $\vec{k}$ is a unit vector in *cms* along the beam direction(let say, beam 1). While σ_{tot} is the unpolarized total cross section (defined earlier) two new observables are related to the standard total cross sections in the pp singlet, σ^s, and triplet states, σ_0^t and σ_+^t (0 and + mean the spin projections) by the relations: $\sigma_1 = \frac{\sigma_0^t - \sigma^s}{4}, \sigma_2 = \frac{\sigma_+^t - \sigma_0^t}{2}$. By measuring at RHIC all three total cross sections: unpolarized σ_{tot}, transversally polarized $\Delta\sigma_T$, and longitudinally polarized $\Delta\sigma_L$ we can reconstruct for the first time all three imaginary parts of helicity amplitudes at t=0. As a result of such measurements we can estimate how much the spin interactions contribute to the total cross section.

The CNI region is very attractive for the measurements of the spin effects. It stems from the fact that in this region two amplitudes interfere: one is the Coulomb spin-flip amplitude which can be calculated in QED very precisely and has specific t-dependence and the second one is the nuclear amplitude, which varies much slower with t and can be taken in the CNI region either as constant or having the exponential dependence with slope parameter B(t). Usually by the measurement in CNI region the parameter ρ is determined. In

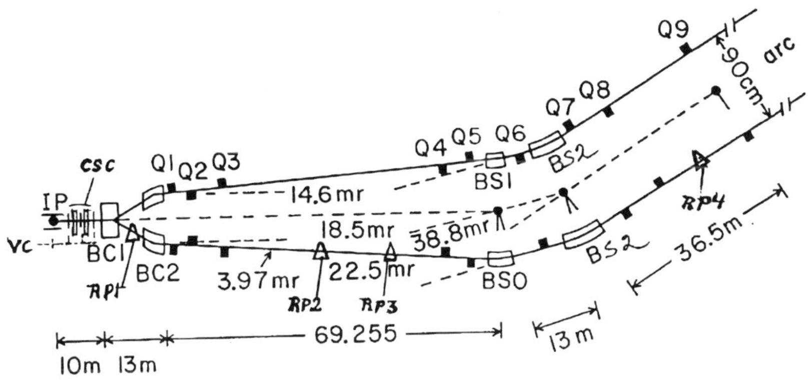

Figure 1: The pp2pp(R7) setup at RHIC.

the case of the polarized beam one can measure the analyzing power [15]

$$A_N(s,t) = A_{max} \cdot \frac{4 \cdot y^{3/2}}{1 + 3 \cdot y^2}, \tag{2}$$

where $y = \frac{|t|}{t_{max}}$, $t_{max} = \sqrt{3} \cdot \frac{8\pi\alpha}{\sigma_{tot}}$ and α is a fine-structure constant. This analyzing power is interesting subject for theory and from other side it can be used for polarimetry of the high energy proton beam. The only measurement of this parameter was done recently by E704 collaboration [16] confirming the expected behaviour. But it was measured at only one energy and with large errors, To be used as a polarimeter one needs more precise measurements(better than 5%) and at several energies. We plan to make such measurements at RHIC by using the same apparatus as it is foreseen for elastic scattering.

We will also measure the spin-correlation parameters in the CNI region, which carry a rich information on the nuclear amplitudes [17].

The region for $|t| > 2(GeV/c)^2$ is essentially unexplored and high precision measurements can shed light on the possible onset of a hard regime spin effects.

4 Experimental setup

We propose to measure the previously outlined aspects of pp elastic scattering in two types of experimental setups, installed around the 2 o'clock Interaction Region(IR) of the RHIC(Fig.1).

For the medium $|t|$- range, we define the following requirements:

- Momentum analysis with $\frac{\Delta p}{p} \leq 2\%$.
- Rejection at the trigger level of events with nearly collinear low- enegy tracks, using the deflection by the magnets.
- Reconstruction of the collision point by extrapolating the observed tracks back to the crossing region, with an accuracy sufficient to reject beam-wall-interactions.
- Use the VC signals for suppressing the elastic triggers accompanied by extra charged particles.
- Acceptance of the large momentum transfer up to $\mid t \mid \simeq 1.5(GeV/c)^2)$.
- Momentum-transfer resolution better than $\Delta t = 0.02(GeV/c)^2$ at $|t| = 1(GeV/c)^2$.

In the standard pp-mode at RHIC, the expected normalized emittance, defined at the 95% level, is $\epsilon = 20\pi \cdot mm \cdot mrad$. The size and angular spread of the beam at the crossing are $\sigma_y = \pm 0.45mm$ and $\sigma_{\theta_y} = \pm 45\mu rad$ respectively.

The scattered protons will be detected by telescopes of detectors in Roman Pots(RP). each containing 2 planes of x and 2 planes of y of the silicon strip detectors(SSD). The size of each plane is $5.5 \times 8.2cm^2$, the pitch is 0.1mm. The detectors will be located in the vertical plane, symmetrically above and below the machine plane. The horizontal bending of BC1 allows an almost complete decoupling between the measurement of the scattering angle, essentially given by the vertical coordinate and the measurement of the momentum, obtained from the horizontal coordinate. Detector RP1 is installed at the distance 20 m from IP and subtends the -t range (0.02 - 1.5) $(GeV/c)^2$, Due to the expected small elastic cross section a set of veto counters(VC) surrounding the IR will be used to suppress the inelastic background at the trigger level. VC and the Cathode Strip Chambers(CSC) will be used also for the determination of the inelastic cross section, σ_{in}, and the measurement of the single diffractive dissociation in pp collisons.

All above consideration can be applied also to the analyzing power, A_N, measurement. If we take the ISR differential cross-section data as an approximate guideline for a 50% coverage, then we have at $|t| = 1(GeV/c)^2$ N $= 0.3$ events/sec for a bin size of $\Delta t = 0.05(GeV/c)^2$. In this $|t|$-region, the measurement errors for the analyzing power and the double- spin asymmetry parameter after a month(about 20 days) of running will be $\delta A_N = 0.2\%, \delta A_{NN} = 0.3\%$.

For very small $\mid t \mid$ measurements(CNI region) the setup will include the detectors RP2. RP3 and RP4(containing the same SSD as above) installed at 57m, 72m and 144m correspondingly from the interaction point(IP). RP2 covers the region $0.002 <\mid t \mid< 0.2)(GeV/c)^2$, RP3 - $0.004 <\mid t \mid< 0.12(GeV/c)^2$, and RP4- $0.0004 <\mid t \mid< 0.03(GeV/c)^2$. At the smallest $\mid t \mid$ the detector resolution will be $\delta t << 10^{-4} GeV^2 (bins) - - > \delta y \approx 0.1 mm$. In order to be able to measure the scattering angle of the protons, their scattering angle θ_y^* has to be larger than the angular spread of the beam $\sigma(y_0')$ at the collision point. As seen from the Table 1 the beam divergence is 4μrad. For the scattering angle of $\Theta = 70\mu$rad one gets the minimum $\mid t \mid= 0.0004(GeV/c)^2$ at the effective distance for RP4, $L_{eff} = 87$m. We are studying a possible way of reaching the smaller $\mid t \mid$ values.

This setup requires a special tune of RHIC, extra power supplies, modifications of the leads to some magnets. A special beam scraping will be required for the run(and during the run) in order to decrease the beam emittance from $\epsilon = 20\pi mm \cdot mrad$ to $5\pi mm \cdot mrad$ to allow detection of scattered protons as close to the beam as possible. A tune change will be needed for the period of data taking to allow large $\beta^* = 195$m at the IP. To accumulate $4 \cdot 10^6$ events, the run will take 4-5 days. From the experience of the UA4/2, we foresee two runs: one engineering run with up to 10^6 events, followed by the data run of $4 \cdot 10^6$ events a few months after the engineering run.

In order to evaluate the performance of the experiment with respect to the possible systematic errors, we performed simulations of small angle scattering. The errors of the elastic scattering parameters σ_{tot}, ρ, B due to uncertainties in the parameters describing the experimental setup were determined. The parameters used in the Monte Carlo simulations are included in Table 1. For comparison the parameters of the UA4/2 experiment [18] are presented too. The advantage of R7 is in the RHIC high luminosity.

The same setup, running with transversally or longitudinally polarized protons, would allow the simultaneous measurements of $\Delta\sigma_T, A_N, A_{NN}$ or $\Delta\sigma_L, A_L$, and A_{LL}.

For $L = 10^{29} cm^{-2} \cdot s^{-1}$, $|t|_{min} = 4 \cdot 10^{-4} (GeV/c)^2$, $\frac{d\sigma_C}{dt}|_{t_{min}} = 1.6 \cdot 10^{-24} cm^2 \cdot (GeV/c)^{-2}$, acceptance =10% R7 counting rate per $10^{-4}(GeV/c)^2$ bin will be 2.5 Hz. In the CNI region, where $|t| \simeq 1.1 \cdot 10^{-3}(GeV/c)^2$, the rate is down to 1.3 Hz at 60% acceptance. For the measurement of ρ and σ_{tot} not to be dominated by statistics, we need the order 10^4 events per $|t|$ bin of $10^{-4}(GeV/c)^2$, leading to run duration of at least 6 hours. The expected precisions in the measurements will be: $\Delta\rho = 0.01, \delta b = 0.05(GeV/c)^{-2}$, and $\delta\sigma_{tot} \simeq 0.5 mb$ for 10^6 events.

In order to measure the beam polarization within $\pm 5\%$, $2.5 \cdot 10^5$ events are needed, assuming that the analyzing power A_N is $\simeq 0.04$. As a rough estimate, the integrated differential cross section is 2.8 mbarn in the range

Table 1: Parameters of the CNI pp2pp setup(expected) and of UA4/2 setup.

Run parameters	pp2pp	UA4-2
Beam momentum, GeV/c	250	270.6
Beam momentum spread, GeV/c	0.25	0.10
Normalized beam emittance, $\pi \cdot mm\cdot$ mrad	5	26
Betatron function at IP, β^*, m	y: 195	x: 2500, y: 87
Beam vertex size in z at IR, σ_Z, cm	15	20
Beam transverse size at IR, σ_x/σ_y, mm	0.4/0.4	5/1
Beam divergence at IR, $\sigma_{x'}/\sigma_{y'}$, μrad	4/4	2.7/14.3
Angle between the beams in x/y direction, μrad	5/5	
Error of angle between the beams in x/y directios, μrad	6/6	4/15
Detector offset, μm	20	25
Detector resolution, μm	100	150
Number of the elastic events, $\times 10^6$	4.0	0.8
Closest to the beam axis position, mm	5	4
RP3 position(distance from IR), m	72	
$L_{eff}(RP3)$, m	y:37	
RP4 position(distance from IR), m	144	48
$L_{eff}(RP4)$, m	y: 87	X: 65, y: 27

$0.002 < |t| < 0.02(GeV/c)^2$. If we assume a fiducial coverage of $\frac{\Delta\phi}{2\pi} = 0.25$, then:

$$N = \text{Ł} \cdot \frac{d\sigma_{el}}{dt} \cdot \frac{\Delta\phi}{2\pi} = \text{Ł} \cdot 1.4 \cdot 10^{-27} cm^2, \quad (3)$$

where N is the number of events per second and Ł is the luminosity. For high β^* tune $\text{Ł} = 2 \cdot 10^{29} cm^{-2} \cdot s^{-1}$, N= 280 events/sec which requires about one hour for a ±5% beam polarization measurement.

5 Conclusion

R7 Collaboration presented the programme of the following measurements at the collider RHIC:

- Total and elastic cross-sections with polarized and unpolarized beams.
- Analyzing power, A_N, and spin-correlation asymmetry, A_{NN}, with transversally polarized beams; with longitudinally polarized beams:A_L and A_{LL}.
- At a later stage we plan to study the single diffraction dissociation by the same apparatus.
- Physics with jet target(polarized and unpolarized)is under development by R7 Collaboration.

Acknowledgments

We would like to express our thanks to Prof.'s M. Islam, A. Martin, B. Nicolescu and O. Selyugin for useful discussions.

1. R7 Collaboration List:

N. Akchurin[1], A. Bogdanov[2], J. Bourotte[3], M. Buenerd[4], B. Chrien[5], M. Conte[6], K. De[7], G. Dudkin[8], R. Giacomich[9], R. Gilb[6], I. Glavanakov[8], W.Guryn[6], M. Haguenauer[3], V. Kaplin[2], A. Karakash[2], Yu. Krechetov[8], D. Lynn[6], G. Naumenko[8], C. Newsom[1], S. Nurushev[2,10], Y. Onel[1], A. Penzo[9], A. Potylitsin[8], M. Pusterla[11], M. Rijssenbeek[12], M. Runtzo[2], A. Rusek[6], M. Sakitt[6], P. Schiavon[9], G. Sette[4], M. Strikhanov[2], C. Tang[12], S. Tepikian[6], A. Ufimtsev[10], A. Vartapetian[7].

References

1. *University of Iowa, Iowa city, USA*
2. *Moscow Engineering Physics Institute(MEPHI), Moscow, Russia*

3. *Ecole Polytechnique/IN2P3-CNRS, Palaiseau, France*
4. *Institut des Sciences Nucleaires, Grenoble, France*
5. *Brookhaven National Laboratory, USA*
6. *Universita di Genova and Sezione INFN, Italy*
7. *University of Texas at Arlington, USA*
8. *Nuclear Physics Institute, Tomsk, Russia*
9. *Universita di Trieste and Sezionne INFN, Italy*
10. *Institute for High Energy Physics, Protvino, Russia*
11. *Universita di Padova, Padova, Italy*
12. *Suny Stony Brook, USA.*

2. P. Gauron, B. Nicolescu, O. Selyugin,*Phys. Lett.* **B 397** , 305 (1997).
3. M. Jakob and G.C. Wick, *Ann. Phys.* (N.Y.) **7** , 404 (1959).
4. L. Puzikov, R. Ryndin, Ya.A. Smorodinsky, *Nucl. Phys.* **3**, 436 (1957).
5. S.M. Bilenky and R.M. Ryndin, *Phys. Lett.* **6**, 217(1963).
6. R.T.N. Philips, *Nucl. Phys.* **43**, 413 (1963).
7. M. Froissart, *Phys. Rev.***123**, 1053 (1963);
   A. Martin, *Nuovo Cimento***42**, 930 (1966).
8. Review of Particle Properties, *Phys. Rev.***D45**, ii.83 (1992).
9. M.J. Menon, A conjecture on Centauro Species, ABSTRACTA C 007-98, IFGW-UNICAMP, 13083-970 Campinas, SP, Brasil.
10. C. Avila et al., *Phys. Lett.***B 445**, 419 (1999).
11. C. Bourrely et al., *Z.Phys.***C 37**, 369(1988).
12. R.J. Glauber and J. Velasco, *Phys. Lett.***B 147**, 380(1984).
13. P. Desgrolar et al., *Z. Phys.***C 63**, 241(1994).
14. A. Donnachie and P.V. Landshoff, *Z. Phys.***C 2**, 55 (1979).
15. L.I. Lapidus and B.Z. Kopeliovich, *Sov. J. Nucl. Phys.***19,** 114 (1974).
16. N. Akchurin et al., *Phys. Rev.***D 48,** 3026 (1993).
17. N.H. Buttimore et al., *Phys. Rev.***D 18,** 694 (1978).
18. C. Augier et al.(UA4/2 Collaboration), "A precise measurement of the real part of the elastic scattering amplitude at Sp$\bar{p}$S", CERN/PPE- 93-115, 1993; M. Haguenauer(UA4/2 Collaboration), *Nucl. Phys.* B(Proc. Suppl.)**B 25,** 266 (1992).

M. M. Islam

High Energy Elastic Scattering and Underlying Effective Field Theory Model of Nucleon Structure

M.M. Islam and E.M. Kubik
Physics Department
University of Connecticut
Storrs, CT 06268
U.S.A.

Abstract

High energy elastic *pp* and $\bar{p}p$ scattering have been analyzed previously in a model where the nucleon has an inner core and an outer cloud. Gauged linear σ-model has been found to provide the underlying effective field theory framework. This model identifies the nucleon core as a soliton arising from topological baryonic charge distribution and attributes the outer cloud to a quark-antiquark condensed ground state. Our present goal is to extend the previous calculations to the LHC c.m. energy of $\sqrt{s} = 14$ TeV. How this is done and the resulting quantitative calculations are presented. Special features of the calculated differential cross section at the LHC energy are pointed out, and the importance of testing them in the proposed TOTEM project is emphasized.

1 Introduction

High energy elastic *pp* and $\bar{p}p$ scattering were measured at the CERN ISR and SPS Collider over the c.m. energy range $\sqrt{s}$ = 23 - 630 GeV and in the momentum transfer range $|t|$ = 0 - 10 GeV^2. These data were analyzed in a model where the nucleon has an inner core and an outer cloud [1]. High energy elastic scattering in this model arises primarily from two processes: *i*) a glancing collision where the outer cloud of one nucleon interacts with that of the other giving rise to diffraction scattering; *ii*) a hard collision where one nucleon core scatters off the other nucleon core via vector meson ω-exchange while their outer clouds overlap and interact independently [2]. In the small momentum transfer region, diffraction dominates, but as the momentum transfer increases, the hard scattering takes over. It was found that such a phenomenological description of the nucleon structure and nucleon-nucleon interaction can be understood in the framework of an underlying effective field theory model — the gauged linear σ-model [3]. In this model, the nucleon core can be identified as a soliton arising from the topological baryonic charge distribution and the outer cloud can be understood as a condensed quark-antiquark ground state analogous to a superconducting ground state. Furthermore, the topological baryonic charge originates from an anomalous action, known as the Weiss-Zumino-Witten action, and ω is coupled to this topological baryonic charge as a gauge boson. As long as this effective model holds, baryonic charge continues to behave as topological charge and ω coupled to it as a gauge boson continues to behave as a gauge boson, i.e., as an elementary spin-1 particle. The $q\bar{q}$ condensed ground state originates from the interaction between left and right quarks via a scalar field. This interaction makes the quarks massive and leads to the spontaneous breaking of chiral symmetry.

2 *pp* Elastic Scattering at LHC

Our present goal is to predict, on the basis of the above core-cloud model, elastic differential cross section at the Large Hadron Collider (LHC) at c.m. energy $\sqrt{s}$ = 14 TeV and momentum transfer range $|t|$ = 0 - 10 GeV^2. The proposed TOTEM project[4,5] will be able to measure elastic differential cross section at LHC throughout this $|t|$ region, and therefore will be able to test stringently various aspects of the model. With this perspective, we briefly outline how the quantitative calculations are carried out in the model[1] and indicate the new features which have been incorporated in the present analysis.

The forward diffraction amplitude is described using the impact parameter representation:

$$T_D(s,t) = ipW \int_0^\infty bdb\ J_0(bq)\Gamma_D^+(s,b) \tag{1}$$

and the profile function

$$\Gamma_D^+(s,b) = g(s)\left[\frac{1}{1+e^{\frac{b-R(s)}{a(s)}}} + \frac{1}{1+e^{-\frac{b+R(s)}{a(s)}}} - 1\right], \tag{2}$$

where $q = \sqrt{|t|}$, $R(s) = R_0 + R_1(\ln s - i\pi/2)$, $a(s) = a_0 + a_1(\ln s - i\pi/2)$, and $g(s)$, which asymptotically goes to a real constant, has the crossing even property: $g^*(se^{i\pi}) = g(s)$. The hard scattering amplitude is given by an amplitude $T_1(s,t)$ due to a single hard collision multiplied by an absorption factor $\left[1 - \Gamma_D^{\bar{p}p}(s,0)\right]$ or $[1 - \Gamma_D^{pp}(s,0)]$. In the absorption factor, the possibility of a C-odd contribution at very small impact parameter is taken into account, so that $\Gamma_D^{\bar{p}p}(s,0) = \Gamma_D^+(s,0) + \Gamma_D^-(s,0)$, $\Gamma_D^{pp}(s,0) = \Gamma_D^+(s,0) - \Gamma_D^-(s,0)$.

The diffraction amplitude obtained from (1) and (2) satisfies the general properties associated with the phenomenon of diffraction:

1. It leads to $\sigma_T(s) \propto (a_0 + a_1 \ln s)^2$, i.e., qualitative saturation of Froissart-Martin bound.

2. It indicates $\rho(s) = \frac{\pi a_1}{a_0 + a_1 \ln s}$ asymptotically, so that the derivative dispersion relation is satisfied ($\rho(s) = \text{Re}T(s,0)/\text{Im}T(s,0)$).

3. It is C-even, and therefore yields equal pp and $\bar{p}p$ total and differential cross sections.

4. It obeys the Auberson-Kinoshita-Martin scaling, i.e., $T_D(s,t) \sim is \ln^2 s \times f(|t| \ln^2 s)$ asymptotically. In addition to the above properties, the diffraction amplitude falls off with increasing $|t|$ faster than the hard amplitude, so that the latter dominates the large $|t|$ region.

To extend previous calculations to the LHC energy of 14 TeV, we face a new problem: $g(s)$ in the diffraction profile function (2) is a complex energy dependent parameter which was determined in previous analyses at each value of s by fitting with the measured differential cross section data. However, we are now trying to predict the differential cross section at a higher energy and therefore $g(s)$ has to be known beforehand. How do we get around this problem? We noticed that accurate model independent analysis by Kundrat and Lokajicek[6] show that absorption due to diffraction at $b = 0$, i.e., $\left|e^{i\chi_D^+(s,0)}\right|$ is finite at high energy and decreases slowly with increasing s. This suggested to us a simple parameterization that we have found satisfactory:

$$e^{i\chi_D^+(s,0)} = \eta_0 + \frac{1}{c_0 + c_1(\ln s - i\pi/2)}. \tag{3}$$

Since $g(s)$ is related with $e^{i\chi_D^+(s,0)} = 1 - \Gamma_D^+(s,0)$, Eq. (3) expresses the energy dependence of $g(s)$ in terms of three energy independent parameters: η_0, c_0, c_1. To take into account the energy dependence of $\Gamma_D^-(s,0)$, we have assumed the following parameterization:

$$\Gamma_D^-(s,0) = i\lambda_0 + \frac{[d_0 + d_1(\ln s - i\pi/2)]^2}{s}, \tag{4}$$

which, of course, has the required C-odd property. The first term on the RHS of (4) shows that, even at asymptotic energy, C-odd contribution to the absorption correction remains finite. The second term on the RHS of (4) is simply a parametrization of the unknown s-dependence of $\Gamma_D^-(s,0)$. In the present analysis, we have ten energy independent parameters: R_0, R_1, a_0, a_1, η_0, c_0, c_1, λ_0, d_0, d_1 and two energy dependent parameters: $\hat{\gamma}(s)$, $\hat{\theta}(s)$, which for simplicity are taken as constants. Compared to the energy 0.546 TeV of the previous analysis[1], the energy range we now cover extends to more than 10 TeV. We can fit well the asymptotic behavior of the total cross section (solid line in Fig. 1). This is to be compared with the parametrization $\sigma_\infty = C_0 + C_2(\ln s)^\gamma$ given by Augier et al.[7] (dotted line in Fig. 1). Also, our present fits of $\rho(s)$ (Fig. 2) and $\frac{d\sigma}{dt}$ at $\sqrt{s} = 546$ GeV (Fig. 3) are quite reasonable given that we have made only visual search of parameters. Experimental data in Fig. 3 (solid circles) are the measured $\frac{d\sigma}{dt}$ for $\bar{p}p$ at 546 GeV by Bozzo et al.[8]. From the fits, we obtain $R_0 = 2.42$, $R_1 = 0.035$, $a_0 = 0.585$, $a_1 = 0.113$, $\eta_0 = 0.001$, $c_0 = 25.0$, $c_1 = 25.0$, $\lambda_0 = 0.35$, $d_0 = 0.5$, $d_1 = 3.2$, $\hat{\gamma} = 8.2/\pi$, $\hat{\theta} = -0.1$. All through our calculations, unit of s is 1 GeV^2. Our predicted $\frac{d\sigma}{dt}$ at the LHC energy of $\sqrt{s} = 14$ TeV is shown in Fig. 3. Also, in Fig. 3, for comparison, we show $\frac{d\sigma}{dt}$ for pp at $\sqrt{s} = 53$ GeV using the parameters in Ref. 1.

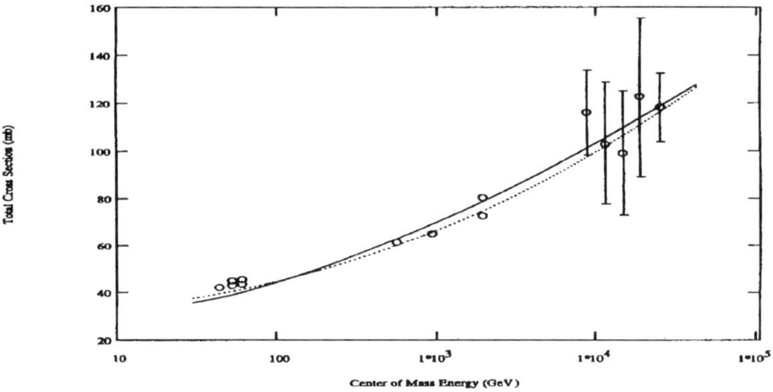

Figure 1: Solid line represents our calculated total cross section as a function of energy using the parameters given in the text. Dotted line represents asymptotic behavior of total cross section given by Augier et. al. (Ref. 7).

3 Special Features of the Asymptotic Elastic Scattering

Examining Fig. 3, we observe several special features:

1. As $\sqrt{s}$ increases from ISR to SPS collider and then to LHC, the forward diffraction cross section increases, the diffraction peak shrinks, and the interference dip moves to smaller values of $|t|$ and rises up. The latter behavior occurs because the shrinkage of the diffraction peak exposes the underlying hard process.

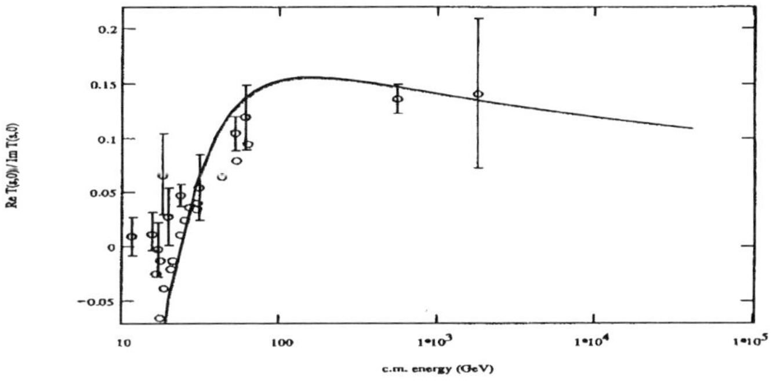

Figure 2: Solid line represents our calculated $\rho(s) = \mathrm{Re}T(s,0)/\mathrm{Im}T(s,0)$ as a function of energy using the parameters given in the text. As in the case of total cross section, $\rho(s)$ for pp and $\bar{p}p$ essentially overlap.

2. For $|t| \gtrsim 1$ GeV^2, we predict a differential cross section that falls off smoothly several decades and behaves as $\frac{d\sigma}{dt} \sim e^{-a\sqrt{|t|}}$ (Orear fall-off). This is in contrast to the impact factor model of Bourrely et al. [9] and Regge pole plus cut model of Desgrolard et al. [10] which include Pomeron, Froissaron, and odderon. The latter two models predict oscillations in the differential cross section at LHC energy in the large $|t|$ region [11].

3. For $|t| > 1$ GeV^2, $\frac{d\sigma}{dt}$ at $\sqrt{s} = 546$ GeV and 14 TeV overlap, i.e., $\frac{d\sigma}{dt}$ has very little energy dependence, only $|t|$ dependence. Such a behavior was predicted by Donnachie and Landshoff [12] on the basis of a triple-gluon exchange model which, in contrast to ours, leads to $\frac{d\sigma}{dt} \sim \frac{1}{t^8}$. There are, however, several issues that raise questions regarding the origin of this behavior [13].

4. $\frac{d\sigma}{dt}$ seems to continue its exponential fall-off all the way to $|t| = 10$ GeV^2 and beyond. In reality, we expect this behavior to change for $|t| \gtrsim 8$ GeV^2. The underlying linear σ-model indicates that the scalar field responsible for making quarks massive must vanish sharply from a constant value f_π (pion decay constant) to zero at some critical distance R. This implies that for momentum transfer q sufficiently large ($1/q < R$), one nucleon probes the other at an impact parameter $b < R$ ($b \simeq 1/q$). The scattering at such a q value will then be due to quark-quark

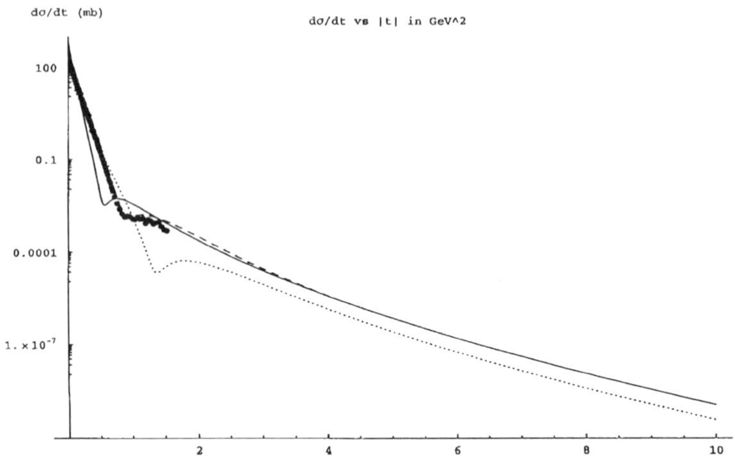

Figure 3: Solid line represents our predicted differential cross section for pp elastic scattering at the Large Hadron Collider at $\sqrt{s} = 14$ TeV. Dashed line is our fit of the measured $\bar{p}p$ differential cross section at $\sqrt{s} = 546$ GeV using the parameters given in the text. Experimental data at 546 GeV (solid circles) are from Ref. 8. Also plotted for comparison is the pp differential cross section at $\sqrt{s} = 53$ GeV using the parameters of Ref. 1 (dotted line).

scattering and is expected to lead to a power fall-off behavior[14]: $\frac{d\sigma}{dt} \sim \frac{1}{t^{10}}$ and consequently to a change in the slope of $\frac{d\sigma}{dt}$ [15]. Such a change of slope was reported by the ISR group[16] at $\sqrt{s} = 53$ GeV.

4 Concluding Remarks

The various features discussed above are connected with the nonperturbative structure of the nucleon, which is yet to be established (even after three decades since the birth of the quark model). It is therefore important to look for these features by accurate measurement of $\frac{d\sigma}{dt}$ at LHC from the very forward direction to momentum transfer greater than 10 GeV^2. The TOTEM project at LHC[4,5] can accomplish this objective. In particular, the large design luminosity of the LHC will allow elastic scattering measurement up to $|t| \simeq 15$ GeV^2. As mentioned above, if a distinct change of slope of $\frac{d\sigma}{dt}$ can be confirmed (say, at $|t| \gtrsim 8$ GeV^2) in the TOTEM project, then it will signal a transition from the nonperturbative regime of soliton and quark-antiquark condensate to a perturbative regime of massless (effective) quarks and will imply a chiral phase transition. Furthermore, if TOTEM verifies the special features of $\frac{d\sigma}{dt}$, then it will mean that the gauged linear σ-model as an effective field theory model can also be applied to study nuclear matter at high temperature and high density[17] even when the c.m. energy of two colliding nucleons is 14 TeV. In particular, it will mean that the model can be applied to lead-lead collisions in CMS, where the c.m. energy of two colliding nucleons will be 5.4 TeV. The model will predict a chiral phase transition for these heavy-ion collisions.

References

1. M.M. Islam, V. Innocente, T. Fearnley, and G. Sanguinetti, EuroPhys. Lett. **4**, 189(1987).
2. M.M. Islam, Nucl. Phys. B (Proc. Suppl.) **25B**, 104(1992).
3. M.M. Islam, Z. Phys. C **53**, 253(1992); Proceedings of the Workshop on Quantum Infrared Physics, ed. by H.M. Fried and B. Müller (World Scientific, 1995) p. 401.
4. TOTEM, CERN/LHCC 97-49, Aug. 1997.
5. TOTEM, CERN/LHCC 99-7, Mar. 1999.
6. V. Kundrat and M. Lokajicek, Z. Phys. C **63**, 619(1994).
7. C. Augier et al., Phys. Lett. B **315**, 503(1993).
8. M. Bozzo et al., Phys. Lett. **147B**, 385(1984); **155B**, 197(1985).
9. C. Bourrely, J. Soffer, and T.T. Wu, Nucl. Phys. B **247**,15 (1984); Z. Phys. C **37**, 369(1988); Frontiers in Strong Interactions, ed. by P. Chiappetta, M. Haguenauer, and J. Tran Thanh Van (Editions Frontieres, 1996) p. 15.
10. P. Desgrolard, M. Giffon, and E. Predazzi, Z. Phys. C 63, 241(1994).
11. M. Buenerd, Frontiers in Strong Interactions, ibid., p. 437.
12. A. Donnachie and P.V. Landshoff, Z. Phys. C **2**, 55(1979).
13. A. Donnachie and P.V. Landshoff, Phys. Lett. B **387**, 637(1996).
14. M.G. Sotiropoulos and G. Sterman, Nucl. Phys. B **425**, 489(1994).
15. M.M. Islam, Frontiers in Strong Interactions, ibid., p. 91.
16. H. De Kerret et al., Phys. Lett. **68B**, 374(1977).
17. M.M. Islam, Proceedings of the Workshop on Quantum Chromodynamics, ed. by H.M. Fried and B. Müller (World Scientific, 1999) p. 119.

Vojtěch Kundrát

HIGH-ENERGY ELASTIC HADRON SCATTERING AND SPACE STRUCTURE OF HADRONS

VOJTĚCH KUNDRÁT, MILOŠ LOKAJÍČEK
Institute of Physics, AS CR, 182 21 Praha 8, Czech Republic

DALIBOR KRUPA
Institute of Physics, SAS, 84228 Bratislava, Slovakia

High-energy elastic hadron scattering has been usually described with the help of amplitude having a dominant imaginary part in a broad interval of momentum transfers t. Perfoming a Fourier-Bessel (FB) transform of such an amplitude one inevitably obtaines a central distribution of elastic hadron collisions in the impact parameter space [1], which contradicts a natural expectation that elastic hadron collisions produced at small $|t|$ values being peripheral in contradistinction to the central character of inelastic processes. Another puzzle concerns then diffractive production processes which are described at the same time as peripheral even if they have very similar characteristics as elastic hadron processes.

The elastic hadron scattering amplitude $h_{el}(s,b)$ at the impact parameter b is given by the FB transform of the elastic hadron amplitude $F^N(s,t)$:

$$h_{el}(s,b) = \frac{1}{4p\sqrt{s}} \int_{t_{min}}^{0} dt\ J_0(b\sqrt{-t}) F^N(s,t); \qquad (1)$$

here s is the square of the total energy in the CMS, p is the corresponding momentum value and $J_0(x)$ is the Bessel function of the zeroth order.

The region of kinematically allowed t values is limited at finite energies. In order to define the FB transform mathematically correctly, the amplitude $F^N(s,t)$ should also be defined in the unphysical region of t [2,3]. It means that the elastic profile $h_{el}(s,b)$ should be considered as a sum of two terms

$$h_1(s,b) = \frac{1}{4p\sqrt{s}} \int_{t_{min}}^{0} dt\ F^N(s,t)\ J_0(b\sqrt{-t}), \qquad (2)$$

and

$$h_2(s,b) = \frac{1}{4p\sqrt{s}} \int_{-\infty}^{t_{min}} dt\ \lambda(s,t)\ J_0(b\sqrt{-t}), \qquad (3)$$

where the unknown function $\lambda(s,t)$, defined in the unphysical region of t, should fulfill two additional mathematical conditions [3].

Similarly one can divide the inelastic overlap function into the sum of two parts $g_1(s,b)$ and $g_2(s,b)$ where the first term represents the FB transform of the inelastic overlap function $G_{inel}(s,t)$ in the physical region of t and the second term is the FB transform of unknown function $\mu(s,t)$ in the unphysical region of t restricted by similar conditions like function $\lambda(s,t)$ [2,3].

The unitarity condition in the impact parameter space has then the form

$$\Im h_1(s,b) = |h_1(s,b)|^2 + g_1(s,b) + K(s,b) \tag{4}$$

where correction function $K(s,b)$ must vanish at $b = 0$ and $b \to 0$ and its integral over all positive impact parameter values should be zero [2,3].

Contrary to general belief the functions $\Im h_1(s,b)$ and $g_1(s,b)$ cannot be non-negative at all finite positive b values. At higher values of b these functions oscillate. It must hold only $\Im h_1(s,0) \geq 0$ and both the functions $\Im h_1(s,b)$ and $g_1(s,b)$ must vanish when b tends to ∞ [2]. Thus, only $|h_1(s,b)|^2$ can be interpreted as the density distribution of elastic hadron scattering in the impact parameter space. Integrals of individual profile functions appearing in Eq. (4) over all positive b values determine then the values of total, elastic and inelastic cross sections.

Let us add now the function $c(s,b)$, fulfilling analogical integral condition as function $K(s,b)$, to the both sides of unitarity equation (4). It is then possible to define the functions $h_{tot}(s,b) = \Im h_1(s,b) + c(s,b)$ and $g_{inel}(s,b) = g_1(s,b)+K(s,b)+c(s,b)$ that can be brought to be non-negative for all positive values of b. These functions can be then shown to represent density distributions of the total and inelastic scatterings in the impact parameter space. One obtains the modified unitarity condition

$$h_{tot}(s,b) = |h_1(s,b)|^2 + g_{inel}(s,b). \tag{5}$$

It is then possible to define the mean-squares of total, elastic and inelastic impact parameters as

$$< b^2(s) >_d \; = \; \frac{\int\limits_0^\infty bdb\, b^2\, d(s,b)}{\int\limits_0^\infty bdb\, d(s,b)}, \tag{6}$$

where the function $d(s,b)$ corresponds to density distributions $h_{tot}(s,b)$, $|h_1(s,b)|^2$ or $g_{inel}(s,b)$ appearing in Eq. (5). All these mean-squares can be expressed with the help of elastic hadron scattering amplitude $F^N(s,t)$ in the t variable.

First, the mean-square of the elastic impact parameter is given [1,4] by

$$< b^2(s) >_{el} \;=\; 4\,\frac{\int\limits_{t_{min}}^{0} dt|t|\,|\frac{d}{dt}F^N(s,t)|^2}{\int\limits_{t_{min}}^{0} dt\,|F^N(s,t)|^2}. \tag{7}$$

Second, the mean-square of the total impact parameter equals [5]

$$< b^2(s) >_{tot} \;= 2B(s,0), \tag{8}$$

where $B(s,0)$ is the diffractive slope in forward direction generally defined as

$$B(s,t) = \frac{2}{|F^N(s,t)|}\frac{d}{dt}|F^N(s,t)|. \tag{9}$$

Finally, the mean-square of inelastic impact parameter can be expressed [5] with the help of Eq. (5)

$$< b^2(s) >_{tot} \;=\; \frac{\sigma_{el}(s)}{\sigma_{tot}(s)} < b^2(s) >_{el} + \frac{\sigma_{inel}(s)}{\sigma_{tot}(s)} < b^2(s) >_{inel}, \tag{10}$$

where $\sigma_{tot}(s)$, $\sigma_{el}(s)$ and $\sigma_{inel}(s)$ are the total, elastic and inelastic cross sections.

The impact parameter profiles are determined, therefore, fully by the complex amplitude $F^N(s,t)$ that can be established from experimental elastic scattering data if the region of very small $|t|$ is involved (interference of Coulomb and nuclear scatterings). However, earlier attempts to derive the given profiles with the help of West and Yennie formula [6] must be regarded as roughly approximative. The problem can be solved in agreement with reality only if all elastic experimental data are described with the help of more exact total scattering amplitude [7]

$$F^{C+N}(s,t) = \pm\frac{\alpha s}{t} f_1(t)f_2(t) + F^N(s,t)\Bigg\{1 \mp i\alpha \int\limits_{t_{min}}^{0} dt'\Bigg[\ln\frac{t'}{t}\Big[f_1(t')f_2(t')\Big]' - \frac{1}{2\pi}\Bigg[\frac{F^N(s,t')}{F^N(s,t)} - 1\Bigg] I(t,t')\Bigg]\Bigg\}, \tag{11}$$

where

$$I(t,t') = \int\limits_0^{2\pi} d\Phi \frac{f_1(t'')f_2(t'')}{t''} \tag{12}$$

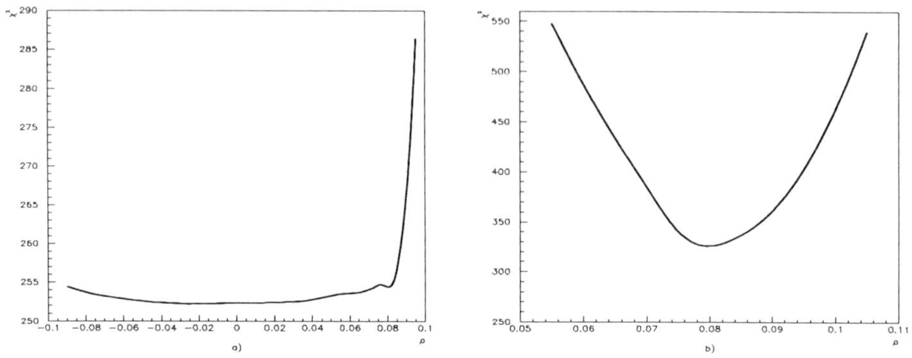

Figure 1: Dependence of the total χ^2 values upon ρ for pp elastic scattering at energy of 53 GeV: a) for peripheral case, b) for central case

Table 1: pp elastic scattering at energy of 53 GeV

Quantity	Peripheral	Central
χ^2/df	252/201	329/204
σ_{tot}	$42.65 \div 42.97$ (± 0.12)	42.69 ± 0.23
B	$13.47 \div 13.67$ (± 0.05)	13.26 ± 0.05
ρ	$-0.12 \div 0.08$ (± 0.009)	0.075 ± 0.005
$\sqrt{< b^2 >_{tot}}$	1.026÷1.033	1.03
$\sqrt{< b^2 >_{el}}$	1.78÷1.80	0.68
$\sqrt{< b^2 >_{inel}}$	0.76÷0.80	1.09

and $t'' = t + t' + 2\sqrt{tt'}\cos\Phi$. α is the fine structure constant and $f_1(t)$ and $f_2(t)$ are the corresponding formfactors. The upper (lower) sign in Eq. (11) corresponds to the scattering of pp $(\bar{p}p)$. The total cross section equals then

$$\sigma_{tot} = \frac{4\pi}{p\sqrt{s}} |F^N(s,0)| \frac{1}{\sqrt{1+\rho^2(s,0)}}, \tag{13}$$

where

$$\rho(s,t) = \tan\zeta^N(s,t) = \frac{\Re F^N(s,t)}{\Im F^N(s,t)}. \tag{14}$$

Based on West and Yennie approach the elastic hadron scattering at given energy is being characterized usually by $\rho(s,0)$ and $B(s,0)$ in addition to $\sigma_{tot}(s)$. Our general formula indicates, however, that the corresponding values cannot provide any reliable description, as peripheral behavior is to be preferred on the basis of corresponding χ^2 values [7]. On the other side, equivalent optimal fits may be obtained belonging to different degrees of peripherality characterized, e.g., by the ratio of mean squares $< b^2(s)_{el} >$ and $< b^2(s)_{inel} >$ (see Eqs. (7)

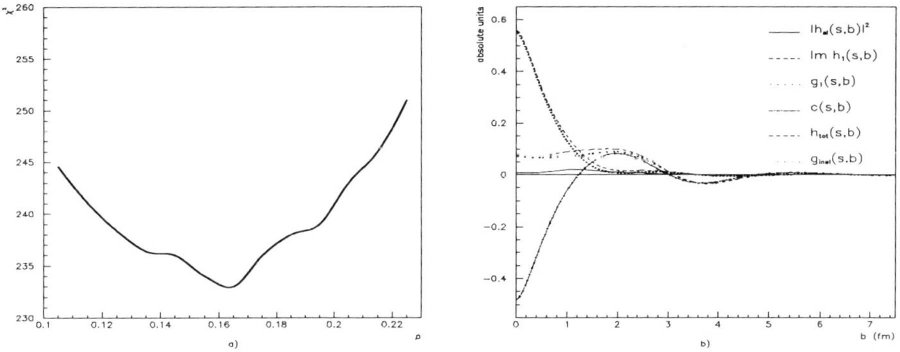

Figure 2: $\bar{p}p$ elastic scattering at energy of 541 GeV: a) dependence of the total χ^2 values upon ρ; b) new profiles constructed with the help of function $c(s,b)$

Table 2: $\bar{p}p$ elastic scattering at energy of 541 GeV

Quantity	Peripheral	Central
χ^2/df	233/213 ÷238/213	354/217
σ_{tot}	60.83 ÷ 63.35 (±1.16)	62.72 ± 0.78
B	16.18 ÷ 16.55 (±0.09)	16.35 ±0.05
ρ	0.11 ÷ 0.18 (±0.022)	0.096 ± 0.013
$\sqrt{< b^2 >_{tot}}$	1.123÷1.135	1.13
$\sqrt{< b^2 >_{el}}$	2.24÷2.35	0.76
$\sqrt{< b^2 >_{inel}}$	0.39÷0.49	1.21

and (10)); and different values of $\rho(s,0)$ should be then regarded as fully admissible. For pp elastic scattering [8] at energy of 53 GeV these admissible values of χ^2 lie in the interval (-0.12, 0.08) see Fig. 1a. If only the central behavior were required the best fit (with a higher χ^2 value) would lead to the admissible $\rho(s,0)$ values lying in a much narrower interval, i.e., (0.075, 0.085) - see Fig. 1b. The admissible values of other quantities are shown in Table 1. The elastic root-mean-square in the peripheral case is always significantly higher than the inelastic one, while in the central case the opposite would be true. The similar analysis was performed also for $\bar{p}p$ elastic scattering [9] at energy of 541 (546) GeV. However, in this case the condition $\sigma_{tot}(1+\rho^2) = 63.3 \pm 1.5$ mb had to be used to normalize experimental data at small $|t|$ [9] which has limited the interval of admissible values also in the peripheral case - see Fig. 2a and Table 2. E.g., admissible values of ρ should lie in the interval (0.11, 0.18) (preliminary results have been published already [10]).

In the latter case ($\bar{p}p$ scattering) we have also established the function

$c(s, b)$ bringing all three profiles to be non-negative values for all $b \geq 0$. The shape of this function as well as resulting profiles (corresponding to the best fit) are shown in Fig. 2b.

The shapes of individual impact parameter profiles were derived for the first time by Miettinen [11] for pp elastic scattering at energy of 53 GeV on the basis of West and Yennie formula. He argued then that protons should be transparent objects allowing rather frequent elastic scattering also in 'head-on' collisions, which was always regarded as a puzzle [12]. Our more exact analysis does not provide any possibility for such conclusions; elastic processes at small b values being practically negligible.

References

1. V. Kundrát, M. Lokajíček Jr. and M. Lokajíček, Czech. J. Phys. B31, 1334 (1981).
2. T. Adachi and T. Kotani, Progr. Theor. Phys. Suppl., Extra Number 310, (1965); 35, 463 (1966); Suppl. 37-38, 297 (1966); 35, 485 (1966).
3. M. M. Islam, Proceedings of the Theoretical Physiscs Institute, University of Colorado, Summer 1967, p.97.
4. F. S. Heney and J. Pumplin, Nucl. Phys. B166, 513 (1979).
5. V. Kundrát, M. Lokajíček and D. Krupa: High-energy elastic hadron collisions and space structure of hadrons; to be published
6. G. B. West and D. R. Yennie, Phys. Rev.172, 1413 (1968).
7. V. Kundrát and M. Lokajíček, Z. Phys. C63, 619 (1994); Mod. Phys. Lett. A11, 2241 (1996).
8. J. Bystricky et al., in Nucleon-nucleon and kaon-nucleon scattering, edited by H. Schopper (Landolt-Börnstein Series, Vol. 1) Springer, Berlin (1980)
9. C. Augier et al., Phys. Lett. B316, 448 (1993); Phys. Lett. B344 (1995) 451. M. Bozzo et al., Phys. Lett. 147B, 385 and 392 (1984); Phys. Lett. 155B, 197 (1985). D. Bernard et al., Phys. Lett. 171B, 142 (1986).
10. V. Kundrát, M. Lokajíček and D. Krupa, "Hadron structure and the ρ value in high-energy elastic hadron collisions", Hadron structure'98 Conference, Sept. 7 - 13, 1998, Stará Lesná, Slovakia, ed. D. Bruncko & P. Striženec, p.389.
11. H.G. Miettinen, in proceedings of the IX th Rencontre de Moriond, Meribel les Allues, Vol.1 (ed. J. Tran Thanh Van), Orsay (1974).
12. G. Giacomelli and M. Jacob, Phys. Rep. 55, 1 (1979).

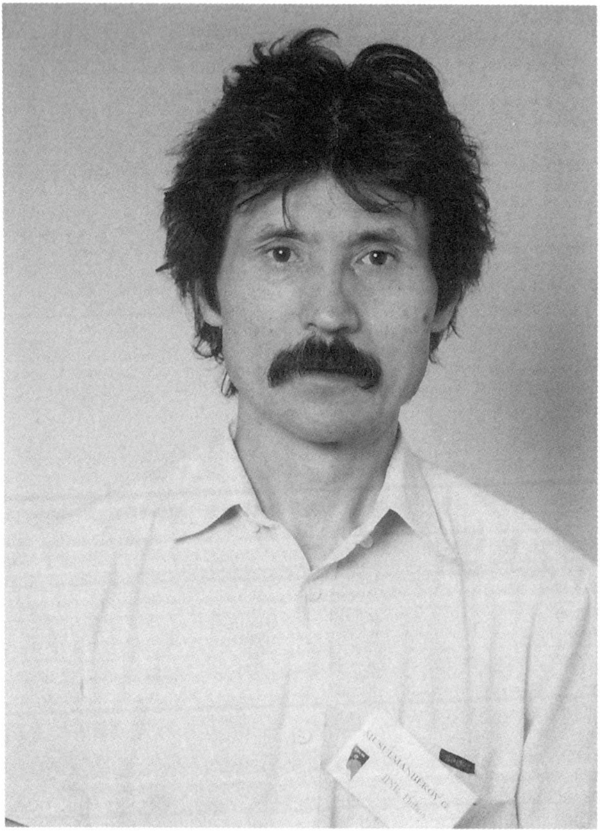

G. Musulmanbekov

DYNAMICAL QUARK STRUCTURE OF HADRONS AND DIFFRACTIVE PROCESSES

G. Musulmanbekov
Joint Institute for Nuclear Research,
141980 Dubna Russia
E-mail: genis@cv.jinr.ru

Diffractive processes are considered in the frame of Strongly Correlated Quark Model (SCQM). It is shown that single diffractive dissociation is a result of fluctuations of hadronic matter distributions inside colliding hadrons.

1 Introduction

Elastic and inelastic pp and $\overline{p}p$ –interactions play an important role in understanding of structure of hadrons. Separation of scattering processes into hard and soft ones, single and double diffraction (hard and soft) and central hadron production tells us that this structure is very complicated. If hard processes are the fields of application of current quark consideration then soft ones are described in constituent quark approach. Interplay between hard and soft interactions in high energy experiments strictly relates to interconnection between small and large size quark configurations inside a hadron. We think that this interplay is a manifestation of fluctuations of hadronic or quark matter distributions inside hadrons and these fluctuations in turn are results of correlated motion of valence quarks inside hadrons.

The question arises: is it possible to construct the dynamical system of quarks which can be observed at one instant of time as constituent (dressed) quarks and at the another one - as current (bare) quarks? Proposed by author the semiclassical model of strongly correlated quarks, SCQM, demonstrated how these configurations were realized inside hadrons[1]. The model possesses the features of both constituent and current quark models. Here we apply this model for description of diffractive processes and show that explicit manifestation of quark correlation inside hadrons is a single diffractive scattering.

2 Strongly Correlated Quark Model

Let us imagine the following hypothetical picture: single, colored quark imbedded into vacuum. Because of vacuum fluctuations one can observe two competing processes: first, polarization (ordering) of $\overline{q}q$ sea by the color field of valence quark (VQ) and second, the tend of vacuum fluctuations to destroy this polarization. As a result, one can say about a vacuum pressure on a single, colored quark. This effect can be interpreted as the "response" of the vacuum on the presence of, say, point defect or dislocation like in solid state physics. What happens if we place in vicinity of this quark the corresponding antiquark? By virtue of opposite color signs their polarization fields interfere in the overlapping region **destructively**. So the pressure of vacuum on quark (antiquark) from outside exceeds that one going from inner space between quark and antiquark. This results in an **attractive** force between quark (dislocation) and antiquark (antidislocation). Intensity of the remaining part of polarization field around quark (antiquark) is hadronic matter distribution. At maximum displacements of $\overline{q}q-$ system, that correspond to small overlapping of polarization fields, hadronic matter distributions have maximum values. Then quark and antiquark start moving towards each other. The closer they each other the larger destructive interference effect and the smaller hadronic matter distributions around VQs and the larger their kinetic energies.

For such interacting $\overline{q}q-$ system the total Hamiltonian is

$$H = \frac{m_{\overline{q}}}{(1-\beta^2)^{1/2}} + \frac{m_q}{(1-\beta^2)^{1/2}} + V_{\overline{q}q}(2x), \tag{1}$$

were $m_{\overline{q}}$, m_q- masses of valence antiquark and quark, $\beta = \beta(x)-$ their velocity depending on displacement x and $V_{\overline{q}q}-$ quark–antiquark potential energy with separation $2x$. It can be rewritten as

$$H = \left[\frac{m_{\overline{q}}}{(1-\beta^2)^{1/2}} + U(x)\right] + \left[\frac{m_q}{(1-\beta^2)^{1/2}} + U(x)\right] = H_{\overline{q}} + H_q, \tag{2}$$

were $2U(x) = V_{\overline{q}q}(2x)$. Therefore, keeping in mind that quark and antiquark are strongly correlated we can consider them separately in 1+1 dimension. Generalization to three–quark system in baryons is performed according to $SU(3)_{color}$ symmetry: an antiquark is replaced by two correspondingly colored quarks to get color singlet baryon and destructive interference takes place between color fields of three valence quarks. Neglecting the mass and charge differences of valence quarks we could say that inside baryon three quarks move along the bisectors of equilateral triangle.

VQ with its polarized surroundings (hadronic matter distribution) forms constituent quark. According to our approach potential energy of valence quark, $U(x)$, corresponds to the energy (mass) M_Q of constituent quark:

$$U(x) = c\int_0^\infty \rho(x,\vec{r}\,')d^3r' \approx M_Q(x) \tag{3}$$

with

$$\rho(d, \vec{r}) = \left|\varphi(d, \vec{r})\right|^2 = \left|\varphi_{\overline{Q}}(-d, \vec{r}) - \varphi_Q(d, \vec{r})\right|^2, \tag{4}$$

where d is a parameter corresponding to VQ's displacement (x). So the knowledge of the mechanism of vacuum polarization around valence quark would give the information about the confining potential. Since we don't know this mechanism we assume that as a first approximation the polarization field can be taken in gaussian form:

$$\varphi_Q(\vec{r}) = \varphi_Q(x, y, z) = \varphi_Q(x_1, x_2, x_3) = \frac{(\det \hat{A})^{1/2}}{(\pi)^{3/2}} \exp\left(-\vec{X}^T \hat{A} \vec{X}\right), \tag{5}$$

where exponent is given in quadratic form. The same is for $\varphi_{\overline{Q}}(\vec{r})$. We fix the mass of constituent quark at maximum displacement

$$M_{Q(\overline{Q})}(x_{\max}) = \frac{1}{3}\left(\frac{m_\Delta + m_N}{2}\right) \approx 360\ MeV, \tag{6}$$

where m_Δ and m_N are masses delta–isobar and nucleon correspondingly. Therefore, the parameters of the model are the masses of VQs, $m_{q(\overline{q})}$, which are chosen to be 5 MeV, maximum displacement, $x_{\max}$, and parameters of gaussian function, $\sigma_{x,y,z}$. As shown below, $x_{\max}$ and $\sigma_{x,y,z}$ are adjusted by comparison of calculated and experimental inelastic cross sections, $\sigma_{in}(s)$, for pp and $\overline{p}p$–collisions.

Using (3)–(5) we can calculate the confining potential $U(x)$ and force $F(x) = -\frac{dU}{dx}$. Their shapes are shown in Fig. 1. As one can see the confining potential is essentially nonlinear. The behavior of potential evidently demonstrates the relationship between constituent and current quark inside a hadron. At maximum displacement quark is nonrelativistic, constituent one (VQ + "polarized sea"), since according to (3) the confining potential corresponds to the mass of constituent quark. At the origin of oscillation, $x = 0$, antiquark–quark in mesons and 3 quarks in baryons, being close each other, have maximum kinetic energy and correspondingly minimum potential energy and mass: they are relativistic, current quarks (bare VQ). Intermediate region corresponds to increasing (decreasing) of quark's mass by dressing (undressing) of quarks due to vacuum polarization. Due to this mechanism of VQs oscillations nucleon runs over the states corresponding to the certain terms of the infinite series of Fock space

$$|\, B\, \rangle = c_1 \,|\, q_1 q_2 q_3 \rangle + c_2 \,|\, q_1 q_2 q_3 \overline{q} q \rangle + \ldots \tag{7}$$

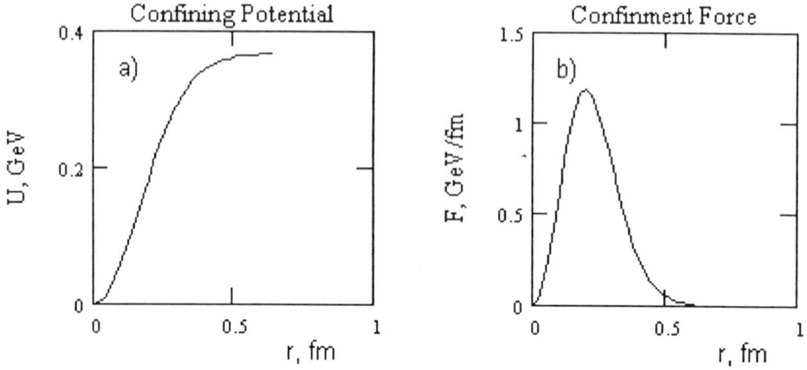

Figure 1: a)Confining potential; b) Confinement force.

Confining force drastically differs from the one given by string models. When VQs are close each other it is very weak and fulfills the "asymptotic freedom" behavior of quarks of QCD. At larger distances between VQs it starts growing rapidly, then reaching maximum value goes down, asymptotically approaching zero. Thus at large distances inside hadrons quarks being in a constituent state are almost free, noninteracting. Hence, it is clear why additive quark model where quarks are treated as massive, unbound and extended objects works well. We must emphasize that interaction between VQs is not direct but a result of polarization of surrounding vacuum combined with destructive interference. Attractive force between VQs in ground state hadrons does not appear as gluon string but goes from **suppression** of vacuum from outside. Hence our approach reflects the features of bag models, as well. The model is in agreement with the experiments (Fig. 2) for description of VQ structure function inside a nucleon

$$F_2^{ep} - F_2^{en} = \frac{x_F}{3}\left[u_v(x_F) - d_v(x_F)\right]. \tag{8}$$

Of cause, our description is classical and we must take into account not only the regular but also chaotic motion of VQs due to vacuum (quantum) fluctuations. It will be the subject of forthcoming papers. Classical consideration of VQ oscillations is justified by E. Schrodinger's paper[2] where he, analyzing the motion of wave packet solution of time dependent Schrodinger equation for harmonic oscillator, demonstrated that this wave packet moves in exactly the same way as corresponding classical oscillator. In our model VQ with its

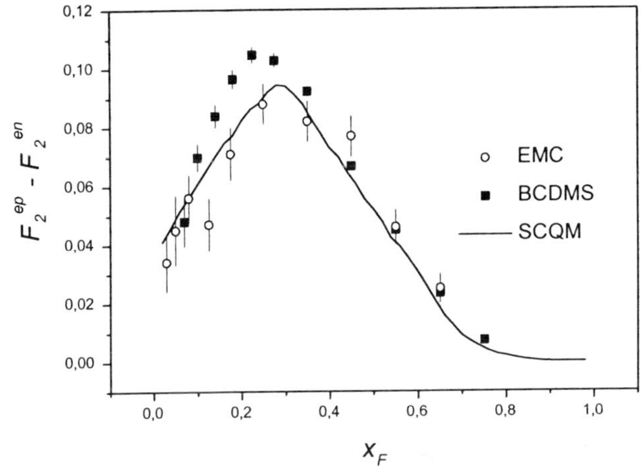

Figure 2: Valence quark structure functions in nucleons; data point are from papers[5,6].

surroundings can be treated as (nonlinear) wave packet. In forthcoming paper, as well, we'll show that this wave packet possesses soliton–like features.

3 Application to Diffractive Processes

Different configurations of quark contents in colliding hadrons realized at the instant of collision result in different types of reactions. The probability of finding any quark configuration inside a hadron is defined by the probability of VQ's displacement in proper frame of a hadron:

$$P(x)dx = \frac{Adx}{\sqrt{1 - m_q^2/(E-V)^2}} \tag{9}$$

with

$$\int P(x)dx = 1. \tag{10}$$

Configurations with nonrelativistic constituent quarks ($x \simeq x_{max}$) in both colliding hadrons lead to multiparticle production in central and fragmentation regions. Hard scattering with jet production and large angle elastic scattering take place when configurations with current VQs ($x \simeq 0$) in both colliding

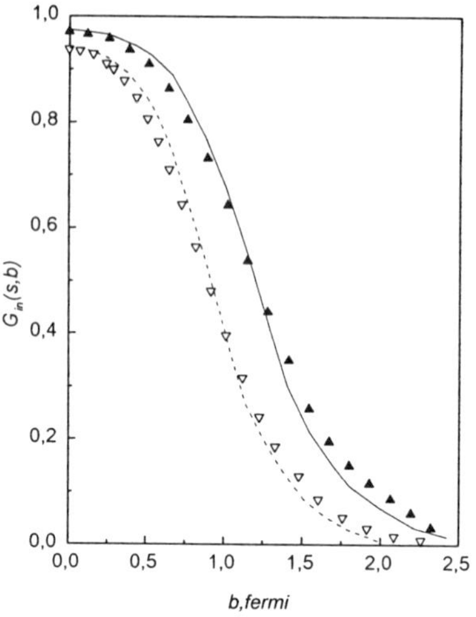

Figure 3: Inelastic overlap function for pp and $\overline{p}p-$ collisions at $\sqrt{s} = 53$ and $540\ GeV$; triangles are Henzi and Valin approsimation[3].

hadrons are realized. Intermediate configurations inside one (both) of colliding hadron are responsible for single (double) diffraction processes and semihard scattering. Using impact parameter representation, namely Inelastic Overlap Function (IOF), we can calculate total, inelastic, elastic and single diffractive cross sections for pp and $\overline{p}p$ collisions. In impact parameter representation IOF can be specified via the unitarity equation

$$2Imf(s,b) = |f(s,b)|^2 + G_{in}(s,b), \tag{11}$$

where $f(s,b)-$elastic scattering amplitude and $G_{in}(s,b)$ is IOF. IOF is connected with inelastic differential cross sections in impact parameter space:

$$\frac{1}{\pi}(d\sigma_{in}/db^2) = G_{in}(s,b). \tag{12}$$

Then inelastic, elastic and total cross sections can be expressed via IOF as

$$\sigma_{in}(s) = \int G_{in}(s,b)d^2\mathbf{b}, \tag{13}$$

$$\sigma_{el}(s) = \int \left[1 - \sqrt{1 - G_{in}(s,b)}\right]^2 d^2\mathbf{b}, \tag{14}$$

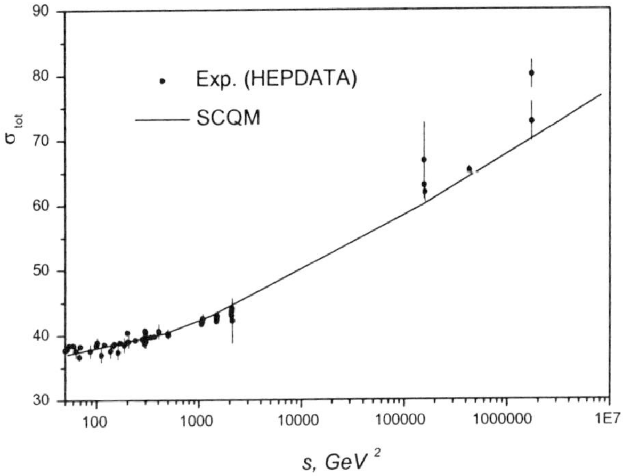

Figure 4: Total cross section for pp and $\overline{p}p-$ collisions; data points are compilations of experimental data taken from electronic data base HEPDATA[9].

$$\sigma_{tot}(s) = 2\int \left[1 - \sqrt{1 - G_{in}(s,b)}\right] d^2\mathbf{b}. \tag{15}$$

Since IOF relates to the probability of inelastic interaction at given impact parameter, (12), we carried out Monte Carlo simulation of inelastic nucleon–nucleon interactions. Inelastic interaction takes place at definite impact parameter b if at least one pion is produced in the region where hadronic matter distributions of colliding protons overlap

$$4M_{q_i}\gamma_{q_i}M_{p_j}\gamma_{p_j}\int \rho_{q_i}(\mathbf{r})\rho_{p_j}(\mathbf{r}-\mathbf{r}')\; d^3\mathbf{r} \geq m^2_{\pi_\perp}, \tag{16}$$

where indices q_i and p_j refer to quarks from different nucleons and $i, j = 1, 2, 3$, M_{q_i}, M_{p_j} – masses of hadronic matter composed in constituent quarks q_i and p_j, γ_{q_i}, γ_{p_j} – their γ-factors; intergrand expression is convolution of hadronic matter density distributions of quarks q_i and p_j. This condition corresponds Hisenberg picture[4] with modified right hand side: m^2_π in the original Heisenberg inequality is replaced by $m^2_{\pi_\perp}$. It is justified by the fact that the average transverse momentum of produced particles increases with energy. Specifying the quark configurations in each colliding nucleons by (9) and (3)–(5) we calculate $G_{in}(s,b)$ for certain values of impact parameter b and then cross

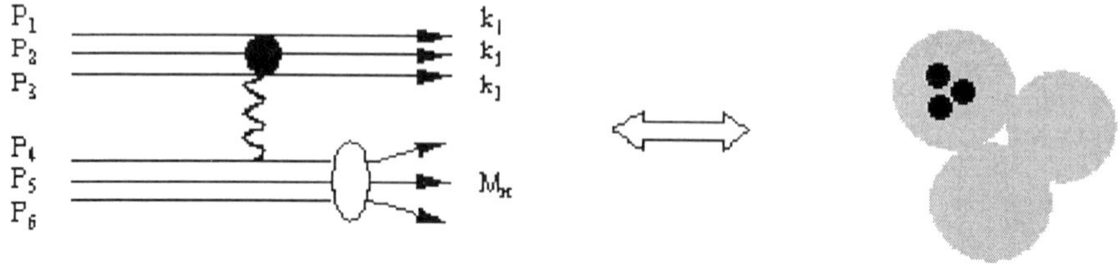

Figure 5: Correspondence between diagram and SCQM description of single diffractive dissociation in $pp \to pX$ processes. Projections of colliding nuclons are drawn on impact parameter plane: big grey circles are constituent quarks of dissociating nucleon, small black circles – (semi)bare quarks of the another nucleon.

sections σ_{in}, σ_{el} and σ_{tot}. The values of adjusted parameters of the model, $x_{\max} = 0.62\ fm$, $\sigma_{x,y} = 0.24\ fm$ and $\sigma_z = 0.12\ fm$, are chosen by comparison of calculated IOF with so called "BEL"–parametrization[3], (Fig. 3), and calculated inelastic cross sections with experimental ones, $\sigma_{in}(s)$, in pp and $\bar{p}p-$ collisions at $\sqrt{s}_{pp} = 540\ GeV$. Fig. 4 shows the result of calculation for pp total cross section at wide range of collision energies. One can see that the model with parameters fixed at one energy ($540\ GeV$) describes the energetic behavior of σ_{tot}. The growth of cross sections with energy is due to the tails of polarization fields around VQs not compensated by destructive interference with fields of two other VQs. With growing collision energy these tails result in the increasing effective size of hadronic matter distribution inside nucleons and correspondingly the increasing radius of interactions. The model gives for cross sections linear logarithmic energy dependence and the curve deviates from data points at very high energies. The reason of this is that in our geometrical approach, ((12) – (16)), we didn't take into account the real part of scattering amplitude. At energies $\sqrt{s} \leqslant 30\ GeV$ calculated cross sections were corrected on contributions of Regge poles exchange by using Donnachie and Landshoff parametrization[7]:$\Delta\sigma_R^{tot} = 56.08 s^{-0.4526}$

An oscillatory motion of VQs appearing in interplay between constituent and bare (current) quark configurations results in fluctuations of hadronic matter distribution inside colliding nucleons. Clear manifestation of these fluctuations is a process of single diffraction (SD). We select SD–events among inelastic $pp \to pX$ events with the criterion $1 - x_F \leqslant 0.1$, where $x_F = \frac{2}{\sqrt{s}}(k_1 + k_2 + k_3)$ (Fig. 5). Here k_1, k_2, k_3 are momenta of quarks forming the final state proton. As one can see, SD–events correspond to constituent quark configuration inside one colliding nucleon and (semi)bare quark configuration inside another one. Fig. 6 shows that calculated SD cross section slightly depends on energy;

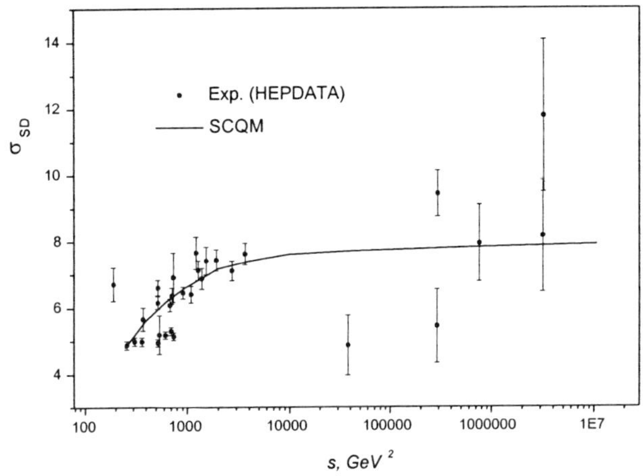

Figure 6: Single diffractive dissociation cross section for pp and $\overline{p}p-$ collisions; data points are compilations of experimental data taken from electronic data base HEPDATA[9].

our calculation for LHC energy gives $\sigma_{SD}^{LHC}(pp) = 8.30 \pm 0.15\ mb$.

Proposed dynamical model of hadron structure, SCQM, possesses some important features. It shows that hadronic matter distribution inside hadrons and in turn the sizes of hadrons are fluctuating quantities. Fluctuations of extended sizes of nucleons allows one to explain why 'black disk" limit is not saturated in pp and $\overline{p}p-$ collisions up to very high energies. Owing to non-compensated tails of polarization fields at outer sides of VQs, it describes the energy dependence of inelastic and total cross sections as a result of increasing effective sizes of colliding hadrons. On the other hand, there is no overlap of hadronic matter distributions in space between VQs of each interacting hadron even at very high energy collisions because destructive interference of their polarization fields reduces resulting hadronic matter at the center of quark system to zero value. So, the model keeps unitarity and does not need of inclusion of such a questionable effect as "antishadowing" [8]. According to our model diffraction dissociation processes are not shadowing effects of nondiffractive inelastic processes – the formers and latters compose inelastic processes, they merely differ by configurations of quarks inside interacting hadrons.

This unified geometrical explanation of diffractive and nondiffractive processes could give a clear answer on long standing question: what is pomeron? Historically the concept of "pomeron", starting from simple Regge pole with intercept one, transformed to a rather complicated object with relatively arbitrary features and smooth meaning. Citing the chair of the conference V. Petrov one could ask: "Is Pomeron

– $j-$ plane singularity?

– particle?

– quasiparticle?

– merely effect of unitarity?"

According to our picture both diffractive and nondiffractive particle production emerge from disturbance (excitation) with succeeding fragmentation of continuous polarization fields of colliding hadrons. The type of interaction depends on quark content configurations of both colliding hadrons occurring at the instant of interaction and the value of impact parameter. So, what we used to call "Pomeron" in $t-$ channel is solely continuum states in $s-$ channel and we claim that Pomeron is unique in elastic, inelastic (diffractive and nondiffractive) and DIS.

This research was partly supported by the Russian Foundation of Basics Research, grants 99-07-90383 and 99-01-01103.

References

1. G. Musulmanbekov, Proc. of XVII Int. Kazimierz Meeting on Particle Theory and Phenomenology, Iowa, 1995, World Scientific, 1996, p. 347-353.; Proc. of XXVI Int.Symp. on Multiparticle Dynamics, Sept. 1-5, Faro, 1996, World Scientific, 1997, p. 357–363; Nucl. Phys. Suppl. **B71**(1999) 117.
2. E. Schrodinger, *"Der stetige Ubergang von der Mikro-zur Makromechanik"*, Naturwissenschaften **14** (1926) 664.
3. R. Henzi and P. Valin, Phys. Lett. **132B** (1983) 443; R. Henzi, Proc. of the 4th Topical Workshop on $\overline{p}p$ Collider Physics, Bern, 1984.
4. W. Heisenberg, Z. Phys. **133** (1952) 65.
5. J.J. Aubert et al., Nucl. Phys. **293B** (1987) 740.
6. A.C. Benvenuti et al., Phys. Lett., **237B** (1990) 599.
7. A.Donnachie and P.V. Landshoff, CERN–TH 6635/92.
8. S.M. Troshin and N.E. Tyurin, hep-ph/9810495.
9. http://durpdg.durham.ac.uk/HEPDATA.

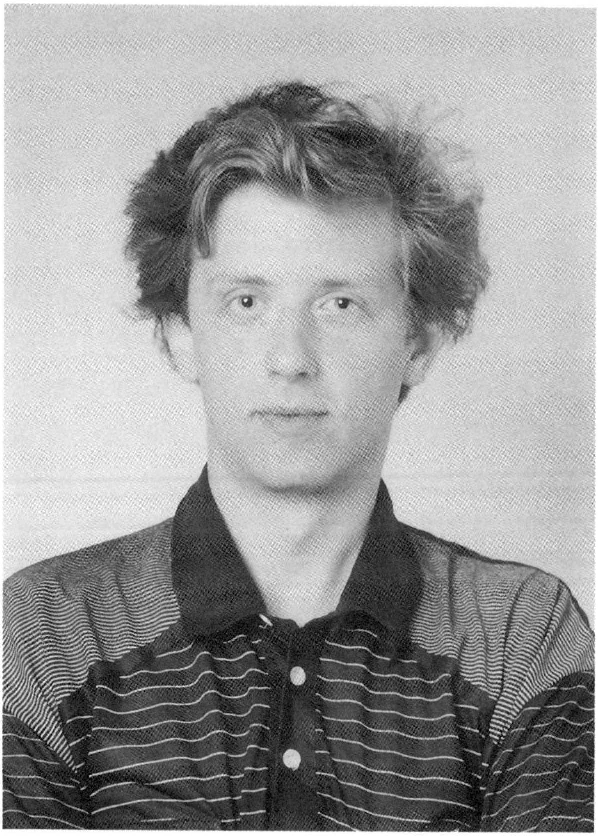

B. V. Polishchuk

REANALYSIS OF E P ELASTIC SCATTERING DATA IN TERMS OF PROTON ELECTROMAGNETIC FORM FACTORS

V. V. EZHELA, B. V. POLISHCHUK
State Research Center of Russia,
Institute for High Energy Physics,
142284, Protvino, Moscow region, Russia
E-mail: ezhela@mx.ihep.su, polishchuk@mx.ihep.su

We have made the comparison of three parametrizations of the proton electromagnetic form factors in the space-like region using the largest data set on $e^{\pm}p$ elastic scattering ever used. All models have correct analytical structure with the minimal set of the well known singularities. For the first time an effective two-pion intermediate coulombic bound state (pionium) was included into GVMD for nucleon formfactors. For the proton charge radius an estimate derived from our best fits is as follows:

$$r_E^p[\text{fm}] = 0.897 \pm 0.002(\text{exp}) \pm 0.001(\text{norm}) \pm 0.003(\text{models}).$$

The traditional "model independent" fit by quadratic polynomial gives in our estimation technology the much crude estimate

$$r^p_{E,polinomial}[\text{fm}] = 0.887 \pm 0.006(\text{exp}) \pm 0.096(\text{norm})$$

Both estimates are closer to the value evaluated from the recent hydrogen Lamb shift measurements.

1. Many attempts were made to model charge and magnetic moment distributions in the nucleons, numerous phenomenological parametrizations of (FFs) have been proposed. Majority of proposed parametrizations are based on various modifications of generalized vector mesons dominance (GVMD) model guided by requirements from general principles of quantum field theories and modest are motivated by QCD features. The common features of the parameterizations are the requirements to give reasonable values of the rms "charge" and "magnetic" radii of the nucleon ($t \to 0$ asymptotic), as well as to reproduce QCD-like behaviour at large $|t|$.

The precise knowledge of the nucleon FFs with as accurate as possible errors treatment is needed in number of phenomenological applications and even in precise tests of well established theories. The analyses of charged hadrons elastic scattering in the Coulomb-nuclear interference regions and of the data on Lamb shifts in atomic hydrogen are the well known examples of such applications.

2. The differential cross section of the elastic $e^{\pm}p$ scattering in the laboratory reference frame is parameterized by the well known Rosenbluth formula[2], based on the one photon exchange approximation QED diagram with modified the proton-photon-proton vertex for extended proton:

$$\Gamma_\mu(q^2) = \gamma_\mu F_1(q^2) + i\sigma_{\mu\nu}\frac{q_\nu}{2m_p}F_2(q^2), \tag{1}$$

with Dirac (F_1) and Pauli (F_2) FFs introduced. These FFs are connected with so called "Sachs" FFs [3] as follows:

$$G_E^{p,n}(q^2) = F_1^{p,n}(q^2) - \frac{q^2}{4m_{p,n}^2}F_2^{p,n}(q^2), \quad G_M^{p,n}(q^2) = F_1^{p,n}(q^2) + F_2^{p,n}(q^2).$$

In the non-relativistic limit they describe an electric charge and magnetic moment distributions within the proton with the normalizations as follows:

$$\begin{aligned} G_E^p(0) &= 1, \qquad & G_M^p(0) &= \mu_p, \\ G_E^n(0) &= 0, \qquad & G_M^n(0) &= \mu_n, \end{aligned}$$

where $\mu_{p,n}$ are the (p,n) magnetic moments and

$$\langle r_E^p\rangle^2 = -6 \cdot \left.\frac{dG_E^p(q^2)}{dq^2}\right|_{q^2=0}$$

is the mean square radius of the proton charge distribution in units $[GeV^{-2}]$. In terms of invariant variables s and t the formula for differential cross section takes the form:

$$\frac{d\sigma}{d(-t)} = \frac{4\pi\alpha^2}{t^2}\left(1+\frac{t}{\tilde{s}}+\frac{tm_p^2}{\tilde{s}^2}\right)\left[\frac{G_E^2+\tau G_M^2}{1+\tau} + \frac{G_M^2}{2}\frac{t^2}{\tilde{s}(\tilde{s}+t)+tm_p^2}\right] \tag{2}$$

where $\tilde{s} = s - m_p^2$, $t = -q^2 = -4E^2\sin^2(\theta/2)\cdot[1+\frac{2E}{m_p}\sin^2(\theta/2)]^{-1}$. Note, that t varies in the range $-\tilde{s}^2/s < t < 0$. When $s \to \infty$, then one has

$$\frac{d\sigma}{d(-t)} \to \frac{4\pi\alpha^2}{t^2}\cdot\left[\frac{G_E^2+\tau G_M^2}{1+\tau}\right], \tag{3}$$

for any fixed value of t. It can be shown that the righthand side of (3) is an upper envelope curve for all $\frac{d\sigma}{d(-t)}(s,t)$.

As it is seen from the Fig. 1. this is confirmed experimentally in the wide ranges of both variables s and t. This is a direct evidence that one

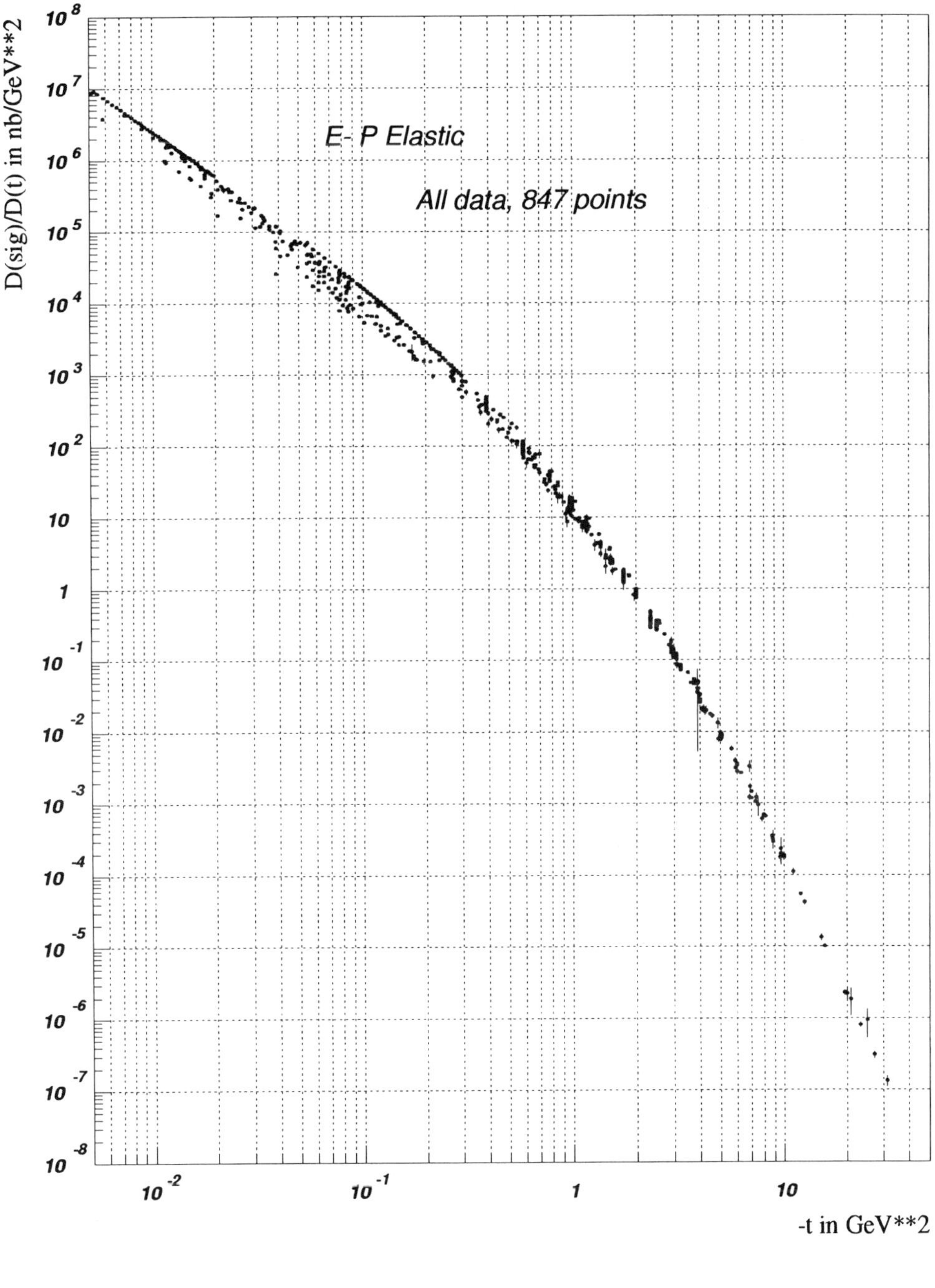

Figure 1: Data atlas.

photon approximation works well at available level of accuracy of the world experimental data.

To model FFs we follow the standard way with introduction of normalized to unity at $t=0$ the isospin FFs, namely $F_{1,2}^{s,v}$. They are connected with the $F_{1,2}^{p.n}$ and $G_{E,M}^{p,n}$ through the relations:

$$F_1^p = \frac{1}{2}\left(F_1^s + F_1^v\right), \quad F_2^p = \frac{\tilde{\mu}_p + \tilde{\mu}_n}{2} F_2^s + \frac{\tilde{\mu}_p - \tilde{\mu}_n}{2} F_2^v$$

$$F_1^n = \frac{1}{2}\left(F_1^s - F_1^v\right), \quad F_2^n = \frac{\tilde{\mu}_p + \tilde{\mu}_n}{2} F_2^s - \frac{\tilde{\mu}_p - \tilde{\mu}_n}{2} F_2^v$$

where $\tilde{\mu}_{p,n}$ are the anomalous magnetic moments of proton and neutron.

3. Isoscalar and isovector FFs[a] are modelled as follows:

$$F_{1,2}^{s,v}(t) = F_{GVMD(1,2)}^{s,v}(t) \times A_{1,2}^{s,v}(t) \tag{1}$$

where $F_{GVMD(1,2)}^{s,v}(t)$ are the generalized vector dominance model factors, and $A_{1,2}^{s,v}$ are the terms which control the asymptotic behavior of FFs at $(-t) \to \infty$, as it predicted by perturbative QCD [11].

$f_{pionium}(t)$ is the term motivated by possible poles of pionium spectrum (corresponding to 1^{--} electromagnetic bound states of the $\pi^+\pi^-$ system).

$$f_{pionium}(t) = \frac{(\sqrt{1 - t/4m_\pi^2}) \cdot (1 + \sqrt{1 - (2m_\pi - \epsilon_\pi)^2/4m_\pi^2})}{\sqrt{1 - t/4m_\pi^2} + \sqrt{1 - (2m_\pi - \epsilon_\pi)^2/4m_\pi^2}} \tag{2}$$

It corresponds to the lowest 1^{--} pionium "bound" state with binding energy $-\epsilon_\pi$, which is unstable under annihilation. We neglect the pionium annihilation width in the form (2). By construction it has the following properties:

$$f_{pionium}(0) = 1, \quad f_{pionium}(t) \to 1 \quad \text{at} \quad |t| \to \infty, \quad f_{pionium}(t)|_{\epsilon_\pi = 0} \equiv 1$$

$f^\rho(t)$ is the model of ρ meson pole contribution to $F_{GVMD(1,2)}^{v}(t)$ in the whole t-plane. It is constructed from the real analytic function $\Phi^\rho(t)$, proposed by Wataghin[8].

$$\Phi^\rho(t) = \frac{m_\rho \Gamma_\rho}{m_\rho^2 - t + m_\rho \Gamma_\rho \cdot \sqrt{\frac{4m_\pi^2 - t}{m_\rho^2 - 4m_\pi^2}}} \tag{3}$$

---

[a]Mathematica and Fortran codes of all models can be obtained by request from: polishchuk@mx.ihep.su.

The pole positions are determined by requirement to reproduce the Breit-Wigner shape in the vicinity of ρ resonance ($t = m_\rho$) at the upper edge of the cut in the small width approximation. The nonzero width ρ meson contributions to the $F^v_{GVMD(1,2)}(t)$ is given by

$$f^\rho(t) = \frac{\Phi^\rho(t)}{\Phi^{\rho)}(0)} \tag{4}$$

with the following properties

$$f^\rho(0) = 1, \qquad f^\rho(t) \sim \frac{const}{t} \quad \text{at} \quad |t| \to \infty$$

$f^\omega_{1,2}(t)$ are the models of ω meson contributions to FFs in the whole t-plane with the righthand cuts. They are constructed as follows[b]:

$$\Phi^\omega_{1,2}(t) = \cos\Theta^\omega_{1,2} \cdot \frac{m_\omega \Gamma^{(3)}_\omega}{m^2_\omega - t + m_\omega \Gamma_\omega/\pi \cdot \ln \frac{9m^2_\pi - t}{m^2_\omega - 9m^2_\pi}}$$
$$+ \sin\Theta^\omega_{1,2} \cdot \frac{m_\omega \Gamma^{(2)}_\omega}{m^2_\omega - t + m_\omega \Gamma_\omega \cdot \sqrt{\frac{4m^2_\pi - t}{m^2_\omega - 4m^2_\pi}}}, \tag{5}$$

$f^\phi_{1,2}(t)$ are the models of ϕ meson contributions to FFs in the whole t-plane with the righthand cuts. They are constructed by a way similar to $f^\omega_{1,2}(t)$.

Another ingredients of our hybrid model are multiplicative factors $A^{s,v}_{1,2}$ which control FFs behaviour at large $|t|$. We use the two forms for it.

$M^{s,v}_{1,2}(t)$ are motivated by parametrization of Mack [4]

$$M^{s,v}_i(t) = \left(\frac{1}{1 - t/t^p_{cut}}\right)^{x^{s,v}_1 + i - 1} \left(\frac{1}{1 - t/t^{s,v}_{cut}}\right)^{x^{s,v}_2 \ln \frac{(t^{s,v}_{cut} - t)}{\lambda^2}}, \quad i = 1, 2 \tag{6}$$

where $t^p_{cut} = 4m^2_p$ is the proton-antiproton threshold, $t^s_{cut} = 9m^2_\pi$, $t^v_{cut} = 4m^2_\pi$, and λ is the scale parameter which is expected to be of the order of the Λ_{QCD}. This model is included in the current analysis in order to test if QCD

---

[b]Here we try to test if differrent branch channels can be identified from the fit. So our form for ω and ϕ contributions differ from that used in [8,9,10]

asymptotic is really prominent at available highest q^2.

$GK_{1,2}^{s,v}(t)$ are motivated by parametrization of Gari and Krümpelman [5] with taking into account the QCD asymptotic logarithms in analogy with parametrizations used in [6,7].

$$GK_i^{s,v}(t) = \frac{(1 - t/t_{cut}^p)^{\nu+1-i}}{\left(1 - \eta_i^{s,v} + \eta_i^{s,v}(1 - t/t_{cut}^{s,v}) \ln\left(\frac{t_{cut}^{s,v} - t}{\lambda^2}\right) / \ln \frac{t_{cut}^{s,v}}{\lambda^2}\right)^{2+\nu}}, \quad i = 1, 2 \tag{7}$$

It normalized by $GK_{1,2}^{s,v}(0) = 1$ at $t = 0$. At large $-t$ they reproduce FFs behaviour at large $-t$ as predicted by pQCD. The exponent ν is given[11] by

$$\nu = \frac{4}{3\beta}, \quad \beta = 11 - \frac{2}{3}N_f, \quad \nu = 0.174 \quad \text{for} \quad N_f = 5$$

Now we are ready to assemble parametrizations for isotopis form factors.

Modified Mack-67 parametrization (Mack-98)

$$\begin{aligned} F_{1,2}^s &= \left((1 - C_{1,2}^\omega) \cdot f_{pionium}(t) + C_{1,2}^\omega \cdot f_{1,2}^\omega(t)\right) \times M_{1,2}^s \\ F_{1,2}^v &= \left((1 - C_{1,2}^\rho) \cdot f_{pionium}(t) + C_{1,2}^\rho \cdot f^\rho(t)\right) \times M_{1,2}^v \end{aligned} \tag{8}$$

with 10 adjustable parameters: $C_{1,2}^{\rho,\omega}$, $\Theta_{1,2}^\omega$, $x_{1,2}^{s,v}$.

Modified Gari-Krümpelman parametrization (GK-98-2V)

$$\begin{aligned} F_{1,2}^s &= \left((1 - C_{1,2}^\omega) \cdot f_{pionium}(t) + C_{1,2}^\omega \cdot f_{1,2}^\omega(t)\right) \times GK_{1,2}^s(t) \\ F_{1,2}^v &= \left((1 - C_{1,2}^\rho) \cdot f_{pionium}(t) + C_{1,2}^\rho \cdot f^\rho(t)\right) \times GK_{1,2}^v \end{aligned} \tag{9}$$

with 10 adjustable parameters: $C_{1,2}^{\rho,\omega}$, $\Theta_{1,2}^\omega$, $\eta_{1,2}^{s,v}$.

Modified Gari-Krümpelman parametrization (GK-98-3V))

$$\begin{aligned} F_{1,2}^s &= \left((1 - C_{1,2}^\omega - C_{1,2}^\phi) \cdot f_{pionium}(t) + C_{1,2}^\omega \cdot f_{1,2}^\omega(t) + C_{1,2}^\phi \cdot f_{1,2}^\phi(t)\right) \times GK_{1,2}^s(t) \\ F_{1,2}^v &= \left((1 - C_{1,2}^\rho) \cdot f_{pionium}(t) + C_{1,2}^\rho \cdot f^\rho(t)\right) \times GK_{1,2}^v \end{aligned} \tag{10}$$

with 14 adjustable parameters: $C_{1,2}^{\rho,\omega,\phi}$, $\Theta_{1,2}^\omega$, $\Theta_{1,2}^\phi$, $\eta_{1,2}^{s,v}$.

4. We use the total sample of data on differential cross-sections for the electron-proton elastic scattering extracted from the PPDS ReacData data base [12] (see also Reaction Data [13]). The complete list of references, full data sample, normalization factors and their covariance matrix are presented elsewhere[14]. The resulting data sample that we have included in the first crude analysis consists of 847 experimental points covered the q^2 range from 0.005 to 31.280 GeV^2 and 1961-1994 time scale (see Fig. 1).

To make data from different experiments mutually consistent on the basis of available models and to determine the areas of applicability of the competitive models, we have developed *the data and models cross assessments (CRAS) technology.*

Main steps of the CRAS technology:

1. Prepare data sample. Select all relevant final data. Make dichotomy: compete data sets – incomplete data sets.

2. Make crude model selection by quality of their fits to the whole data sample with weights constructed from all uncertainties quadratically combined.

3. Filter out incomplete data sets. Refit survived at Step.2 models to compete data sets. Further model selection by quality of the fits.

4. Refit survived at Step.3 models to the complete data sets with weights constructed as on Step.2 but without admixtures of the normalization systematic uncertainties and with correction (renormalization) factors and corresponding covariance matrix derived from the fits with penalty functions for each model.

5. Construct unique (averaged over models) array of correction factors and corresponding covariance matrix.

6. Refit all survived models to the complete data sets with averaged correction factors. Construct areas of applicability for each model and area of applicability common for *all* models. We'll refer further the data from this area as "assessed data".

7. Refit all models to the assessed data sets and compare results with the results of the previous step.

8. Select the best model for extrapolations.

9. Evaluatc of thc common for all competitive models derived physical parameters from each model and construct current recommended from the "point of view" of all competitive models value.

The results of fits to the assessed data sets are presented in the Table 1.

5. From the fit results of all models to the assessed data sample we have derived our best estimate of the proton charge radius:

$$r_E^p = 0.897 \pm 0.002(exp) \pm 0.001(norm) \pm 0.003(models) \quad [\text{fm}]$$

given by modified Gary-Krümpelman model (GK-98-2V).

Acknowledgments

We give our thanks to S.R. Slabospitsky for his carefully reading the manuscript and suggestions to make it more readable. This work have been supported in parts by Russian Foundation for Basic Research by Grants: RFBR-93-02-3733, RFBR-96-07-89230, and RFBR-99-07-90356.

References

1. R. Hofstadter and R. McAlister, Phys. Rev. 98 (1955) 217;
   R. McAlister and R. Hofstadter, Phys. Rev. 102 (1955) 851;
2. M. Rosenbluth, Phys. Rev. 79 (1950) 615;
3. Ernst E.J.,Sach R.G.,Wali K.C., Phys. Rev. 119, (1960) 1105
4. G. Mack Phys. Rev. 154 (1967) 1617;
5. M. Gari, W. Krümpelman, Zeit. Phys. A322 (1985) 689;
   M. Gari, W. Krümpelman, Phys. Lett. B173 (1986) 10;
   M. Gari, W. Krümpelman, Phys. Lett. B274 (1992) 159;
6. S. Furuichi and K. Watanabe, Progr. Theor. Phys 82 (1989) 581
   S. Furuichi and K. Watanabe, Progr. Theor. Phys 83 (1989) 565
7. P. Mergell, U.-G. Meissner, D. Drechsel, Nucl. Phys. A596 (1996) 367
8. V. Wataghin, Nucl. Phys. B10 (1969) 107;
9. V.A. Meshcheryakov, G.V. Meshcheryakov, Mod. Phys. Lett. A9 (1994) 1603
10. V.A. Meshcheryakov, G.V. Meshcheryakov, Yad. Fiz. 60 (1997) 1400
11. S.J. Brodsky and G. Farrar, Phys. Rev. D11 (1975) 1309
    S.J. Brodsky and G.P. Lepage, Phys. Scr. 23 (1981) 945
    G.P. Lepage and S.J. Brodsky Phys. Rev. D22 (1980) 2175
12. `http://dbserv/BAFIZ/ppds.html`
13. `http://durpdg.dur.ac.uk/HEPDATA/REAC`
14. `http://wwwppds.ihep.su:8001/ep2ep_tot`

Table 1: The results of fits to the assessed data sets. Upper error is quadrature sum of statistical and non-normalization systematic uncertainties, lower one is the error induced by the renormalization procedure.

	Mack-98	GK-98-2V	GK-98-3V
χ^2/NDP	510.3/685	493.9/685	494.8/681
x_1^s	$1.59^{\pm 0.17}_{\pm 0.039}$	–	–
x_1^v	$1.806^{\pm 0.020}_{\pm 0.0056}$	–	–
x_2^s	$0.0953^{\pm 0.0057}_{\pm 0.0018}$	–	–
x_2^v	$0.0284^{\pm 0.0014}_{\pm 0.0004}$	–	–
η_1^v	–	$0.005652^{\pm 0.000032}_{\pm 0.000021}$	$0.005844^{\pm 0.000032}_{\pm 0.000021}$
η_2^v	–	$0.004407^{\pm 0.000077}_{\pm 0.000027}$	$0.004495^{\pm 0.000073}_{\pm 0.000024}$
η_1^s	–	$0.03275^{\pm 0.00015}_{\pm 0.00011}$	$0.03358^{\pm 0.00015}_{\pm 0.00010}$
η_2^s	–	$0.01253^{\pm 0.00055}_{\pm 0.00019}$	$0.01253^{\pm 0.00050}_{\pm 0.00016}$
C_1^ρ	$-0.213^{\pm 0.011}_{\pm 0.003}$	$9.25^{\pm 0.19}_{\pm 0.08}$	$9.877^{\pm 0.077}_{\pm 0.036}$
C_2^ρ	$0.6722^{\pm 0.0078}_{\pm 0.0019}$	$0.457^{\pm 0.010}_{\pm 0.003}$	$0.4161^{\pm 0.0089}_{\pm 0.0028}$
C_1^ω	$-1.725^{\pm 0.020}_{\pm 0.007}$	$-7.98^{\pm 0.20}_{\pm 0.08}$	$-8.798^{\pm 0.082}_{\pm 0.036}$
C_2^ω	$-9.07^{\pm 0.30}_{\pm 0.09}$	$1.594^{\pm 0.030}_{\pm 0.011}$	$1.688^{\pm 0.022}_{\pm 0.009}$
C_1^ϕ	–	–	$0.1937^{\pm 0.0053}_{\pm 0.0034}$
C_2^ϕ	–	–	$-0.137^{\pm 0.072}_{\pm 0.020}$
θ_1^ω[deg]	$91.521^{\pm 0.014}_{\pm 0.009}$	$91.50427^{\pm 0.00094}_{\pm 0.00068}$	$91.50409^{\pm 0.00085}_{\pm 0.00062}$
θ_2^ω[deg]	$91.414116^{\pm 2.1e-5}_{\pm 0.8e-5}$	$91.415123^{\pm 0.1e-5}_{\pm 0.1e-5}$	$91.415123^{\pm 0.1e-5}_{\pm 0.1e-5}$
θ_1^ϕ[deg]	–	–	$157.8^{\pm 2.6}_{\pm 1.9}$
θ_2^ϕ[deg]	–	–	$159.9^{\pm 13.0}_{\pm 8.0}$
λ[GeV]	$0.0035^{\pm 0.0009}_{\pm 0.0003}$	$0.21991^{\pm 0.00023}_{\pm 0.00008}$	$0.219171^{\pm 0.000084}_{\pm 0.000038}$
r_E^p[fm]	**0.8937**	**0.8971**	**0.9005**

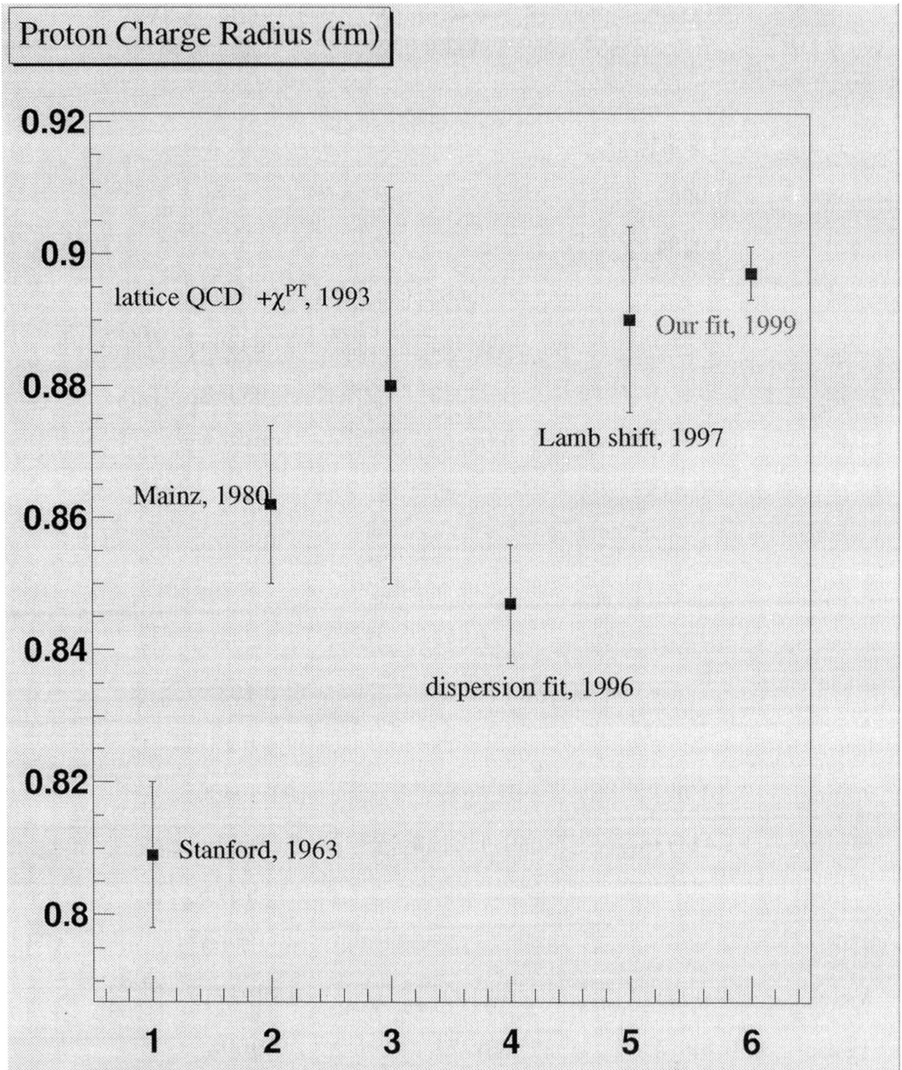

Figure 2: The proton charge radius as it is usually known. (The original picture was in the article of S.G.Karshenboim, see hep-ph/9712347)

A. V. Kisselev

CHARGED HADRON MULTIPLICITY IN DIFFRACTIVE DEEP INELASTIC SCATTERING

A. V. KISSELEV
Institute for High Energy Physics,
142284 Protvino, Russia
E-mail: kisselev@mx.ihep.su

The calculations of average multiplicity of charged hadrons in diffractive deep inelastic scattering have been presented. The quark-gluon content of the pomeron is studied and a comparison of theoretical formulas with the H1 data is made.

1 Introduction

The study of hadronic spectra and multiplicities in deep inelastic scattering (DIS) is of great interest. The reason is that it enables us to investigate a dependence of an efficiency of hadron production on both energy and large momentum scale (Q^2) which defines the size of interaction region.

As was first pointed out in [1], one can expect that for larger values of the momentum transfer, the inelastic process favours larger multiplicity of hadrons. This idea has been realized in a number of papers[2] in the framework of perturbative QCD.

Diffractive DIS (DDIS) observed at HERA [3] give a new insite into the physical picture of the large rapidity gap events and enables us to measure diffractive parton densities. The presence of rapidity gap in DDIS means that the internal structure of a colourless object with the vacuum quantum numbers, namely pomeron, is being probed.

A lot of efforts has been done to describe DDIS in the framework of QCD (see, for instance, [4]). It was shown in [5] that up to power corrections, the inclusive diffractive structure functions factorize into a hard photon-parton scattering part and diffractive parton distributions.

Applying the ideas of Regge approach to diffractive quark and gluon densities, one arrives at the Ingelman-Schlein proposal[6]. Both data on DDIS and experimental evidences of hard diffraction from CDF and D0 Collaborations[7] indicate that Ingelman-Schlein picture right at least quantitively. However, this model overestimates the diffractive production rates of jets and W's [8]. This descrepancy can be attributed to the so-called pomeron flux factor [9]. In particular, pomeron structure function as extracted from HERA data is dependent on the flux factor[10].

It has been proposed in [11] to study characteristics of final hadrons in

DDIS which, on one hand, are very sensitive to the quark-gluon structure of the pomeron, and, on the other hand, do not depend on a particular expression of the pomeron flux factor.

2 Hadron Multiplicity in Diffractive DIS

Recently, the data from H1 Collaboration [12] on hadronic final states in diffractive process of the type

$$ep \to e'XY \tag{1}$$

have appeared. The hadronic systems X and Y in (1) are separated by the rapidity gap. The system Y is closest to the proton beam and its invariant mass is limited to $M_Y < 1.6$ GeV. Invariant mass of X system, M_X, varies from 3 to 36 GeV. The square momentum transfer at the pY vertex, t, is limited to $|t| < 1$ GeV^2, and the variable Q^2 lies within the limits $21 \div 27$ GeV^2 [12].

The multiplicity structure of the hadronic system X has been studied as a function of its invariant mass M_X. Hadronic system in (1) is a result of the virtual photon-pomeron interaction

$$e\mathbb{P} \to \mathrm{e'X}. \tag{2}$$

In addition to W and Q^2, the following kinematical variables are used to describe DDIS:

$$x_{\mathbb{P}} \simeq \frac{M_X^2 + Q^2}{W^2 + Q^2}, \qquad \beta \simeq \frac{Q^2}{M_X^2 + Q^2}. \tag{3}$$

It was obtained that for low masses ($M_X < 10$ GeV) the average multiplicity, $\langle n\rangle^{DDIS}(M_X, Q^2)$, is similar to that in DIS and e^+e^- annihilation. For $M_X > 10$ GeV it is larger than hadron multiplicity in DIS, $\langle n\rangle^{DIS}(W, Q^2)$, and hadron multiplicity in e^+e^- annihilation, $\langle n\rangle^{e^+e^-}(W)$, at comparable values of energy $W = \langle M_X\rangle$. The ratio $\langle n\rangle^{DDIS}(M_X, Q^2)/\langle n\rangle^{e^+e^-}(M_X)$ grows in M_X at fixed Q^2.

To describe the features of the H1 data qualitatively and quantitatively, a formalism previously developed for non-diffractive DIS [2] was generalized for DDIS [11]. Namely, it was obtained in the framework of perturbative QCD:

$$\langle n\rangle^{DDIS}(M_X, Q^2) = \langle n\rangle^{e^+e^-}(M_X) \times \\ \times \left[\Delta_{\mathbb{P}} + \left(1 + \frac{C_A}{2C_F}\varepsilon(M_X^2, \beta)\left(1 - \frac{3}{2}\varepsilon(M_X^2, \beta)\right)\right)(1 - \Delta_{\mathbb{P}})\right], \tag{4}$$

where

$$\varepsilon(M_X^2, \beta) = \sqrt{\frac{\alpha_s(M_X^2)}{\pi}}\left[\sqrt{\frac{\ln(1/\beta)}{\xi(M_X^2)}} - \frac{d}{\sqrt{2N_c}}\right], \tag{5}$$

$d = (1/6)(11N_c + 2N_f/N_c^2)$, $C_A = N_c$ is a number of colours, $C_F = (N_c^2 - 1)/2N_c$, N_f is a number of flavours and ξ is QCD evolution parameter.

Thus, average hadronic multiplicity in DDIS, $\langle n\rangle^{DDIS}(M_X, Q^2)$, is realted to average hadronic multiplicity in e^+e^- annihilation taken at $W = M_X$.

The quantity Δ_{IP} in (4) defines quark/gluon ratio inside the pomeron:

$$\Delta_{IP}(\beta, Q^2) = \frac{D^q_{IP}(\beta, Q^2)}{D^g_{IP}(\beta, Q^2) + D^q_{IP}(\beta, Q^2)}. \qquad (6)$$

Here D^g_{IP} and D^q_{IP} are respectively the gluon and the singlet quark diffractive distributions. They obey DGLAP evolution equations.

Equation (4) explaines the observed excess of charged hadron multiplicity in DDIS with respect to those in non-diffractive DIS and e^+e^- annihilation, as well as its more rapid growth in diffractive mass (see above).

The result of the fit of the H1 data with the use of formula (4) is shown in Fig. 1. The dotted (dashed) curve corresponds to a pure quark (gluon) content of the pomeron, while the solid line is the best fit which corresponds to a mixing of quark and gluon components in approximately equal shares [11].

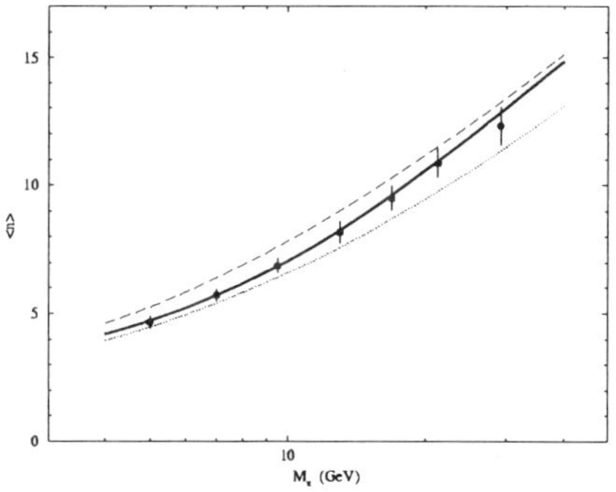

Figure 1: Average multiplicity of charged hadrons in DDIS. Data obtained by the H1 Collaboration. Theoretically calculated curves are described in the text.

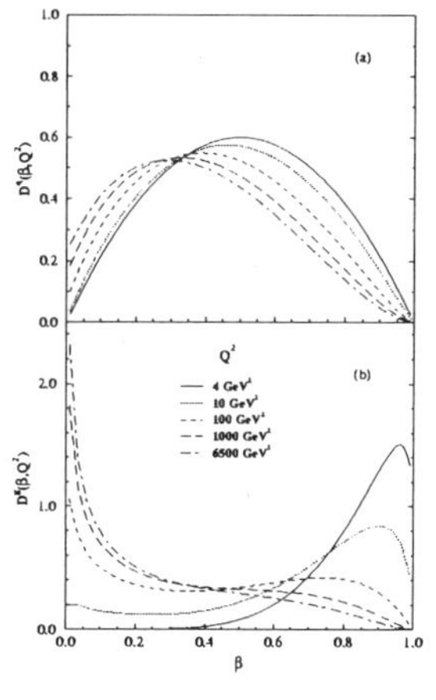

Figure 2: Quark singlet (a) and gluon (b) distributions inside the pomeron obtained by the fitting procedure and their evolution in Q^2.

The initial parton distributions inside the pomeron (at $Q^2 = 4$ GeV^2) and their evolution with Q^2 are presented in Fig. 2. As one can see, the gluon distribution has a super-hard profile, in agreement with H1 analysis [13]. The gluon momentum fraction in the pomeron, f_g, appeared to be approximately 50% at the initial scale [11]. It is closed to the value $f_g = 0.54^{+.16}_{-.14}$ obtained recently by CDF Collaboration [14].

3 Conclusion

The theoretical description of the hadronic multiplicity in diffractive DIS based on perturbative QCD calculations is given. The corresponding formulas have been obtained in [11]. They successfully explain the excess of particle production in diffractive DIS relative to that in non-diffractive DIS and e^+e^- annihilation seen by the H1 Collaboration [12].

It is shown that the data are sensitive to the quark-gluon content of the pomeron and favour the super-hard profile of the gluon distribution inside the pomeron.

To analize Q^2 dependence of average multiplicity of charged hadrons at fixed values of diffractive mass M_X, more data in extended kinematical ranges of diffractive DIS are needed.

References

1. J. Benecke, T.T. Chou, C.N. Yang and E. Yen, *Phys. Rev.* **188**, 2159 (1969).
2. G.Sh. Dzhaparidze, A.V. Kisselev and V.A. Petrov, *Sov. J. Nucl. Phys.* **35**, 927 (1982); A. Bassetto, *Nucl. Phys.* B **202**, 493 (1982); A.V. Kisselev and V.A. Petrov, *Phys. Lett.* B **131**, 468 (1983); V.A. Petrov, A.V. Kisselev and O.P. Yushchenko, *Phys. Lett.* B **193**, 357 (1987); A.V. Kisselev and V.A. Petrov, *Particle & Nuclei* **19**, 21 (1988); A.V. Kisselev and V.A. Petrov, *Phys. Lett.* B **214**, 434 (1988).
3. ZEUS Coll., M. Derrick *et al., Phys. Lett.* B **315**, 481 (1993); H1 Coll., T. Ahmed *et al., Nucl. Phys.* B **429**, 477 (1994).
4. M. Diehl, Talk given at the 7-th International Workshop on Deep Inelastic Scattering and QCD (DIS 99), Zeuthen, Germany, 1999.
5. J.C. Collins, *Phys. Rev.* D **57**, 305 (1998).
6. G. Ingelman and P.E. Schlein, *Phys. Lett.* B **152**, 256 (1985).
7. CDF Coll. F. Abe *et al., Phys. Rev. Lett.* **78**, 2698 (1997); D0 Coll. S. Abachi *et al.,* in Proc. of the 28-th Int. Conference on High Energy Physics (ICHEP 96), eds. Z. Ajduk and A.K. Wroblewsky (World Scientific, Singapore, 1996).
8. L. Alvero *et al., Phys. Rev.* D **59**, 074022 (1999); R.J.M. Covolan and M.S. Soares, in Proc. of the LAFEX Workshop on Diffractive Physics (LISHEP 98), Rio de Janeiro, Brazil, 1998.

9. K. Goulianos, *Phys. Lett.* B **358**, 379 (1995).
10. R.J.M. Covolan and M.S. Soares, *Phys. Rev.* D **57**, 180 (1998).
11. R.J.M. Covolan and A.V. Kisselev, *Phys. Rev.* D **60**, 34003 (1999).
12. H1 Coll., C. Adloff *et al., Eur. Phys. J.* C **5**, 439 (1998) (see also, A. Bunyatyan, These Proceedings).
13. H1 Coll., C. Adloff *et al., Z. Phys.* C **76**, 613 (1997).
14. A. Solodsky, These Proceedings.

M. J. Menon

HADRONIC AND ELEMENTARY MULTIPLICITY DISTRIBUTIONS IN A GEOMETRICAL APPROACH

P. C. BEGGIO, M. J. MENON
Instituto de Física "Gleb Wataghin",
Universidade Estadual de Campinas, Unicamp,
13083-970 Campinas, SP, Brazil
E-mail: menon@ifi.unicamp.br

P. VALIN
Dépt. de Physique, Université de Montréal,
CP. 6128, Succ. Centre-Ville
Montréal, Québec, Canada H3C 3J7
E-mail: pierre.valin@lmco.com

We construct the hadronic multiplicity distribution in terms of an elementary distribution (at a given impact parameter) and the inelastic overlap function characterized by the observed BEL (Blacker-Edgier-Larger) behaviour. With suitable parametrizations for the elementary quantities, based on some geometrical arguments and the most recent data on e^+e^- annihilation, an excellent description of pp and $\overline{p}p$ inelastic multiplicity distributions at the highest energies is obtained.

1 Introduction

It is well known that the underlying theory of hadronic interactions, the quantum chromodynamics (QCD) is not yet able, by its own, to describe the bulk of experimental data associated with soft (long range) processess, in particular elastic scattering. At present, phenomenological approaches are very important as a source of information for adequate theoretical developments. On the other hand, from the experimental point of view, a renewed interest on both pp and $\overline{p}p$ elastic and inelastic scattering is expected with the advent of the next accelerator generation, the RHIC and LHC [1]. At this stage, due to our limited theoretical understanding of elastic scattering and, on the contrary, the success of QCD in treating hard (inelastic) processes, it may be important to investigate possible *connections between elastic and inelastic channels*, even from a phenomenological point of view. In this work we shall treat this subject in the contexts of the impact parameter picture, unitarity and the eikonal approximation.

Following other authors [2], we express the "complex" (overall) hadron-p multiplicity distributions (inelastic channel) in terms of an "elementary" distribution associated with an elementary process taking place at given impact parameter and the inelastic overlap function (which is constructed from elastic channel data). The novel aspects concern: (a) quantitative correlation between the *violations* of the KNO scaling [3] (inelastic channel) and geometrical scaling [4] (elastic channel); (b) introduction of suitable parametrizations for the elementary quantities, based on some geometrical arguments and the most recent data on e^+e^- annihilation. With this formalism, the hadronic multiplicity distribution may be evaluated without any free parameter and an excellent reproduction of the experimental data on pp (ISR) and $\bar{p}p$ ($S\bar{p}pS$) inelastic multiplicities is achieved. A detailed discussion on all the results presented here and also a complete list of references to the experimental data may be found in a recent paper [5].

2 Impact Parameter Formalism

In the geometrical picture, unitarity correlates the elastic scattering amplitude in the the impact parameter b space, $\Gamma(b,s)$, with the inelastic overlap function, $G_{in}(b,s)$, by $2Re\Gamma(b,s) = |\Gamma(b,s)|^2 + G_{in}(b,s)$. For a purely imaginary elastic amplitude in momentum transfer space the profile function $\Gamma(b,s)$ is real and in the eikonal approximation is expressed by $\Gamma(b,s) = 1 - exp[-\Omega(b,s)]$, so that

$$G_{in}(b,s) = 1 - exp[-2\ \Omega(b,s)] \equiv \sigma_{in}(b,s) \tag{1}$$

is the probability for an inelastic event to take place at b and s, namely, $\sigma_{in}(s) = \int d^2\mathbf{b}\ G_{in}(b,s)$. In this picture the topological cross section for producting an even number N of charged particles at $\sqrt{s}$ may be expressed by $\sigma_N(s) = \int d^2\mathbf{b}\ \sigma_N(b,s) = \int d^2\mathbf{b}\ \sigma_{in}(b,s)\sigma_N(b,s)/\sigma_{in}(b,s)$.

In this context, the formal connection between hadronic and elementary multiplicity distributions is obtained as follows. Let φ be the elementary multiplicity distribution, $<n>(b,s)$ the average number of particles produced at b and s and $z = N(s)/<n>(b,s)$ a KNO variable associated with the elementary process taking place at b (and s). Then, in general, $\varphi =<n>(b,s)\sigma_N(b,s)/\sigma_{in}(b,s) = \varphi(z,s)$. Representing the overall multiplicity distribution by Φ and the corresponding KNO variable by $Z = N(s)/<N>(s)$, where $<N>(s)$ is the average multiplicity at $\sqrt{s}$, we have in general $\Phi =<N>(s)\sigma_N(s)/\sigma_{in}(s) = \Phi(Z,s)$. Both distributions are normalized by the usual conditions [2]

$$\int_0^\infty \varphi(z)dz = 2 = \int_0^\infty \varphi(z)zdz, \qquad \int_0^\infty \Phi(Z)dZ = 2 = \int_0^\infty \Phi(Z)ZdZ. \tag{2}$$

Now, introducting a *multiplicity function* as the ratio

$$\frac{<n>(b,s)}{<N>(s)} \equiv m(b,s), \tag{3}$$

the relationship between Φ and φ follows from the above equations:

$$\Phi = \frac{\int d^2\mathbf{b}\,\frac{G_{in}(b,s)}{m(b,s)}\varphi(\frac{Z}{m(b,s)})}{\int d^2\mathbf{b}\; G_{in}(b,s)} = \Phi(Z,s). \tag{4}$$

This result means that, once one has parametrizations for $G_{in}(b,s)$ and the elementary quantities φ (multiplicity distribution) and $m(b,s)$ (multiplicity function) the overall hadronic multiplicity distribution may be evaluated. In this work we consider $G_{in}(b,s)$ from analyses of elastic pp and $\overline{p}p$ scattering data (taking account of geometrical scaling violation) and infer the elementary quantities based on geometric arguments and experimental data on e^+e^- annihilation, as explained in what follows.

In the elastic channel, the breaking of Geometrical scaling is quite well described by the BEL behaviour, analytically expressed by the Short Range Expansion[6], $G_{in}(b,s) = P(s)exp\{-b^2/4B(s)\}k(x,s)$, with k expanded in terms of a short-range variable $x = b\ exp\{-(\gamma b)^2/4B(s)\}$. With suitable parametrizations for $P(s)$ and $B(s)$ an excellent agreement with experimental data on pp and $\overline{p}p$ elastic scattering is achieved and we shall use this well known result[6] as the input from the elastic channel.

3 Elementary Hadronic Process

We now turn to the discussion of the elementary hadronic processes, characterized by φ and m in Eq. (4). At a given impact parameter, the most elementary process is an e^+e^- collision, which is a process occuring in a unique angular momentum state and therefore also at a given impact parameter (zero in this case). Although e^+e^- annihilations cannot be exactly the same as collisions between hadron constituents, it is reasonable, even from a geometric point of view, to think that some characteristics of both processes may be similar. The point is to find out or infer what they could be.

First, let us consider that the average multiplicity at given impact parameter depends on the center-of-mass energy in the form of a general power law $<n>(b\ fixed, s) \propto E_{CM}^{\gamma}$. Now, from Eq. (1), $exp\{-2\ \Omega(b,s)\}$ is the transmission coefficient, i.e. the probability of having no interaction at a given impact parameter, and therefore Ω should be proportional to the thickness of the target, or the energy E_{CM} that can be deposited at b for particle production at a given s. Then, we can express $<n>(b,s) \propto \Omega^{\gamma}(b,s)$ and comparison with Eq. (3), allows us to infer the multiplicity function:

$$m(b,s) = \xi\Omega^{\gamma}(b,s), \qquad (5)$$

with ξ being calculated by the normalization condition of the overall multiplicity distribution, Eq. (2). With this, Eq. (4) becomes

$$\Phi = \frac{\int d^2\mathbf{b}\ \frac{G_{in}(b,s)}{\xi\Omega^{\gamma}(b,s)}\ \varphi(\frac{Z}{\xi\Omega^{\gamma}(b,s)})}{\int d^2\mathbf{b}\ G_{in}(b,s)} = \Phi(Z,s), \qquad (6)$$

where

$$\xi = \frac{\int db^2 G_{in}(b,s)}{\int db^2\ G_{in}(b,s)\Omega^{\gamma}(b,s)} = \xi(s). \qquad (7)$$

We proceed with the determination of the elementary distribution φ and the power coefficient γ, through quantitative analyses of e^+e^- data and under the following arguments. Because the elementary process occurs at a given impact parameter, its elementary structure suggests that it should scale in the KNO sense. Now, since experimental information on e^+e^- multiplicity distributions shows agreement with this scaling, we shall base our parametrization for φ just on these data. In particular, it is sufficient to assume a gamma distribution, normalized according to Eq. (2), $\varphi(z) = 2K^K z^{K-1} exp\{-Kz\}/\Gamma(K)$. Fit to the most recent data, covering the interval 22.0 $GeV \leq \sqrt{s} \leq 161$ GeV furnished $K = 10.775 \pm 0.064$ with χ^2/N_{DF}=508/195.

Finally, we consider fits to the e^+e^- average multiplicity through a general power law $< n >_{e^+e^-} = \alpha(\sqrt{s})^{\gamma}$. We collected the most recent experimental data at the highest energies, covering the interval 5.1 GeV $\leq \sqrt{s} \leq 183$ GeV. Fitting to this equation yields $\alpha = 2.06 \pm 0.02$ and $\gamma = 0.522 \pm 0.002$, with χ^2/N_{DF}=354/45. The above parametrization deviates from the data above $\sqrt{s}$ $\sim$ 100 GeV and this contributes to the high χ^2 value. However, as commented before, we do not expect that e^+e^- annihilation exactly represent the collisions between hadrons constituents.

4 Results and Conclusions

With the above results we are now able to predict the hadronic multiplicity distribution $\Phi(Z,s)$, Eq. (6), without free parameters: $G_{in}(b,s)$ (and $\Omega(b,s)$) comes from analysis of the elastic scattering data and $\varphi(z)$ and γ from fits to e^+e^- data. We express Φ in terms of the scaling variable $Z' = N'/ < N' >$ where $< N' > = N(s) - N_0$ with $N_0 = 0.9$ leading charges removed. The predictions for pp scattering at 52.6 GeV and $\overline{p}p$ at 546 GeV are shown in Figs. 1 and 2, together with the experimental data. The behaviour of the

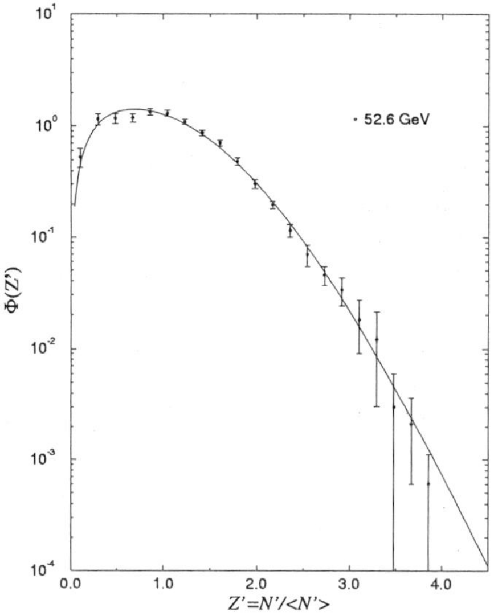

Figure 1: Scaled multiplicity distribution for inelastic pp data at ISR energies compared to theoretical expectations using Eq. (6).

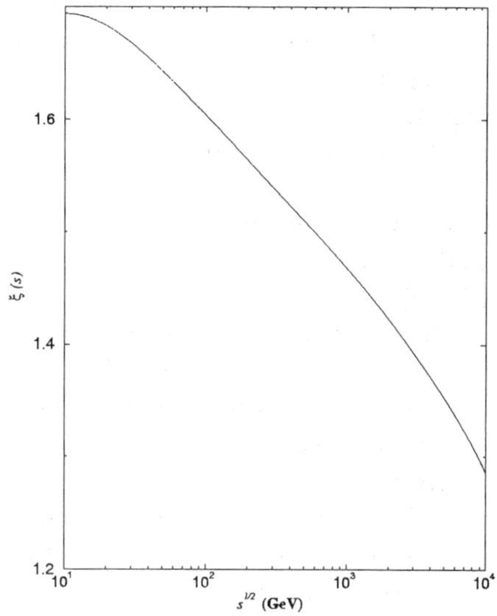

Figure 3: Normalization factor from Eq. (7) as function of the energy.

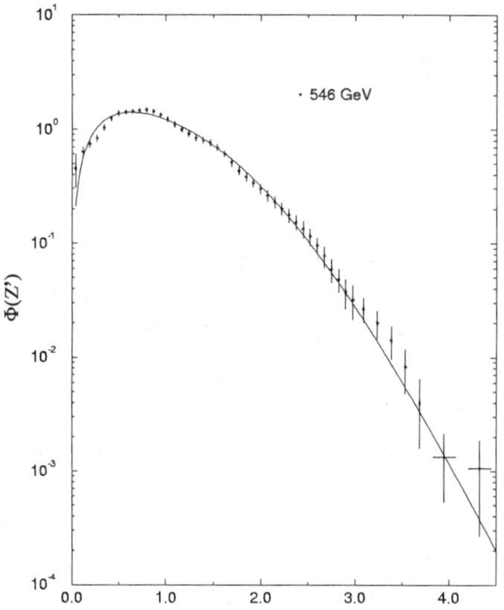

Figure 2: Same as Fig. 1 for inelastic $\overline{p}p$ data at collider energies.

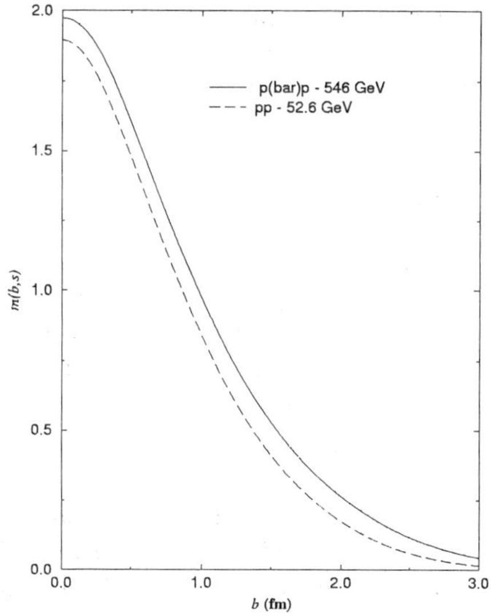

Figure 4: Multiplicity function from Eq. (5) for pp at 52.6 GeV and $\overline{p}p$ at 546 GeV.

normalization factor $\xi(s)$, as determined from Eq. (7), is shown in Fig. (3) in a wide interval of energy. The multiplicity functions $m(b,s)$, Eqs. (1) and (5), corresponding to the energy and reactions of Figs. 1 and 2 are displayed in Fig. 4.

From Figs. 1 and 2, we conclude that, in the context of the geometrical approach and based on the BEL behaviour for $G_{in}(b,s)$, our guess for the elementary hadronic quantities (based on geometrical arguments and quantitative analysis of e^+e^- data) yields excellent predictions for the hadronic distribution $\Phi(Z,s)$.

The fact that the average multiplicity associated with hadron constituents, increases faster then e^+e^- data above $\sim$ 100 GeV seems quite reasonable, since at these energies we expect additional contributions from gluons/quarks interactions which are not present in lepton-lepton collisions.

Anyway, despite the success of the geometrical approach in the phenomenological context, it is in general difficult to obtain direct connections with a microscopic theory. Some aspects of these connections have been recently discussed [5] and other results, along these lines, were also presented at this Conference [7].

Acknowledgments

M.J.M. is thankful to Professor V. A. Petrov and the whole Organizing Committee of the Conference for the kind hospitality and the excellente atmosphere of the meeting. Thanks are also due to FAPESP and CNPq for financial support.

References

1. S. Nurushev, talk at this Conference; S. Weisz, talk at this Conference.
2. S. Barshay, *Phys. Lett.* B **116**, 193 (1982); *Phys. Rev. Lett.* **49**, 1609 (1982); C.S. Lam and P.S. Yeung, *Phys. Lett.* B **119**, 445 (1082).
3. Z. Koba, H.B. Nielsen and P. Olesen, *Nucl. Phys.* B **40**, 317 (1972).
4. J. Dias de Deus, *Nucl. Phys.* B **59**, 231 (1973).
5. P.C. Beggio, M.J. Menon and P. Valin, *Phys. Rev. D* (to be published), preprint hep-ph/9908389.
6. R. Henzi and P. Valin, *Nucl. Phys.* B **148**, 513 (1979); *Phys. Lett.* B **164**, 411 (1985); *Z. Phys.* C **27**, 351 (1985).
7. N. Tyurin, talk at this Conference; G. Musulmanbekov, talk at this Conference.

V. A. Petrov

GRAND TOTAL

V.A. PETROV
Division of Theoretical Physics, IHEP,
Protvino, Russia

This is a great pleasure to state that our Workshop has finished without loss of enthusiasm.

As the previous talk concerned partly history I dare to remind you recent celebration of the 200-th anniversary of great Russian poet, Alexandre Pushkin. It seems to me that present situation in our field is nicely summarized with the following lines (in my imperfect translation from Russian):

"Oh, how many wonderful discoveries
To us prepare spirit of enlightment,
Experience, son of hard mistakes,
And Genius, friend of paradoxes,
And Chance, god-inventor."

Coming to physics I would not like to characterize or to briefly retell every talk. I listened to all of them and all of them are interesting (at least in some specific sense). So I only could spoil your first impressions. What I do want to draw your attention to is a somewhat higher level of criticism and a healthy trend to really understand what are the real problems in diffractive particle physics. I cannot say that we have succeeded very much in this respect, even at our special discussion session 29 June.

Nonetheless some vague feeling of the necessity of such discussions is evident. This necessity is strongly backed by a splendid bouquet of experimental talks from ZEUS and H1, L3, ALEPH and DELPHI, CDF and E866 which displayed an impressive panorama of new and intriguing phenomena. Phenomena that still cannot find their fully satisfactory (if any) explanation and description.

New projects both near (RHIC) and far (TOTEM) (cf also A. Krish's talk on plans on future polarized beams experiments), they also appeal to our move towards greater understanding of what do we do. Are all these titanic efforts - intellectual, technological, financial - justified?

According to the title of the Blois Workshops our main concern is diffractive phenomena at high energies.

This is an old subject. What have we learnt about that after 30-40 years of quite thorough investigation? Or at least at our meeting here?

At this conference 10 (of total 48) talks had in their titles the word "diffraction" (or "diffractive"). Due mostly to HERA we have got a new promising direction: "hard diffraction". But where is the very diffraction picture? Or when it will be seen? Maybe that is why the entry "diffraction" in the capital Oxford English Dictionary is described as a phenomenon inherent to optics and acoustics but not to high-energy physics. These old questions look even more acute now.

6 talks had in their titles the word "Pomeron". Factually the Pomeron was discussed in the bulk of the talks. What is (or are) the Pomeron? We have at the moment the "hard" Pomeron (or even many of them) which is non-universal and depends in general on colliding particles species. We have "ITEP-Orsay" Pomeron, we have "BFKL" Pomeron(s), we have "dipole" Pomeron etc. Some people prefer to call the "Pomeron" just the leading j-plane singularity of the elastic scattering amplitude. Certainly, the problem is not terminological and in all cases we search for some agent responsible for leading high-energy behaviour. We could ask: if the experimental data can decide in favour of one, "true", theory (or model)? One can hardly state that at present this is (or will be soon) the case. In this respect many discussions acquire evident metaphysical features.

QCD, the patent fundamental theory of strong interaction, in the best case "motivates" the models, but nobody can at the moment extract from it clear and firm predictions which can be confirmed or falsified at least in principle.

In such situation it is natural to keep closer to the experimental data as an objective message from Nature.

But what do we see? Take, e.g., the summary of some very interesting and instructive talk. There we meet such subjective, model-dependent conclusions as, e.g.: "DGLAP analysis of F_2^P shows substantial gluon component". In this respect I would like to remind you a very spectacular case of the repeated measurement of the real part of the $p\bar{p}$-scattering amplitude by UA4 Collaboration. The result clearly depends on the type of unknown strong interaction amplitude (its t-dependence) and is, generally, model-dependent. So all conclusions on the discovery or ruling out of so-called "maximal Odderon" (or any other) have no objective meaning if they are based on a unique observable (differential cross-section).

Another point which is worth mentioning is a precocious use of ill-defined concepts and preliminary results on them.

We have seen in the extremely exciting report on gamma-gamma interactions at LEP that recent variants of "hard" (or BFKL) Pomerons do not fit the data. What could it mean? Is QCD wrong? Certainly, we hardly make such a conclusion. On the other hand recent publication results on gamma-gamma total cross-sections are compared with the model of Bourelly-Soffer-Wu. Argeement is quite satisfactory. Does it mean that we got a proof that

QCD describes these phenomena without problems? Certainly, it would be wrong in spite of high heuristic value of such models.

I hope that I am not too pessimistic and this is not my intention. Contrary to that, I believe that these several days we spent here in discussions and disputes give us a proof that hadronic diffraction physics is very actual, interesting and promising discipline. And I am sure that we will continue our discussions in two years, a t the IX-th Blois Workshop, where, no doubt, new effects and new ideas will be reported. Thank you and good-bye!

List of participants of the International Conference on Elastic and Diffractive Scattering (VIIIth "Blois Workshop")

1. Ado Yu.M. *IHEP, Protvino*
2. Ageev A.I. *IHEP, Protvino*
3. Alekseev A.I. *IHEP, Protvino*
4. Amelung C. *ZEUS and H1, DESY, Hamburg*
5. Arbuzov B.A. *IHEP, Protvino*
6. Arkhipov A.A. *IHEP, Protvino*
7. Asseev A. *IHEP, Protvino*
8. Borodulin V.I. *IHEP, Protvino*
9. Berger E. *Institute of Theoretical Physics, Heidelberg*
10. Bondarenko N. *Kharkov University*
11. Buniatian A.*H1 and ZEUS, DESY, Hamburg*
12. Dakhno L. *St.Petersburg Nuclear Physics Institute*
13. Denisov S.P. *IHEP, Protvino*
14. Ezhela V. *IHEP, Protvino*
15. Dmitrovskii V.Yu. *IHEP, Protvino*
16. Gershtein S.S. *IHEP, Protvino*
17. Iannotti L. *ZEUS and H1, DESY, Hamburg*
18. Islam M. *University of Connecticut*
19. Jenkovszky L. *Bogolyubov Instutute for Theoretical Physics, NASU, Kiev*
20. Kaidalov A. *ITEP, Moscow*
21. Khrustalev O.A. *Bogolyubov ITMP, Moscow*

22. Kang K. *Brown University, Providence*

23. Kisselev A.V. *IHEP, Protvino*

24. Kienzle M. *CERN and Geneva University*

25. Kim V. *St.Petersburg Nuclear Physics Institute*

26. Klishevich S.M. *IHEP, Protvino*

27. Kotov V.I. *IHEP, Protvino*

28. Kotikov A. *JINR, Dubna*

29. Komachenko Yu.Ya. *IHEP, Protvino*

30. Krisch A. *University of Michigan*

31. Kuraev E. *JINR, Dubna*

32. Kundrat V. *Institute of Physics, ASCR, Prague*

33. Landsberg L.G. *IHEP, Protvino*

34. Lehto M. *Sheffield University*

35. Logunov A.A. *IHEP, Protvino*

36. Martin A. *CERN, Geneva*

37. Martynov E.*Bogolyubov Instutite for Theoretical Physics, NASU, Kiev*

38. Menon M. *Gleb Wataghin Institute of Physics, Campinas*

39. Mestvirishvili M.A. *Bogolyubov ITPM, Moscow*

40. Mnatsakanova M.N. *Skobeltsyn NPI, Moscow*

41. Moiseev A.M. *IHEP, Protvino*

42. Musulmanbekov G. *JINR, Dubna*

43. Myznikov K.P. *IHEP, Protvino*

44. Nicolescu B. *CNRS, University of Paris VI*

45. Nikitina O. *Dashkova school, Kremenki*

46. **Nikolaev N.** *L.D.Landau Institute for Theoretical Physics, Chernogolovka and Institute for Nuclear Physics, Research Centre Juelich Gmbh*

47. Nurushev S. *IHEP, Protvino*

48. Obraztsov V.F. *IHEP, Protvino*

49. Olsson J. *H1 and ZEUS, DESY, Hamburg*

50. Petrov V.A. *IHEP, Protvino*

51. Polishchuk B. *IHEP, Protvino*

52. Predazzi E. *University of Torino*

53. Prokudin A.V. *IHEP, Protvino*

54. Rogalyov R.N. *IHEP, Protvino*

55. Ryabov Yu.G. *IHEP, Protvino*

56. Ryutin R.A. *IHEP, Protvino*

57. Samokhin A.P. *IHEP, Protvino*

58. Saponov P.A. *IHEP, Protvino*

59. Serebryakov B.A. ***IHEP, Protvino***

60. Selyugin O. *JINR, **Dubna***

61. Sokolov S.N. *IHEP, **Protvino***

62. Solodsky A. *Rockefeller University, New-York*

63. Soloviev L.D. *IHEP, Protvino*

64. Soloviev V.O. *IHEP, Protvino*

65. Solov'yanov V.L. *IHEP, Protvino*

66. Stoyanova D.A. *IHEP, Protvino*

67. Teplyakov V.A. *IHEP, Protvino*

68. Terazawa H. *KEK, Tokyo*

69. Troshin S.M. *IHEP, Protvino*

70. Tsokur V.A. *IHEP, Protvino*

71. Tyapkin A.A. *JINR, Dubna*

72. Tyapkin I. *CERN, Geneva and JINR, Dubna*

73. Tyurin N.E. *IHEP, Protvino*

74. Ugarov V.P. *IHEP, Protvino*

75. Vacca G. P. *II Institute of Theoretical Physics, University of Hamburg*

76. Vasiliev M. *Texas A&M University, College Station and RRC "Kurchatov Institute", Moscow*

77. Vernov Yu.S. *INR RAS, Moscow*

78. Zaitsev A.M. *IHEP, Protvino*

79. Zinov'ev Yu.M. *IHEP, Protvino*

80. Jsembery J. *CEA/DSM/DAPHNIA/SPP Gif-sur-Yvette*

81. Weisz S. *CERN, Geneva*